The Institute of Mathematics and its Applications Conference Series

Previous volumes in this series were published by Academic Press to whom all enquiries should be addressed. Forthcoming volumes will be published by Oxford University Press throughout the world.

NEW SERIES
1. *Supercomputers and parallel computation* Edited by D. J. Paddon
2. *The mathematical basis of finite element methods*
   Edited by David F. Griffiths
3. *Multigrid methods for integral and differential equations*
   Edited by D. J. Paddon and H. Holstein
4. *Turbulence and diffusion in stable environments* Edited by J. C. R. Hunt
5. *Wave propagation and scattering* Edited by B. J. Uscinski
6. *The mathematics of surfaces* Edited by J. A. Gregory
7. *Numerical methods for fluid dynamics II*
   Edited by K. W. Morton and M. J. Baines
8. *Analysing conflict and its resolution* Edited by P. G. Bennett
9. *The state of the art in numerical analysis*
   Edited by A. Iserles and M. J. D. Powell
10. *Algorithms for approximation* Edited by J. C. Mason and M. G. Cox
11. *The mathematics of surfaces II* Edited by R. R. Martin
12. *Mathematics in signal processing*
    Edited by T. S. Durrani, J. B. Abbiss, J. E. Hudson, R. N. Madan, J. G. McWhirter, and T. A. Moore
13. *Simulation and optimization of large systems*
    Edited by Andrzej J. Osiadacz
14. *Computers in mathematical research*
    Edited by N. M. Stephens and M. P. Thorne
15. *Stably stratified flow and dense gas dispersion*
    Edited by J. S. Puttock
16. *Mathematical modelling in non-destructive testing*
    Edited by Michael Blakemore and George A. Georgiou
17. *Numerical methods for fluid dynamics III*
    Edited by K. W. Morton and M. J. Baines
18. *Mathematics in oil production*
    Edited by Sir Sam Edwards and P. R. King
19. *Mathematics in major accident risk assessment*
    Edited by R. A. Cox
20. *Cryptography and coding*
    Edited by Henry J. Beker and F. C. Piper
21. *Mathematics in remote sensing*
    Edited by S. R. Brooks
22. *Applications of matrix theory*
    Edited by M. J. C. Gover and S. Barnett

Continued overleaf

23. *The mathematics of surfaces III*
    Edited by D. C. Handscomb
24. *The interface of mathematics and particle physics*
    Edited by D. G. Quillen, G. B. Segal, and Tsou S. T.
25. *Computational methods in aeronautical fluid dynamics*
    Edited by P. Stow
26. *Mathematics in signal processing II*
    Edited by J. G. McWhirter
27. *Mathematical structures for software engineering*
    Edited by Bernard de Neumann, Dan Simpson, and Gil Slater
28. *Computer modelling in the environmental sciences*
    Edited by D. G. Farmer and M. J. Rycroft
29. *Statistics in medicine*
    Edited by F. Dunstan and J. Pickles
30. *The mathematical revolution inspired by computing*
    Edited by J. H. Johnson and M. J. Loomes

The Mathematical Revolution Inspired by Computing

Based on the proceedings of a conference on The Mathematical Revolution Inspired by Computing organized by The Institute of Mathematics and its Applications and held at Brighton Polytechnic in April 1989.

J. H. JOHNSON
The Open University, Walton Hall

M. J. LOOMES
Hatfield Polytechnic

CLARENDON PRESS · OXFORD · 1991

Oxford University Press, Walton Street, Oxford OX2 6DP
Oxford New York Toronto
Delhi Bombay Calcutta Madras Karachi
Petaling Jaya Singapore Hong Kong Tokyo
Nairobi Dar es Salaam Cape Town
Melbourne Auckland
and associated companies in
Berlin Ibadan

Oxford is a trade mark of Oxford University Press

Published in the United States
by Oxford University Press, New York

© The Institute of Mathematics and its Applications 1991

All rights reserved. No part of this publication may be reproduced, stored in a retrieval system, or transmitted, in any form or by any means, electronic, mechanical, photocopying, recording, or otherwise, without the prior permission of Oxford University Press

A catalogue record for this book is available from the British Library
ISBN 0–19–853658–5

Library of Congress Cataloging in Publication Data
The Mathematical revolution inspired by computing : based on the proceedings of a conference on the mathematical revolution inspired by computing, organized by the Institute of Mathematics and Its Applications and held at Brighton Polytechnic in April 1989 / [edited by] J.H. Johnson, M.J. Loomes.
(The Institute of Mathematics and Its Applications conference series ; new ser., 30)
Includes bibliographical references.
1. Computer science—Mathematics—Congresses.   I. Johnson, J. H. (Jeffrey H.)   II. Loomes, Martin.   III. Institute of Mathematics and Its Applications.   IV. Series.
$QA76.9.M35M387$ 1991   91–12272 $004'.01'51$—dc20
ISBN 0–19–853658–5

Keyed by Dr J. Johnson
at the Open University and
printed in Great Britain by
Courier International Ltd
Tiptree, Essex

Preface

The mathematical revolution inspired by computing! You may think there is no revolution taking place in mathematics, and if so you will be in the company of the majority of those attending the IMA conference with this deliberately provocative title. This book is the proceedings of that conference, which was held in Brighton in 1989. Revolution or not, we hope you will agree that some very exciting things are happening in mathematics and that computers are having a significant impact.

We have divided the book into four parts. The first is an introduction, which gives an overview of the impact of computers on mathematics and sets the context for the other papers in this volume. Parts two and three consider the impact of computers on mathematics and the impact of mathematics on computing respectively, while the fourth part considers issues relating to the proposition that computers are revolutionising mathematics. Of course, these divisions are artificial and many of the papers could have been put in any or all of the parts. We hope that the authors will forgive us if they feel their paper has been misplaced.

Most of the participants at the conference found it a highly stimulating three days, and most of us learnt of things about which we previously knew nothing. Certainly, editing these proceedings has been an education for us. We hope this means that the book will help to make a wide variety of ideas available to the general mathematical audience, while acting as a useful reference for specialists in the various areas.

The IMA has taken a great interest in the impact of computers on mathematics over the years, and the Institute's Bulletin frequently carries articles on the subject. One of many is an excellent introductory account given in Professor Churchhouse's *Presidential Address: Mathematics and Computing* [2]. Also, an earlier volume in this series has the title *Computers in Mathematical Research* [4]. The impact of computers in mathematics has also stimulated great interest in other countries, especially the USA where, for example, the *Notices of the American Mathematical Society* began a column in 1988 [1]. Also, there are frequent conferences on the subject in the USA, for example, [3].

Fractals and chaos are perhaps the best known areas in which computers are impinging on mathematics. They give mathematics a colourful face, and anything which can make mathematics more attractive to the layman must be good for the discipline. However, as this book shows, the impact of computers is much wider than that. Arguably, through computer algebra and computer proof, computers are changing the way mathematics work or even the way mathematicians think! But this will not be developed here, since the arguments for and against are well rehearsed in these proceedings.

A significant, if mundane, effect of computers on mathematics is the possibility of 'composing mathematics at the keyboard'. Barwise remarks that 'As a writer, I have completely switched over from writing mathematics papers in longhand and having them typed, to using $\mathrm{L\!A\!T\!E\!X}$, a version of Knuth's mathematical text processing system $\mathrm{T\!E\!X}$.' [1]. We think this must be a widespread experience for mathematicians: composing mathematics at the keyboard is quicker, less frustrating, and gives far better results than using a typewriter.

We received a lot of help when setting this book in $\mathrm{L\!^AT_EX}$, at the Open University. Special thanks go to Steve Daniels, Marilyn Moffat, and Andy Boddington of the ACS Research Advisory Service, and also to Phil Thomas, for their support. Thanks also go to Jian Peng of the Programming Research Group at Oxford University, who was extremely helpful in providing the $\mathrm{L\!^AT_EX}$ macros for setting the book and explaining how to use them.

The conference was generously sponsored by Apple Computers (UK) Ltd, Barclays Bank plc., British Telecom plc., The Buxton-Douglas Partnership, Computer Weekly, Digital Equipment Ltd, Gould-Johnson Research Associates, ICL, The Open University, Oxford University Press, Praxis Systems plc., SD-Scicon, South Bank Polytechnic, Sun Microsystems Ltd, Unilever Research plc., and Vision Scientific Ltd. We are very grateful for the flexibility this gave us.

Behind this conference were many people who worked hard to contribute to its success. In particular, we would like to thank the officers of the IMA who worked so efficiently in the organisation of the conference. Also, we were especially fortunate in the composition of the organising committee, and our thanks go to Richard Bosworth, Tim Denvir, Norman Fenton, Chic Rattray, and Robin Whitty.

Our colleagues have also been very supportive, including Sally Boyle, Judith Daniels, Nigel Cross, Rodney Ranzetta, Derek Richards and Phil Steadman. Thanks also go to the staff of Oxford University Press for their help in getting the book into your hands. Most of all, we would like to thank those who attended the conference, and those who contributed to this book.

Milton Keynes and Hatfield   J.H.J.
January 1991   M.J.L.

References

[1] Barwise, J., 'Computers and Mathematics', *Notices of the American Mathematical Society,* **35**, 5, 693–694, 1988.

[2] Churchhouse, R. F., 'Presidential Address: Mathematics and Computers', *Bulletin of the I.M.A.,* **25**, 3/5, 40–49, 1989.

[3] Kaltofen, E., Watt, S. M., (eds), *Computers and Mathematics,* Springer-Verlag, (New York), 1989.

[4] Stephens, N. M., Thorne, M.P., (eds) *Computers in Mathematical Research,* Clarendon Press, (Oxford), 1989.

Contents

List of Contributors	ix

PART I   AN INTRODUCTION TO THE MATHEMATICAL REVOLUTION INSPIRED BY COMPUTING

An Introduction to the Mathematical Revolution Inspired by Computing *Jeffrey Johnson*	3

PART II   COMPUTERS IN MATHEMATICS

Cryptography—The Catalyst *Fred Piper*	21
The Return of the Visual *R. V. Evans*	33
The Mathematics of Chaos *D. K. Arrowsmith*	47
Computing The Unpredictable: Deterministic Chaos and The Nervous System *Arun V. Holden*	75
Word Processing Algorithms, Rewrite Rules and Group Theory *D. B. A. Epstein*	87
Computer Assisted Proof for Mathematics: an Introduction Using the LEGO Proof System *Rod Burstall*	101
A New Method of Automated Theorem Proving *Yang Lu*	115
Making Discrete Mathematics Executable on a Computer *R. D. Knott*	127
The Wider Uses of the Z Specification Language in Mathematical Modelling *Allan Norcliffe*	145
Scene Analysis via Galois Lattices *M. Andrew, D. Bose, and S. Cosby*	157
The Mathematics of Complex Systems *Jeffrey Johnson*	165

PART III MATHEMATICS IN COMPUTING

The Mathematics of Complex Computational Systems 189
Stephen B. Seidman

A Euclidean Basis for Computation 205
Dan Simpson

An Extension of Turing Machines 211
Claudio Sossai

Algorithmic Languages and the Computability of Functions 221
Newcomb Greenleaf

The Parallel Computation Hypothesis and its Applications to Computer Science 233
V. J. Rayward-Smith

The Mathematics of Complexity in Computing and Software Engineering 243
N. E. Fenton

The Mathematics of Calibration 257
K. L. Tse and R. W. Whitty

PART IV IS THERE A MATHEMATICAL REVOLUTION INSPIRED BY COMPUTING?

Computing and Foundations 269
J. M. E. Hyland

The Development and Use of Variables in Mathematics and Computer Science 285
Meurig Beynon and Steve Russ

The End of the Defensive Era of Mathematics 297
C. Ormell

Revolution, Evolution or Renaissance? 307
D. J. Cooke

The Superfluous Paradigm 323
Daniel I. A. Cohen

Contributors

M. ANDREW *Royal Sussex County Hospital, Brighton.*

D. K. ARROWSMITH *School of Mathematical Sciences, Queen Mary and Westfield College, University of London, Mile End Road, London, E1 4NS.*

MEURIG BEYNON *Department of Computer Science, University of Warwick, CV4 7AL.*

D. K. BOSE *IT Research Institute, Brighton Polytechnic, BN2 4GJ.*

ROD BURSTALL *Laboratory for Foundations of Computer Science, King's Building, University of Edinburgh, Edinburgh, EH9 3JZ.*

DANIEL I. A. COHEN *Hunter College, City University of New York, 695 Park Avenue, New York 10022, USA.*

D. J. COOKE *Department of Computer Studies, Loughborough University of Technology, Leics, LE11 3TU.*

S. COSBY *IT Research Institute, Brighton Polytechnic, BN2 4GJ.*

D. B. A. EPSTEIN *Mathematics Institute, University of Warwick, CV4 7AL.*

R. V. EVANS *ICL, Kings House, 33 Kings Road, Reading, RG1 3PX.*

N. E. FENTON *Centre for Software Reliability, City University, Northampton Square, London EC1V 0HB.*

NEWCOMB GREENLEAF *Department of Computer Science, Columbia University, New York, NY 10027, USA, newcomb@cs.columbia.edu.*

ARUN V. HOLDEN *Department of Physiology and Centre for Nonlinear Studies, The University, Leeds, LS2 9NQ.*

J. M. E. HYLAND *Department of Pure Mathematics, University of Cambridge, 16 Mill Lane, Cambridge, CB2 1SB.*

JEFFREY JOHNSON *Centre for Configurational Studies, The Open University, Milton Keynes, MK7 6AA.*

R. D. KNOTT *Department of Mathematics, University of Surrey, Guildford, Surrey, GU2 5XH.*

ALLAN NORCLIFFE *Department of Mathematical Sciences, Sheffield City Polytechnic, Pond Street, Sheffield, S1 1WB*

C. ORMELL *School of Education, University of East Anglia, Norwich, NR4 7TJ.*

FRED PIPER *Department of Mathematics, Royal Holloway and Bedford New College, University of London, Egham Hill, Surrey, TW20 0EX.*

V. J. RAYWARD-SMITH *School of Information Systems, University of East Anglia, Norwich, NR4 7TJ.*

STEVE RUSS *Department of Computer Science, University of Warwick, CV4 7AL.*

STEPHEN B. SEIDMAN *Department of Computer Science and Engineering, Auburn University, Alabama 36830, USA.*

DAN SIMPSON *Department of Computing, Brighton Polytechnic, Moulsecoomb, Brighton, BN2 4GJ.*

CLAUDIO SOSSAI *Via Vlacovich 20-35124 Padova, Italy.*

K. L. TSE *Nijenrode, The Netherlands School of Business, Straatweg 25, 3621 BG Breukelen, The Netherlands.*

R. W. WHITTY *Department of Mathematics, Goldsmiths' College, University of London, New Cross, London, SE14 6NW.*

YANG LU *Institute of Mathematical Sciences, Chengdu Branch, Academia Sinica, 610015 Chengdu, Sichuan, China.*

PART I

An Introduction to the

The Mathematical Revolution Inspired by Computing

An Introduction to The Mathematical Revolution Inspired by Computing

Jeffrey Johnson
Centre for Configurational Studies,
The Open University, Milton Keynes, MK7 6AA

Abstract

This paper introduces the general reader to the considerable impact of computers on mathematics, and thereby to the proceedings of a conference on this subject (references to papers appearing in this volume will be marked by a dagger symbol, e.g. [1]†). The relationship between mathematics and calculating machines goes back at least to Pascal and Leibniz and remains a very active research area to this day. Whereas the original motivation was to facilitate numerical calculation, the interplay between computers and mathematics is much more complex today in the context of the 'information revolution' which affects all aspects of society. In order to impose some structure, three main (overlapping) areas are identified: mathematics in computing, computing in mathematics, and the changes that computers have precipitated in mathematics, its methods, and its philosophy. The first considers the common concerns at the foundations of mathematics and computer science, and the use of mathematics to model computation and computing machines. The second concerns the 'facilitating' affect of computers on creative work in general, and on mathematics in particular: we can do things now that we could not do before. The third concerns the proposition that mathematics is undergoing momentous and qualitative change. Some professional mathematicians are unaware of the changes that the future may bring, and others are hostile to computers invading their privileged universe. Despite this, it can be argued that we are living through some of the most exciting times in the whole history of mathematics.

*If you do not think about the future,
you cannot have one.*
John Galsworthy

1 Introduction

The twentieth century has seen the dawn of the information age with almost every aspect of life being affected by computers. Science and its methods have undergone major changes as computers have become standard equipment in every branch: from modelling molecules in chemistry and biology, to controlling particle accelerators in physics, to analysing data in sociology and political science, and so on.

And of course, computers have made a considerable impact on mathematics and mathematicians.

Mathematics and many great mathematicians have played a major part in the development of computing. Long before the electronics era, mathematicians devised algorithms[1] for calculation and symbolic manipulation. The nineteenth century mathematician Babbage was but one of many in a long line working on the design of practical calculating engines which includes Pascal (addition) and Leibniz (multiplication).

In the twentieth century the development of computing has been closely intertwined with that of mathematics. At the end of the nineteenth century mathematicians were becoming increasingly concerned with the foundations of their subject, there being three main schools of thought: Logicist, Intuitionist, and Formalist. Hollingdale [10] summarises them thus: " following [13], we can describe these three mathematical philosophies very briefly as follows. The Logicists believed that mathematics can be derived from logic, and is indeed an extension of logic. The Intuitionists adopted a radically different approach: they conceived of a fundamental mathematical intuition based on our ability to perceive a sequence of events ordered in time (so giving us our notion of the natural numbers, as an obvious example). According to this view, mathematics is a constructive process that builds its own universe, independent of the world of everyday experience; mathematical ideas are embedded in the human mind, outside which they have no separate existence. The Formalists took something of a middle position. For them, mathematics consists of several branches, each with its own axioms, rules, concepts and theorems. The objects of mathematical thought reduce to symbolic elements which no longer stand for idealized physical entities".

A hypothetical consequence of the Formalist position proposed by Hilbert is that a computer could be programmed to do all of mathematics. Cohen [29]† writes "Hilbert's vision of a universal algorithm to solve mathematical theorems required a unification of Logic, Set Theory, and Number Theory. This project was initiated by Frege, rerouted by Russell, repaired by Whitehead, derailed by Gödel, restored by Zermelo, Frankel, Bernays and von Neumann, shaken by Church and finally demolished by Turing". [2]

Mathematics and computing have more deliberate relationships than this early history, because of the rigour and precision required for the latter. In particular, computer scientists are interested in languages which are capable of returning compound propositions such as the general answer to a general question, rather than values such as a number. This requires a theory in which functions can be vari-

---

[1] The word algorithm comes from the ninth century Persian mathematician Abu Ja'far Mohammed ibn Mûsâ al-Khowârizmî.

[2] Benacerraf and Putnam comment "Yet it should not be forgotten that if today it seems somewhat arbitrary just where one draws the line between logic and mathematics, this is a victory for Frege, Russell, and Whitehead: before their work, the gulf between the two subjects seemed absolute" [1],and Kreisel writes "Hilbert's programme ... is a rich line of research in foundations. ... As far as piecemeal understanding is concerned, its importance consists in having led to the fruitful study of the constructive aspects of axiomatic systems" [14].

ables, and Church's lambda calculus has been taken up for this purpose. However the questions at the foundations of computer science remain very close to those at the foundations of mathematics. In particular both mathematical logicians and theoretical computer scientists are very interested in category theory because of the primacy it gives the notion of functions, and the natural way that it can be used to handle complex logical concepts [6] [15] [36]†.

In the 1940's the phenomenally successful computer architecture of today's sequential computers was developed by von Neumann and others, notably Atanasoff, Eckert, and Mauchly [8]. It is interesting to note that even in those early days, the questions of parallel computation and machine learning were receiving the attention of mathematicians including von Neumann himself [23].

The close inter-relationship between the foundations of mathematics and computing will be discussed in more detail in Section 2.

Since mathematics has had such a clear impact on the theory and development of computing machines, it is sometimes difficult to appreciate the wider impact that computers are having on mathematics and the day to day work of mathematicians. Mathematics involves having new ideas and working through their consequences, this reflecting the usual "1% inspiration and 99% perspiration" of intellectual inquiry. Computers have already proved to be useful tools for reducing the manual labour of repetitive calculation, including the symbolic manipulation of general mathematical entities. Also, there can be no doubt that computers have encouraged mathematical creativity in some areas, and that in some cases they have directly stimulated the discovery of new mathematical systems. These aspect of the impact of computers on mathematics will be investigated in Section 3.

However, the impact of computers on mathematics goes far beyond a common interest in foundations, the source of inspiration for new mathematical structures, the practical possibility of investigating new mathematical structures, and the new intuitions and stimuli offered by computer graphics. In 1976 Appel, Haken, and Koch upset the mathematical world by presenting their computer proof of the four colour problem [29]†. Some mathematicians rejected the idea of a proof that could not be checked by one man, while others concluded that the established views on what constitutes a mathematical proof require revision. There is as yet no consensus on the impact of this development, but it may have startling consequences.

2 Mathematics in Computing

Mathematics has always involved some degree of calculation and symbolic manipulation, which many mathematicians consider to be necessary evils. Little intellectual satisfaction can be gained from working through an established algorithm, and errors frequently frustrate progress. Making machines to take over this drudgery is very attractive to mathematicians, who are also well equipped to specify and design such machines. One such, Leibniz, wrote in the seventeenth century "... it is unworthy of excellent men to lose hours like slaves in the labour of calculation, which could safely be relegated to anyone else if the machine were used" [4] [8].

2.1 Early Calculating Machines

Some of the earliest calculating machines were invented by the fifteenth century Persian astronomer, al-Kāshī (1393-1449). These simplified the calculations of astrologers and astronomers [8]. Following John Napier's invention of logarithms in 1614, Edmund Gunter devised in 1620 a forerunner to the slide rule, which was subsequently invented by William Oughtred in 1632.

Goldstine [8] attributes the invention of the first automatic digital calculating machine to Wilhem Schickard, who was professor of astronomy, mathematics, and Hebrew at Tübingen. In a letter to Kepler in 1623 he wrote of his machine that it "immediately computes the given numbers automatically, adds, subtracts, multiplies, and divides. Surely you will beam when you see how [it] accumulates left carries of tens and hundreds by itself or while subtracting takes something away from them". Unfortunately Schickard's work suffered a setback due to a fire in 1624, and he died of plague in 1634 before his invention became widely known [8].

About the year 1641, the eighteen year old Blaise Pascal invented and constructed one of the earliest calculating machines. It performed addition with automatic mechanical 'carry', the feature which makes it a true calculating machine. In 1672 Gottfried Wilhelm Leibniz made a prototype calculating machine which mechanised multiplication using the 'Leibniz stepped wheel', a device still in use in the 1940's [10].

The nineteenth century mathematician Charles Babbage designed a number of computing machines including his Difference and Analytical Engines. The Analytical Engine was *digital*, separated the *store* from the *mill* (which actually did the calculations), and used three types of card: *numerical constant cards, variable cards,* and *operation cards* [4] [11]. Whereas Babbage never actually finished building any of his machines, Pehr Georg Scheutz of Stockholm built a difference engine in 1853 which saw service in London and New York.

Herman Hollerith invented an electromechanical machine for the 1890 census in the U.S.A. which stored population data on punch cards. In 1911 Hollerith's Tabulating Machinary Company became the Computer-Tabulating-Recording Company, which became the International Business Machines Corporation under Thomas J. Watson in 1924.

For a more complete history of early computing machines see [8].

2.2 Foundations: Decidability, Gödel's Theorems, Church's Lambda Calculus, and the Turing Machine

At the turn of the century, Hilbert devised a programme to formalise all of mathematics and to show by purely syntactic means that finitary methods could never lead to contradiction. This is equivalent to finding a *decision algorithm* for all of mathematics, which Gödel's theorems of 1931 showed to be impossible. Goldschlager and Lister [7] give the following concise summary: "Deciding the truth of a given statement in a formal system was called the *Entscheidungsproblem* by Hilbert, who considered it to be a fundamental open problem in mathematics. Unfortunately for Hilbert's objective, the 1930's brought a wave of research which showed that the Entscheidungsproblem is not computable. That is, no algorithm of the type for which Hilbert longed, exists. A cynic might say that mathematicians could heave a sigh of relief, for if such an algorithm did exist, they would be out of a job just as soon as the algorithm was found. In fact mathematicians were stunned by this remarkable discovery."

Following this, attention turned to the class of functions that can be evaluated systematically. Neale and Neale [17] write that " in 1936 Alonzo Church, following a suggestion made by Gödel in conversation, put forward the thesis that every calculable function is recursive. ... Church had conceived independently the plan of identifying effective calculability with a feature which he called λ-definability, but he discovered that this was equivalent to recursiveness, and a year later A. M. Turing proved the same of *computability-by-a-machine* which he had offered as another analysis of the notion of effective calculability. The importance of this thesis is that it makes possible a new approach to the decision problem; and Church took advantage of this to prove first the impossibility of a universal decision procedure for a certain fragment of number theory and then, as a consequence, the impossibility of such a procedure for general logic."

Turing's used his abstract 'universal computing machine' (for more details see [31]†) to investigate the *halting problem* of knowing which inputs do or do not eventually cause a computer to stop (for a readable account see page 216 [21]). Turing proved the Halting Problem is *undecidable*, that is, there exists no algorithm to solve it. This mathematical result places practical constraints on what computers can be expected to do.

What is an algorithm? Goldschlager and Lister write "Gödel defined an algorithm as a sequence of rules for forming complicated mathematical functions out of simpler mathematical functions. Church used a formalism called the lambda calculus, whilst Turing used a hypothetical machine we call a Turing Machine. Turing defined an algorithm as any set of instructions for his simple machine." ([7], page 69). Church's hypothesis, also known as the *Turing-Church Thesis*, can be paraphrased as saying that these and all other 'reasonable' definitions of algorithm are equivalent. Simpson's paper in this volume approaches the Church-Turing thesis from an intuitionist viewpoint [44]†. For more details of λ-calculus see [19][30]†[36]†.

Computer science has highlighted the need to make the concept of the term *variable* more precise than is normal in mathematics, especially when the *value* of a variable may be a *function* or *predicate*. There is clearly a difference between $x = y$ in mathematics and the assignment $x := y$ in a programming language. Furthermore variables and their values have dynamic properties in programming languages which are not commonly met in conventional mathematics [27]†. These ideas are directly relevant in Computer Algebra when one may want an expression returned rather than a particular value of that expression [3].

As shown in this volume, lambda calculus and combinators are attracting great interest in computer science. Revesz prefaces his book on the subject with the comments "they seem to have successfully captured the most general formal properties of the notion of a mathematical function, which in turn, is one of the most powerful concepts of modern mathematics. ... An explicit and systematic use of lambda-calculus in computer science was initiated by Peter Landin, Christopher Strachey, and a few others who started the development of a formal theory of semantics for programming languages based directly on lambda-calculus. Their approach is now called *denotational semantics* and it is widely accepted as a relevant theory. At first, however, denotational semantics was thought to be flawed by some experts because of its naive use of the type-free lambda-calculus, which did not seem to have any consistent mathematical model at that time. The first mathematical model for the type-free lambda-calculus was discovered by Dana Scott only in 1969 when he was trying to prove the nonexistence of such models." [19].

2.3 Category Theory and Topos Theory

Although category theory was developed in the nineteen forties to establish certain natural equivalences in algebraic topology, its tremendous generality has been increasingly recognised by mathematicians and logicians. In the foreword to the book *Computational Category Theory*, John Gray writes "The fact of the matter is that category theory is an intensely computational subject, as all its practitioners well know. Categories themselves are the models of an essentially algebraic theory and nearly all the derived concepts are finitary and algorithmic in nature" [20].

Opinions on category theory vary among mathematicians, from it being "abstract nonsense" to the position put forward by Goldblatt that " ... set theory provides a general conceptual framework for mathematics. Now, since category theory, through the notion of topos, has succeeded in axiomatising set-theory, the outcome is an entirely new *categorial foundation of mathematics*! The category-theorists attitude that "function" rather than "set membership" can be seen as the fundamental mathematical concept has been entirely vindicated. The pre-eminent role of set theory in contemporary mathematics is suddenly challenged. A revolution has occurred in the history of mathematical ideas (albeit a peaceful one) that will undoubtedly influence the direction of the path to the future" [6].

The kind of category called a *topos* is particularly interesting as it has a rich internal logic determined by the lattices of subobjects of objects of the category. In general these lattices are not Boolean algebras but Heyting algebras and so the logic is not Boolean but intuitionistic (and hence constructive). The close relationship between Category Theory and lambda-calculus is explained in [15]. Hyland [36]† discusses these ideas which he argues are fundamental to the foundations of mathematics.

2.4 Measurement, Complexity and NP-Completeness

Of those problems with theoretical algorithmic solutions, some cannot be implemented practically because they would take too long to execute or require too much memory. Often these things depend on the number of elements in the input data. The *Big-O* notation is used to estimate the order of the complexity of algorithms. For example, an algorithm which ran in a time proportional to the square of the size of the input, n, would be said to have *complexity* of the *order* n^2, written $O(n^2)$. For more details of this see [31]† or [33]†. Algorithms whose complexity is a polynomial are said to be *polynomial*, and if n^k is the highest order term of the polynomial the complexity is said to be $O(n^k)$ since the lower order terms become relatively insignificant as n increases.

Although the Big-O notation gives some indication of a program's complexity and possible performance, it is rather a blunt measuring instrument for some purposes. Thus the mathematics of *measurement* is of interest in assessing the behaviour of computer programs. Furthermore, the relevant features may not map onto a totally ordered scale, and may have algebraic characterisation. For a discussion of these issues see [33]†, [37]†, and [46]†.

Of those algorithms which have no known polynomial time algorithm, some have the property that a proposed solution can be verified in polynomial time. The class of such problems is called NP and it contains all 'feasible' and polynomial time problems, and a number of 'open problems' which are believed to be non-polynomial, e.g. the travelling salesman problem. Some well known open problems in NP are *NP-complete*, that is, if any one such has a polynomial time algorithm, then *every* problem in NP has a polynomial time algorithm [42]†. It is widely believed that all the NP-problems are infeasible, but [34]† presents another view on this.

2.5 The Mathematics of Programming

In his inaugural lecture at Oxford, Hoare [9] begins with the principles:

1. *Computers are mathematical machines.* Every aspect of their behaviour can be defined with mathematical precision, and every detail can be deduced from this definition with mathematical certainty by the laws of pure logic.

2. *Computer programs are mathematical expressions.* They describe with unprecedented precision and in every minutest detail the behaviour, intended or unintended, of the computer on which they are executed.

3. *A programming language is a mathematical theory.* It includes concepts, notations, definitions, axioms and theorems, which help a programmer to develop a program which meets its specification, and to prove that it does so.

4. *Programming is a mathematical activity.* Like other branches of applied mathematics and engineering, its successful practice requires determined and meticulous application of traditional methods of mathematical understanding, calculation and proof.

Hoare's lecture covers many issues but one of the most important is the production of error-free programs. The previous sections show that computing is widely accepted as essentially mathematical in the study of machines and languages, as

Hoare's principles indicate. Also, many in mathematics and computer science share the view that computer programs are mathematical expressions, and that it is therefore possible to *prove* them correct by mathematical means. Although this is only part of the story, *specification* languages based on set theory such as Z and VDM are becoming used increasingly to specify what a program is supposed to do, and to ensure that the final result will meet the specification. For examples of the use of specification languages see [33]†, [39]†, [43]†.

3   Computing in Mathematics

Mathematicians may be less defensive about the impact of computers on their subject if they consider the changes that have taken place in the world of design with the introduction of Computer Aided Design. Twenty years ago academic architects became interested in the computer's ability to draw pictures and (apparently) solve hard problems. After some initial over-optimism, it was realised that computers could become a design *aid* by helping with drafting, keeping information well organised in a database (part lists, costs, etc.), and doing some elementary calculations (quantities, estimates, etc.). The commercial arguments for such systems were that they would save drafting time when changes were made to the design, that they could give a qualitative improvement by providing realistic perspective views of buildings for clients quickly and inexpensively, and that there would be less errors. The saving in time would allow *more alternatives* to be investigated, and the graphic output would improve communication between the designer and the client.

Most traditional designers hated the idea of machines entering what they thought to be an essentially creative human process. Some engineering designers such as those in aeronautics had used computers for years, but it is only in the last ten years that even electronics design has moved away from the drawing board to the computer. Despite early resistance, there is now hardly any area of design from clothes fashions to civil engineering which does not depend on computers. What did the designers gain, and what did they lose? The promised gains in terms of trying more alternatives, being more productive, being more creative, avoiding dreary calculations, making fewer errors, and keeping information better organised have certainly been delivered. The losses tend to involve regret at losing traditional crafts such as drawing, and the disruption experienced while adapting to the new (and not always perfected) technology. Although computers are ubiquitous in design, nobody seriously expects the computer to become the designer, and human creativity and ingenuity remain highly valued.

I am sure that a similar history will occur (is occurring) in mathematics, with mathematicians becoming more creative and productive (in their own terms) through the use of computers. And mathematicians need not worry about going out of business, since they have already *proved* that mathematics cannot be reduced to an algorithm.

3.1 Computer Algebra and Automatic Theorem Proving

It is a standing joke that mathematicians are hopeless at arithmetic. Whether or not this is true, it is certain that mathematicians do make errors as they work through algorithms to obtain results needed for more rewarding things. And even if they did not make errors, we can recall Leibniz's strictures on losing hours like slaves in the labour of calculation.

In fact there now exist a number of computer packages which help mathematicians in their day to day work, for example MACSYMA, REDUCE, CAYLEY, and MATHEMATICA. The book by Davenport, Siret and Tourner [3] is a standard reference which discusses the major theoretical issues in computer algebra, and begins with a list of the kind of features possessed by computer algebra systems: operations on integers, on rational, real and complex numbers *with unlimited accuracy*; operations on polynomials in one or more variables and on rational fractions, calculating greatest common divisors, factorising over the integers; calculations on matrices with numerical and/or symbolic elements; simple analysis including differentiation, integration, expansion in series; manipulation of formulae including substitutions, selection of coefficients and parts of formulae, numerical evaluation, pattern recognition, controlled simplifications; and so on. Most of the computer algebra systems mentioned above are written in LISP: here we can see some of the fruits of mathematical computer science being returned to a wider mathematical community.

Epstein [31]† gives an insight into how mathematicians are adapting computer proof to attack some of the traditional problems in mathematics. First he takes us back to to an event in 1954 which "should have startled the mathematical world": Wolfgang Haken's constructive algorithm to solve the Knot problem which determines if a knot can be deformed into an unknotted circle. "This was the first result in a development which has now become a major industry, and is, in my opinion, one of the most important strands of modern mathematics, namely the attempt to produce algorithmic processes which can answer all the important questions about compact 3-dimensional manifolds". In this paper Epstein goes on to show how string matching can be an important tool for group theorists. He reviews the practical problems of reducing the complexity of string matching algorithms (which may run for hours on supercomputers), and discusses the application of these algorithms for attacking mathematical problems.

Knott [38]† shows how mathematics and computing can come full circle by making discrete mathematics executable on a computer using a library of Prolog routines. Given the close relationship between discrete mathematics and the theory of computer science which underlies logic programming languages, this might be regarded as a very natural thing.

Of particular interest in mathematics is the possibility of having a computer actually *prove* a proposed theorem. Burstall [28]† gives an interesting discussion of the issues of computers helping us to *find* proofs as opposed to *verifying* them. Burstall explains how the *Calculus of Constructions*, a form of *Type Theory*, is a logical language with proof rules suitable for formulating constructive mathematics, and illustrates these ideas with the LEGO proof system developed at Edinburgh.

It may come as a surprise to some readers that a great deal of work on computer proof has been going on in China for many years. This started in the mid nineteen seventies with the work of Wu Wen-tsün [24] and among a growing literature, the book by Shang Ching Chou [2] summarises the method and presents 512 geometry theorems. Yang Lu [47]† presents a new method of automatic theorem proving based on the Zhang-Yang theorem, and illustrates this with a remarkable new theorem in spherical geometry: "If the area of a spherical triangle equals one quarter of the area of the sphere, then the midpoints of the three sides form an equilateral triangle with sides of length $\pi r/2$, where r is the radius of the sphere". Trivially, this has the equally astonishing corollary that the internal triangle has area one eighth that of the sphere, and one half that of the original triangle! The reader can verify this theorem by implementing Yang's simple computer program. This theorem has subsequently been proved by conventional methods [47]†.

3.2 The Impact of Graphical Feedback

Two areas of mathematics are currently enjoying intense research activity due to computer graphics. They are of course the inter-related fields of *fractals* and *chaos*. In both cases the research began with very simple iterated equations, and in both cases viewing these equations the right way produces some surprising results.

As Evans [32]† explains, Mandelbrot [16] began with the equation $z_{t+1} = z_t^2 + c$ where z_k and c are complex numbers and t is an integer. The *Mandelbrot set* is the set of values c for which z does not diverge with t. This set forms the beetle-like shape shown on page 45 of this volume. This appears to be a large shape surrounded by 'dust'. This prompted the question as to which specks of dust are actually connected to the main shape. In 1982 Douaday and Hubbard demonstrated a remarkable mathematical result: they are *all* connected to the main body by what Mandelbrot is said to have called "devil's polymers" [5]. The proof is summarised in [18]. In this respect it is worth noting that the computer has stimulated questions which have been answered in a way acceptable to conventional mathematicians.

Evans [32]† discusses how the Mandelbrot set is related to the work of Fatou and Julia at the turn of the century. Of course they did not have computer graphics to stimulate and guide their work, and it lay dormant for many years before the recent computer-inspired revival of interest in the subject following Mandelbrot's work.

The recent interest in chaos began in the early nineteen sixties when Lorentz was studying weather systems using the equations $dx/dt = \sigma(y-x)$, $dy/dt = rx-y-xz$, and $dz/dt = xy-bz$ where t, x, y, and z are real variables, and σ, r, and b are three real parameters. In the literature they are often set to the values $\sigma = 10$, $b = 8/3$, and $0 < r < \infty$. Lorentz discovered by accident that his system of equations diverged rapidly from what appeared to be the same starting values (in fact they were slightly different). The interesting features of the *Lorentz Equations* are (a) that they are very sensitive to the initial conditions, and (b) that they are bounded. In other words, if one plots the points (x, y, z) corresponding to increasing values of t they form a *trajectory* in three dimensional space. The trajectories passing through any two very close points (x, y, z) and (x', y', z') at t begin to diverge very rapidly as t increases. However, all the trajectories stay within a bounded region of space called a *strange attractor*. This mixture of divergence with boundedness is only possible by the trajectories repeatedly 'folding' over themselves. These ideas and their mathematics are explained and developed in more detail by Arrowsmith [26]†.

Although fractals and chaos have caught the popular imagination, computer graphics is having a profound affect on other areas of mathematics. Not surprisingly the mathematics underlying computer graphics has been considerably developed over the last the last twenty years. This includes B-spline and Bézier curves and surfaces used in CAD and surface modelling [32]†.

Pictures have always been used to explore and communicate mathematical concepts. Since the theory underlying computer graphics is essentially mathematical, computers are excellent at producing pictures of mathematical objects. For this reason alone one would expect computers to have a major impact on mathematics. The interaction with computers is often improved by graphical interfaces, which give the possibility of seeing and manipulating graphical representations of the objects and spaces under investigation. Certainly commercial packages to assist mathematicians in their work increasingly put a great deal of effort into their graphic display capabilities. Also we know that *teaching* mathematics is greatly assisted by students being able to interact with images of the objects they are studying [12].

There can be little doubt that the ability of computers to communicate ideas and results by graphical means is a major dimension of the impact they are having on mathematics.

3.3 Mathematical Modelling, Simulation, and Complex Systems

Twenty years ago I worked as a graduate student attempting to build models of complex social systems using combinatorial concepts from algebraic topology. Collecting, *storing*, and *analysing data* were considered to be essential in that enterprise, and remain so today [37]†. The mathematics itself progressed in a fairly traditional way, but the *motivating force* which drove the enterprise forward was the *feedback* from the computer analysis of particular systems. Arguably, that research enterprise reflects a traditional mixture of applied and pure mathematics in the context of scientific investigation of physical and social systems. This mathematical activity would have been stillborn without computers to store data, to search for particular combinatorial structures, perform numerical calculations, and present the results in comprehensible ways. This particular anecdotal experience generalises: interaction with a computer allows one to make tentative hypotheses, test them on the machine, and either modify them or take them forward to the next tentative hypothesis. At some stage, the resulting sequence of unrejected hypotheses is ripe for synthesis and development into mathematical theory.

Seidman writes that "The process of model development first involves the use of existing mathematics, followed by the construction of new mathematical tools, which are finally abstracted and incorporated into mathematics. Although this process is beyond observation for physical systems, and is difficult to observe for social systems, it can be directly examined for computational systems." and illustrates his point by considering the mathematics of complex parallel computing machines [43]†.

If one believes that pure mathematics and applied mathematics are essentially complementary, then the impact of computers on our ability to construct useful mathematical descriptions of complex systems can be seen to be dramatic and significant.

In physical science and engineering, systems of equations can be investigated in ways that were impossible without the fantastic computational power of modern computers. That this can lead to new mathematics is amply illustrated by the theory of chaos arising out of attempts to model weather systems. Not only does a new pure mathematical theory of chaos arise out of applications, but that mathematics then finds new applications. This is illustrated by Holden's application of chaos to investigate phenomena in neurology [35]†.

Computers are very widely used to *simulate* the hypothesised behaviour of complex systems. For example, Finite Element Analysis is now a standard tool in engineering which, it can be argued [37]†, *extends* a conventional continuous mathematics of very limited physical generality to a piecewise-continuous mathematics of great physical generality and *empirical validity*.

Research into discrete combinatorial mathematics has benefited particularly from computing, both from algorithms which can be used to investigate particular structures, and from the fact that many algorithms are conceived in graph-theoretic or other structural terms. Examples of applications of such structures include [25]† and [37]†.

3.4 Computers in Mathematics Teaching

Used skilfully as an aid to human teachers, computers and graphics can be very powerful pedagogic aids in mathematics. Carefully designed colour screens and animations can motivate study and give deep intuitions into complex ideas and structures in a way that is almost impossible otherwise. At any stage in learning it sometimes helps to internalise ideas by *practice with feedback*. Human teachers tend to get rather bored with this, but computers can provide interactive exercises at a level appropriate to the student and give feedback on performance. In some cases computers can 'explain' what the student has done wrong, or how to do something correctly [12].

4 Is there a mathematical revolution inspired by computing?

There are many ways of interpreting this question, but really it comes down to this: is there a revolution underway in mathematics, and if so, is it due to computers?

Cohen [29]† presents a vigorous argument that there has been no revolution in mathematics, and certainly not one that is computer inspired. As seen in Section 2, much of computer science is inspired by mathematics, and a lot of that mathematics predates computers. This point is also made by Cooke [30]†, and is generally accepted.

Ormell [40]† argues that often the scientific and technological 'consumers' of applied mathematics have been disappointed, and this has led to a 'defensive era' in mathematics. The reason, he argues, is that successful applications depend on successful computation which was a hit-or-miss affair in the pre-computer days. "The arrival of copious computing power together with a cogent account of the final *purpose* of an application enables us at last to throw off this defensive attitude and the associated pure-foremost conception of mathematics". Thus Ormell sees a new role for the mathematician in society and the need for a new form of education to prepare them for this new role. Certainly the impact of computers on mathematics is such that the *content* and the *form* of a good mathematics education thirty years ago are likely to be different from what we would consider to be a good mathematics education today. My own view is that a remarkable amount of the content will remain common to both, while the form will change dramatically, and there is the possibility of a corresponding increase in *quality*.

Cohen [29]† writes that "a mediocre mathematician with a computer might be able to simulate the creative powers of a top notch mathematician with pencil and paper". If that were true by itself it would reflect an important contribution to a subject in which *creative* ideas and *understanding* are everything. However, there is no extra intellectual virtue in results obtained the hard way, nor is there any reason to believe that the best mathematicians cannot enjoy amplification of their creative powers by using computers. It can be argued that computing is changing mathematics into an *empirical* enterprise, one in which the mathematician has ideas which are tested interactively by computer. It may be that mathematics has *always* been such an activity, but that the experiments were less explicit because they were hidden inside the individual's head as ideas were tried, tested, refined or rejected. Be that as it may, computers give mathematicians the possibility of trying things out, and give rapid feedback. Although they may not make an intellectual contribution to the dialogue, they can act as a interactive sounding board for ideas. In this way they can assist in the creative process, the 1% inspiration, by allowing unpromising ideas to be discarded early and saving some of the 99% perspiration.

But many of the papers in this volume implicitly or explicitly accept that mathematics is undergoing change as a result of computers. In our conference debate it was stated that the computer has been no more revolutionary than the invention of the light bulb. I cannot agree. The light bulb gave the mathematician up to twice as much time to work in established ways, and that was the extent of its impact. It had absolutely no impact on the nature of mathematical thought, it did not speed up the creative mathematical process or amplify the deductive process, it did not provoke a debate on the acceptable nature of proof in mathematics, and it did not make mathematical ideas more transparent or easy to assimilate. The papers contributed to this volume show that computers have had all these affects on mathematics and the ways mathematicians can think. It seems that computers are changing mathematics rather than just the social world within which mathematics exists. I think the *extent* of this change will be seen to be dramatic. Whether the *speed* of change is great enough for it to be called revolutionary or evolutionary depends on one's historical perspective. And of course it does not matter which words are used to describe this change, as long as you know about it and its implications.

Acknowledgements Many people suggested corrections and improvements to an earlier version of this paper. My thanks to them all.

References

[1] Benacerraf, P., Putnam, H., *Philosophy of Mathematics*, Cambridge University Press (Cambridge), 1985 (first published 1964, Prentice-Hall)

[2] Chou, S. C., *Mechanical Theorem Proving*, Reidel Publishing Company, (Dordrecht, Holland), 1988.

[3] Davenport, J.H., Siret, Y., Tournier, E. *Computer Algebra*, Academic Press (London), 1988.

[4] Fauvel, J., Gray, J., *The History of Mathematics: a reader*, Macmillan Press (Basingstoke), 1988.

[5] Gleick, G., *Chaos*, Heinemann (London), 1988.

[6] Goldblatt, R. *Topoi: The Categorial Analysis of Logic*, Studies in Logic, Volume 98, North Holland (Amsterdam), 1984.

[7] Goldschlager, L., Lister, A., *Computer Science*, Prentice Hall International Series in Computer Science (New York), 1988.

[8] Goldstine, H. H., *The computer from Pascal to von Neumann*, Princeton University Press, 1972.

[9] Hoare, C. A. R., 'The Mathematics of Programming', in *Essays in Computer Science*, C. A. R. Hoare, C. B. Jones (editor), Prentice Hall International Series in Computer Science, 1989.

[10] Hollingdale, S., *Makers Of Mathematics*, Pelican Books, (London), 1989.

[11] Hyman, A., *Charles Babbage: Pioneer of the Computer*, Oxford University Press, 1982.

[12] Johnson, J. H., 'Computers in education: principles from practice', *Computers in Adult Education*, **2**, 7-17, January 1991.

[13] Kline, M., *Mathematical Thought from Ancient to Modern Times*, Oxford University Press, (Oxford), 1972.

[14] Kreisel, G., 'Hilbert's Programme', *Dialectica* **12** 346-72, 1958.

[15] Lambek, J. L., Scott, P. J., *Introduction To Higher Order Categorial Logic*, Cambridge University Press (Cambridge), 1986.

[16] Mandelbrot B. *The Fractal Geometry Of Nature*, W. H. Freeman & Co. (New York), 1983 (first edition 1977).

[17] Neale, W., Neale, M., *The Deveolpment of Logic*, Clarendon Press (Oxford), 1984 (first edition 1962).

[18] Peitgen, H. O., Saupe, D., *The Science of Fractal Images*, Springer Verlag (New York), 1988.

[19] Revesz, G., *Lambda Calculus, Combinators, and Functional programming*, Cambridge University Press (Cambridge), 1988.

[20] Rydeheard, D. E., Burstall, R. M., *Computational Category Theory*, Prentice Hall International Series in Computer Science (London), 1988.

[21] Stewart, I., *The Problems of Mathematics*, Oxford University Press (Oxford), 1988 (first edition 1987).

[22] Turing, A. M., 'On Computable Numbers with an Application to the Entscheidungsproblem', *Proc. London Math. Soc. (2)*, **42**, 230-265, 1937.

[23] von Neumann, J. 'The General and Logical Theory of Automata', *Cerebral Mechanisms in Behaviour*, reproduced in *The World of Mathematics*, J. R. Newman (compiler and commentator), Tempus Books, (Redmond, Washington), 1988 (first edition 1956).

[24] Wu Wen-Tsün 'On The Decision Problem And The Mechanization of Theorem-proving In Elementary Geometry', *Scientia Sinica*, **XXI**, 2, 159-172, March-April 1978.

References included in this volume

[25]† Andrew, M., Bose, D. K., Cosby, S., 'Scene Analysis via Galois Lattices', 157-164

[26]† Arrowsmith, D. K., 'The Mathematics of Chaos', 47-74

[27]† Benyon, M., Russ, S., 'Variables in Mathematics and Computer Science', 284-295

[28]† Burstall R. M., 'Computer Assisted Proof For Mathematics: An Introduction Using The LEGO Proof System', 101-114

[29]† Cohen, D. I. A., 'The Superfluous Paradigm', 323-329

[30]† Cooke. J., 'Revolution, Evolution, or Renaissance', 307-322

[31]† Epstein, D. B. A., 'Word Processing Algorithms, Rewrite Rules And Group Theory', 87-100

[32]† Evans, R. V., 'The Return Of The Visual', 33-46

[33]† Fenton, N., E., 'The Mathematics of Complexity in Computing and Software Engineering', 243-256

[34]† Greenleaf. N., 'Mathematics and Informal Languages for Algorithms', 221-232

[35]† Holden, A. V., 'Computing The Unpredictable: Deterministic Chaos And The Nervous System', 75-86

[36]† Hyland, J. M. E., 'Computing and Foundations', 269-284

[37]† Johnson, J. H., 'The Mathematics Of Complex Systems', 165-186

[38]† Knott, R. D., 'Making Discrete Mathematics Executable On A Computer', 127-144

[39]† Norcliff A. 'The Wider Uses Of The Z specification Language In Mathematical Modelling', 145-156

[40]† Ormell C. 'The End Of The Defensive Era In Mathematics', 297-306

[41]† Piper, F., 'Cryptography - The Catalyst', 21-32

[42]† Rayward-Smith, V. J., 'The parallel computation hypothesis and its application to computer science', 233-242

[43]† Seidman, S. B., 'The Mathematics Of Complex Systems', 189-203

[44]† Simpson, D., 'A Euclidean Basis For Computation', 205-210

[45]† Sossai, C., 'The Necessity of Uncertainty', 211-220

[46]† Tse, K. L., Whitty, R., 'The Mathematics of Calibration', 257-266

[47]† Yang Lu 'A New Method Of Automated Theorem Proving', 115-126

PART II

Computers in Mathematics

Cryptography - The Catalyst

Fred Piper
Department of Mathematics, Royal Holloway & Bedford New College
University of London, Egham Hill, Surrey, TW20 OEX

1 Introduction

In this paper we discuss some applications of mathematics to cryptography. It was, after all, cryptographic requirements which inspired the revolution in computing.

The most commonly recognised application of cryptography is the use of encryption to 'scramble' messages prior to transmission or storage. The effect of this encryption is to make the resultant data unintelligible to any eavesdropper. However the provision of secrecy is no longer the sole application of cryptographic techniques. Indeed it is probably not even the most common. In addition to the 'traditional' military and government usage, most financial institutions now use cryptographic techniques to provide at least one of the following:

1. Message integrity

2. User verification

3. Digital signatures

4. Access control.

The provision of message integrity enables the recipient of a transmitted message to be able to detect any (deliberate) alteration to it during the transmission. It is required when, although secrecy is not important, it is crucial that the contents of a message cannot be altered by a fraudster. (An obvious example is a routine financial transaction between two parties where, although the amount may not be confidential, it is crucial that no-one can alter the value of the transaction). Clearly a standard parity check will not provide the necessary protection since anyone who deliberately changes the contents of the message needs only alter the parity check appropriately to avoid detection.

Figure 1 illustrates a typical cipher system. Thus the enciphering algorithm may be regarded as a family of functions and the enciphering key $k(E)$ as a method of selecting one of them. If we let $f_{k(E)}$ denote the enciphering function determined by $k(E)$ then we have $\mathbf{c} = f_{k(E)}(\mathbf{m})$. Similarly the deciphering algorithm may be regarded as the family of inverse functions and, if $k(D)$ is the deciphering key corresponding to $k(E)$, $\mathbf{m} = f_{k(D)}(\mathbf{c})$.

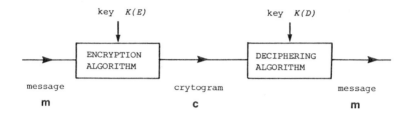

Figure 1. A typical cipher system

The fundamental requirement of this type of system is that knowledge of c without $k(D)$ should not enable an eavesdropper to determine m. A further requirement is that knowledge of one or more pairs c, m with $c = f_{k(E)}(m)$ should not enable an attacker to determine $k(D)$. If this latter requirement is not satisfied then, once he has determined $k(D)$, the eavesdropper will be able to decipher all further cryptograms which are enciphered using the key $k(E)$.

There are two fundamentally different types of cipher system:

1. *Conventional* or *Symmetric* systems where $k(D)$ is easily obtained from $k(E)$

2. *Public* or *Asymmetric* systems where it is computationally infeasible to determine $k(D)$ from $k(E)$

In any cipher system where the deciphering algorithm is known, it is clearly crucial that $k(D)$ be kept secret. In fact in this situation it is only the secrecy of $k(D)$ which is protecting the message. Thus an immediate consequence of the definition is that the security of conventional systems relies on *both* $k(E)$ *and* $k(D)$ being kept secret. However the transmitter and receiver must agree on the keys to be used before they begin a conversation. The distribution of the keys throughout a secure network is often a complex issue and, in general, the management of the keys is one of the major 'headaches' for the designer of a secure conventional system.

This is probably the appropriate place to point out that cryptography is not a branch of mathematics. It is a subject which is of great interest to many mathematicians and there are many areas of mathematics which are relevant to it. It is, however, a multi-discipline subject. Classical cryptographic procedures include:

1. Designing a cipher algorithm;

2. Deciding how it is to be used;

3. Incorporating it into the existing communications system;

4. Devising a key management scheme.

Of these four procedures, only the first is mathematics.

2 Exhaustive Key Searches

We have already observed that, if it is assumed that the deciphering algorithm is known, the 'protection' of a message m relies solely on the secrecy of the deciphering key $k(D)$. Although there are obvious advantages to keeping the algorithm secret, it is generally accepted that the security of a system should not depend on the secrecy of the algorithm. (We will say more about this when we discuss the Data Encryption Standard). If it is accepted that the algorithm is known then an interceptor is presented with an obvious form of attack. He can try all possible keys and attempt to determine the correct one. This form of attack is called an exhaustive key search and is normally achieved by some process of elimination. Of course, in order to launch an exhaustive key search attack, the cryptanalyst needs some way of eliminating incorrect keys. This is usually either the redundancy in the language or knowledge of some known plaintext/ciphertext pairs.

One obvious consequence of the existence of exhaustive key searches is that the number of keys must be large enough to withstand this type of brute force attack. In order to show the effect that increasing computer power is having on the size required for a key space we include some examples.

There are about 3×10^7 seconds in a year, i.e. about 2^{24} seconds. The following table shows approximate key size needed to withstand an exhaustive key search for a year. Note that we are assuming that the key is a bit string and that, if the length is k, then all possible k-bit strings are potential keys, i.e. there are 2^k keys.

Type of Attack	Key Size
A human effort at 1 key/sec	25 bits
A processor at 10^6 keys/sec	45 bits
1000 such processors	55 bits
10^6 processors/chips at 10^6 keys /sec	65 bits

The calculations involved in obtaining this table are, of course, very simple. However it emphasises the fact that, given today's computing power, systems need very large key spaces. It is no longer appropriate to rely on mathematical subtlety to defeat the cryptanalyst. The time taken for an exhaustive key search gives an upper bound for the security of a system. If the number of keys is sufficiently large then it might be agreed that a 'good' algorithm is one for which the fastest form of attack is the exhaustive key search.

3 Simple Substitution Ciphers

It is, of course, important to observe that, although necessary, a large key space does not guarantee security. The classical counter example is provided by simple substitution ciphers of the English language, i.e. ones in which permutations of the alphabet are used. Here the number of keys is 26! which is about 10^{26}. However, as the following discussion shows, such ciphers are well known to be totally insecure.

First a simple exercise to illustrate their weakness and to wake you up. Find the message in the following cryptogram which is the result of using a single substitution cipher on a sentence written in English:

C XU X UXT

This example is, of course, artificially chosen. However the English language has well known statistical properties which can be exploited by a cryptanalyst. The following table shows the expected frequency distribution of the letters in standard English text.

LETTER	NUMBER	LETTER	NUMBER
A	8.167	N	6.749
B	1.492	O	7.507
C	2.782	P	1.929
D	4.253	Q	0.095
E	12.702	R	5.987
F	2.228	S	6.327
G	2.015	T	9.056
H	6.094	U	2.758
I	6.966	V	0.978
J	0.153	W	2.360
K	0.772	X	0.150
L	4.025	Y	1.974
M	2.406	Z	0.074

If a simple substitution cipher is used then each letter is replaced by a unique one and the frequency of a letter in the text will be identical to the frequency of the replacement in the cryptogram. Thus, provided the cryptogram is sufficiently long, a simple frequency analysis will go a long way towards breaking the cipher. Figure 2 shows a typical histogram obtained from a cryptogram where a simple substitution cipher was used.

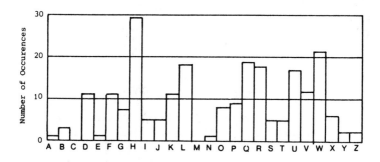

Figure 2. Letter Frequency Table

From this histogram it would be reasonable to guess that H represents E and W represents T. This would then be confirmed' if there were a popular triple W ? H, since this would represent the word THE. Simple reasoning (plus luck) is usually sufficient to deduce the message.

4 The One-Time-Pad

There is (essentially) only one unbreakable cipher system; the *one-time-pad*. For this discussion we will assume that the message is in binary form, although it must be emphasised that this is not necessary. The key is now a random bit-sequence whose length is the same as that of the message and the mixer could be an exclusive-or gate, i.e. the sequences are mixed by modulo-2 addition of the corresponding bits. (Although we do not want to give a formal definition of either 'random' or 'unbreakable' we note that knowledge of some corresponding plaintext/ciphertext bits give knowledge of the corresponding bits in the random sequences. However since a random sequence is unpredictable this does not enable the attacker to deduce any further bits in the random sequence and thus he cannot determine any more of the message).

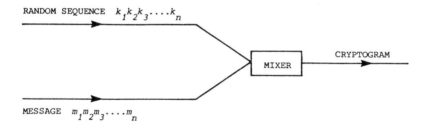

Figure 3. The One-Time-Pad

Unfortunately the key management problems make this type of system unusable in most applications. Note, for instance, that the receiver needs the same random sequence as the transmitter. Clearly he cannot generate it independently of the transmitter and thus the transmitter must send it to him. In most situations protecting the random sequence is as hard as protecting the message!

5 Theoretical Versus Practical Security

If a system does not use the one-time-pad then there is a sense in which it can be broken. However in practice breaking it may be extremely difficult. This simple observation leaves us to consider various aspects of theoretical and practical security. The most relevant remarks to make here are:

1. A theoretically secure system may not be secure in practice, e.g. the one-time-pad (because of the key distribution problem)

2. A theoretically breakable system may be secure in practice, e.g if breaking it requires sufficient memory or operations to render the process computationally infeasible.

The mention of the term computationally infeasible needs some qualification. We begin by stressing that most systems have a cover time, i.e. a maximum time for which secrecy is required. Thus the term 'computationally infeasible' can frequently be replaced by 'computationally infeasible within the cover time'.

In order to assess the time needed to break a system it is necessary to have some idea of the mathematical problem which the attacker must solve to break it and to know the time complexity function of that problem. (Of course other factors are also relevant, but we will concentrate on the time complexity) The following simple table illustrates the well known fact that exponential functions increase significantly faster than polynomials and, without any need for explanation, shows the meaning of computationally infeasible! (Note: to obtain the figures it was assumed that the computer performed 10^6 operations per second).

TIME COMPLEXITY FUNCTION	INPUT LENGTH		
	10	30	50
r	10^{-5} secs	3×10^{-5} secs	5×10^{-5} secs
r^3	10^{-3} secs	2.7×10^{-2} secs	1.25×10^{-1} secs
2^r	10^{-3} secs	17.9 mins	37.7 years
3^r	5.9×10^{-2} secs	6.5 years	2×10^8 centuries

6 Stream and Block Ciphers

In an attempt to mimic the one-time-pad, cryptographers introduced the concept of a stream cipher. From the diagram it can be seen that the random sequence of the one-time-pad has been replaced by a sequence which is systematically generated. This has the disadvantage that the keystream sequence is definitely not random. However this is offset by the considerable practical advantage that the transmitter and receiver can simultaneously generate the same sequence.

Stream ciphers are widely used and the design of 'good' keystream generators is of great interest to mathematicians (see [12] for an excellent treatment). We do not intend to discuss this topic here, but merely to answer the following simple question: How secure is a stream cipher?

If a stream cipher is used to protect a message then there is a positive integer n such that as soon as an interceptor knows $2n$ consecutive bits of message/cryptogram he can determine the entire message. Furthermore he achieves

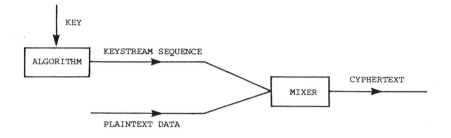

Figure 4. A Stream Cipher

this by inverting an $n \times n$ matrix.

Since the one-time-pad is not being used it is not surprising that such a statement is possible. It means that the user must either

- make n so large that it will take too long to invert the matrix, or
- change the key before the interceptor has discovered $2n$ consecutive bits, or
- both of these.

Note that the first appears to be an imprecise statement (or, more precisely, it is an imprecise statement!). However the term 'too long' can be given a precise value if the cover time of the system is known. Once this is done then knowledge of the complexity of matrix inversion, plus an estimate of the attacker's computing power, will enable us to attach a numerical value to the term 'so large'.

An alternative to the stream cipher approach is to divide the message into 'blocks' of s bits and then encipher blocks. The system is then called a *block cipher*. A typical value for s is 64. Thus a key for a block cipher may be regarded as determining a permutation of the integers $\{0, 1, ...2^s - 1\}$, or a permutation of the vectors of $V(s, 2)$ etc. This enables the cryptographer to utilise many different branches of mathematics.

If K is the key space, M the message space and C the cryptogram space, then a cipher system is a set of mappings $M \times K \to C$. In order to be able to analyse the system it is tempting to choose mappings from well studied mathematical families. However attempts to choose mathematically elegant systems usually result in insecurity and the most popular block cipher is probably the Feistel cipher (see Figure 5).

A typical Feistel cipher (see [1]) will have 16 rounds by which time it should be clear that, for almost any 'reasonable' choice of f, the cryptogram will not resemble the message. The effectiveness of such a cipher is (essentially) assessed by statistical tests to check for the lack of (obvious) relationships between input, output and key.

Public domain cryptography did not really exist until the 1970's and it was during that decade that financial institutions began to appreciate the need for data

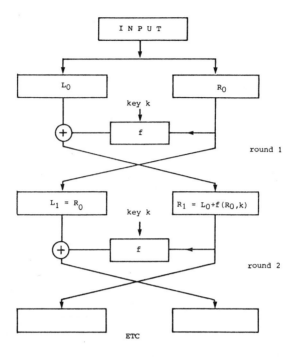

Figure 5. A Feistel Cipher

security. During this same period the academic community also began to take an interest. Two of the most significant cryptography publications in the 1970's were:

1. Federal Information Processing Standards, Publication 46 (1977)

2. New Directions in Cryptography: Diffie & Hellman, Trans IEEE Information Theory, **67** (1976)

(1) contains details of the Data Encryption Standard while (2), (possibly the most appropriately named paper ever written!) , contained two brilliant innovative ideas which were to have a long lasting effect on the later development of cryptography.

7 The Data Encryption Algorithm

This is a Feistel cipher with 16 rounds. It has block size 64 with a 56-bit key. The FIPS publication contains complete details of the cipher. It became an ANSI standard algorithm and is widely used by financial institutions.

DES has been under public scrutiny for nearly 15 years and many academics have tried (without any obvious success) to determine the design criteria behind its components and to find an attack which is faster than an exhaustive key search. Clearly the users of DES have accepted the general principle that the security of a system should not rely on the secrecy of the algorithm!

Despite its continued use by the financial industry, there is general acceptance of the fact that current computing capabilities make a key size of only 56 bits seem small. Thus the use of DES has to be accompanied by strict key management procedures to insure against the exposure of keys which are either currently being used or whose value might be exploited by an attacker.

8 New Directions in Cryptography

In this classic paper Diffie and Hellman:

1. Proposed a key exchange scheme based on modular exponentiation;

2. Suggested the use of systems where knowledge of a key k does not imply knowledge of its inverse.

8.1 The Diffie-Hellman Key Exchange Scheme

This scheme is based on the following assumption:

If F is any finite field and a is a primitive element of F then, provided F is large enough, $x \rightarrow a^x$ is not feasibly invertible.

The protocol by which two parties A and B agree on a common key is:

They agree on a primitive element a of a large finite field F.

A thinks of a (large) integer x and send a^x to B.

B thinks of a (large) integer y and send a^y to A.

NOTE: A and B do not disclose the values of x or y to anybody and, in particular, not to each other.

A raises a^y to the power x while B raises a^x to the power y. Clearly both A and B now have the same value, i.e. a^{xy}. Furthermore, provided our initial assumption is correct, no interceptor will know this value.

Note that since A does not know y and B does not know x, neither party could predict the agreed value in advance. Thus this is definitely a key exchange scheme and not a method for securing messages. Note also that A and B need to be able to authenticate each other.

8.2 Public Key Systems

The other innovative idea in the Diffie-Hellman paper was the concept of a *public key system*. For a public key system an enciphering algorithm is agreed and each would-be receiver publishes the key which *anyone* may use to send a message to him. Thus for a public key system to be secure it must not be possible to deduce the message from a knowledge of the cryptogram and the enciphering key. Once such a system is set up, a directory of all receivers plus their enciphering keys is published. However the only person to know a given receiver's deciphering key is the receiver himself.

For a public key system, encipherment must be a 'one-way function' which has a 'trapdoor'. The trapdoor must be a secret known only to the receiver. A one-way function is one which is easy to perform but very difficult to reverse. A trapdoor is a trick on another function which makes it easy to reverse the function. Of course as soon as a trapdoor is introduced the attacker may decide to concentrate on finding the one-way function rather than trying to invert the one-way function. There are three main types of public key system (see [14]):

1. Diffie-Hellman type system

2. RSA systems and their derivatives

3. Systems based on NP-complete problems.

9 The RSA System

As an example of a public key system we list the celebrated RSA system. To implement an RSA system we publish integers n and h where $n = pq$ (p and q large primes) and h is chosen so that $(h, (p-1)(q-1)) = 1$. If the message is an integer m with $0 < m < n$ then the cryptogram is $c = m^h \pmod{n}$. The primes p and q are 'secret' (i.e. known only to the receiver), and the system's security depends on the fact that knowledge of n will not enable the interceptor to work out p and q.

Since $(h, (p-1)(q-1)) = 1$ there is an integer d such that $hd = 1(mod(p-1)(q-1))$.

(NOTE: without knowing p and q it is 'impossible' to determine d. This is because the modulus, i.e. $(p-1)(q-1)$ will be unknown)

To decipher raise c to the power d. Then $m \equiv c^d \ (\equiv m^{hd}) \bmod n$. The system works because if $n = pq$, $a^{(p-1)(q-1)+1} \equiv a(mod\ n)$ for all a.

One of the main drawbacks of using RSA is that it requires considerable processing power and is slow to implement. However it should be noted that the computation of a^x requires no more than $2\ log_2 x$ multiplications. There are many ways of trying to minimise the number of multiplications. The following example illustrates a method where the expected number is about $1.5(log_2 x)$. However this can be improved.

Example

$$23 = 10111$$

$$a^{23} = (((a^2)^2 a)^2 a)^2 a$$

This requires 7 multiplications, In general the number of multiplications depends on the number of 1's in the binary expansion of x.

Nevertheless the time needed for an RSA encryption makes it unsuitable for data encryption and its two main applications are to provide digital signatures and to provide key encryption to protect keys during their distribution through a network which uses a conventional cipher system.

10 Identification

We end this paper with a brief discussion of the *Fiat-Shamir Identification Scheme* (see [6]).

Here the idea is for a prover A to establish his identity to a verifier B in such a way that, no matter how often the procedure is followed, no one (not even B) obtains any knowledge which will enable him to impersonate A. In order to identify himself, A establishes knowledge of some secret information. The security of this particular scheme relies on the following:

If we have an algorithm for computing square roots modulo n then we can factorize n.

We will illustrate the scheme by giving a protocol whereby A proves that he knows the value of a specific square root modulo n without revealing the actual value.

If the protocol is repeated k times the probability of an 'impersonation' is 2^{-k}. So if $k = 20$ the chances of an impostor 'passing' is about one in a million. In order to illustrate the scheme we will assume the use of a smart card.

10.1 Setting up the simplified scheme

[Step 1] Express identity I in binary.
[Step 2] Using a suitable hashing function f find a small integer c such that $f(I, c) = v$ is a square modulo n.
[Step 3] Compute u such that $u^2 \equiv v (mod\ n)$
[Step 4] Store u securely on the card.

NOTE: I, c, f (and hence v) are public. n is also public. However knowledge of the factors of n is needed to compute u.

10.2 The Identification Protocol

[Step 1] Card transmits I, c, n to terminal.
[Step 2] Card generates arbitrary r and computes $x \equiv r^2 (mod\ n)$
[Step 3] Card computes $y \equiv v/x (mod\ n)$
[Step 4] Card transmits x and y to terminal.
[Step 5] Terminal selects either x or y and asks for its square root $(mod\ n)$.
[Step 6] Card provides answer.

NOTE: If the card does not know u there is a 50% chance that it will provide correct answer. Thus if this protocol is carried out k times, the chances of a fraudster passing is 2^{-k}.

References

[1] Beker, H. and Piper, F., *Cipher Systems*, Van Nostrand Reinhold

[2] Beker, H. and Piper, F., *Secure Speech Communications*, Academic Press.

[3] Davies, D. W. and Price, W. L., *Security for Computer Networks*, Wiley.

[4] Denning, D., *Cryptography and Data Security*, Addison Wesley.

[5] Diffie, W. and Hellman, M. E., (1976), New Directions in Cryptography, *A Trans. Inform. Theory*, IT-22, Vol. 6, 644-654.

[6] Fiat, A. and Shamir, A., (1987), How to Prove Yourself: Practical Solutions to Identification and Signature Problems, *Advances in Cryptology. CRYPTO 86.* Lecture Notes in Computer Science 263, 186-198.

[7] Koblitz, N., *A Course in Number Theory and Cryptography*, Springer-Verlag (New York)

[8] Konheim, A. G., *Cryptography: A Primer*, Wiley Interscience.

[9] Kranakis, E., *Primality and Cryptography*, Wiley & Teubner.

[10] Meyer, C. H. and Matyas, S. H., *Cryptography*, Wiley Interscience.

[11] National Bureau of Standards, (1977), *Data Encryption Standard*, Federal Information Processing Standards Publication 46.

[12] Rueppel, R. A., *Analysis and Design of Stream Ciphers*, Springer-Verlag (Berlin).

[13] Seberry, J. and Pieprzyk, J., *Cryptography: An Introduction to Computer Security*, Prentice Hall.

[14] Welsh, D., *Codes and Cryptography*, Oxford University Press.

The Return of the Visual

R. V. Evans

ICL, Kings House, 33 Kings Road

Reading, RG1 3PX

Abstract

The rapidly growing field of computer graphics has an obvious impact on computing, but is having an increasing effect on mathematics. The roots of computer graphics are clearly mathematical, involving simple Euclidean geometry, but the current striving for visual realism is having a far wider mathematical impact. The requirements placed on computer graphics by its users are causing it to turn increasingly to mathematics for answers to its problems, providing new impetus. Whereas the power of the graphical computer, as a tool for experimentation, is enabling striking progress in hitherto dormant areas.

1 Introduction

In the early days of mathematics, much of it was used to solve problems people saw in the world about them: problems about the shapes and sizes of objects, about distances and navigation, about time and the stars. As mathematics grew and matured, mathematicians were forced to dig deeper, and to be more abstract, in order to understand the foundations and the general principles of their subject. But this took them, slowly at first, away from the plane of the world we know, through non-Euclidean geometry, finite arithmetic and transcendentals, but ultimately to the unimaginable, where intuition is all but blind. Pure mathematics was becoming very rarified, and to the outsider, uninteresting. The beauty that mathematics has always had was becoming hard to see and often austere.

Whole areas of mathematics almost seemed to be in danger of dying from neglect, geometry had all but disappeared from school curricula, when a change worthy of Catastrophe Theory came over the whole process, short-circuiting the gaps between branches of mathematics, and reversing the trends. That change is a consequence of the emergence of the field of computer graphics. The computer industry as a whole has been painfully slow to realise either the inevitablility or the power of graphics, but I like to think that mathematicians as a whole have not been so foolish.

Computers are good at doing exactly what they are told, the fact that computers so often do stupid things has a simple and well known explanation. If I tell my non-graphical computer to do a million calculations and print out the result, it will obediently do so. Unfortunately the result is likely to be either a list of a million answers, or a single answer. If the answer to my ultimate question is 42, then I shall not be much the wiser. If the answers take me a year to read, then I shall probably lose patience, or at best have forgotten the start long before I reach the end.

Now graphics changes this; it gives you information in a form that you can take in at your own rate, and to your own limits. Graphics supports a quick glance, or a detailed scrutiny. My million answers may be in the form of a graph, or a contour map, or something else, but they will be in a form that can take advantage of the power of the eye and the brain, and be rapidly and flexibly assimilated. How does this involve mathematics? In two ways; by providing *impetus*, and by *enabling*.

Impetus is provided by the problems that must be solved to allow computer graphics to satisfy the demands of people, who have seen what it can do, and want a lot more. Enablers are provided by the ability of computer graphics to combine powerful computation and display capabilities, providing a powerful experimentation engine.

Computer graphics is therefore producing a supply of problems of a highly visual nature, and at the same time providing a versatile visualisation engine, capable of turning the previously unimaginable into pictures. This combined effect is revolutionising mathematics by causing a return to the visual, through a whole new set of problems and methods.

2 Curves

The world about us is perceived largely through boundaries, either surfaces that we see or feel, that reflect, refract and have texture, or curves representing the edges of objects, their joins and intersections.

I remember when first introduced to differential geometry, what a feeling of power it gave. The ability to find the slope, the curvature, inflexions, maxima and minima of a curve; and all those lovely shapes, lemniscates, cardioids and cycloids, things with loops and cusps. I felt I could have created any shape I wanted, but I had not learned that analytic ability does not necessarily imply synthetic ability.

Now Computer Graphics had to solve the problem of making curves 'to order' in earnest. Draughtsmen liked the power that early CAD systems gave, but were

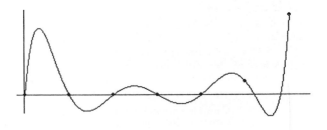

Figure 1. A Polynomial of degree seven

The Return of the Visual

not going go be satisfied with straight lines, or even conics. They were used to nice bendy bits of metal, wood or plastic that they could anchor at chosen *control points* and use to produce smooth curves.

The naïve first thought would be this: "given a set of points the curve should go through, find a nice polynomial of appropriate degree, with the right number of coefficients to solve for". Sadly, polynomials through n-points can oscillate quite wildly, and the more points you apply to control them, the more they tend to oscillate between those points. Figure 1 illustrates this. The seven dots are the *control points*, but the curve $y = x(x - .5)(x - 1)(x - 1.5)(x - 2)(x - 2.6)(x - 2.9)$ oscillates more than might be expected. Also, the form $Y = f(x)$ is hopelessly restrictive, depending on the orientation of the axes; and so the parametric form $x = x(t)$, $y = y(t)$ is more useful.

One clue as to how to proceed comes from the draughtsmen. The physics of their bendy strip is amenable to analysis and can be represented by a parametric cubic between control points, with adjacent cubic segments continuous in first and second derivatives. This leaves a couple of degrees of freedom at the ends, which can be specified by any of:

- curvature = 0 at ends;
- slope defined at ends;
- curvature the same at an end and the next-to-end point.

These piecewise cubics, called 'splines' after the draughtsman's tools, give quite good curves, but have the problem that any change in a control point has side effects throughout the curve, meaning that the whole curve will have to be recalculated, and that the movement of a point at one end may have undesirable effects at the other end.

2.1 Approximations

Another approach with useful consequences is to relax the constraint that the curve must pass through the control points, while still ensuring they exert control. Bézier curves do this using the ideas of *blending* or *basis function*, and making a general point on the curve a weighted average of the *control points*.

A Bézier curve based on four control points (P, Q, R, S) provides a good example; it is a cubic which passes through the first and last points P and S, and is tangential at P to PQ, and at S to RS.

In this case the general point is given by:

$$Z(u) = (1-u)^3 P + 3u(1-u)^2 Q + 3u^2(1-u)R + u^3 S \qquad (2.1)$$

Note the blending functions, the coefficients of the control points, are binomial coefficients. This is true for the general Bézier curve on n control points.

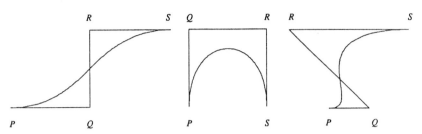

Figure 2. Bézier curves on four control points

Bézier curves with 4 control points have several useful features:

- The 4 points form a *convex hull*, or *control polygon*, within which the curve always lies.
- The tangents are easy to manipulate and match.
- They never oscillate wildly.
- They are defined in terms of points, not tangents, and convenient for interactive use.

The next step is the B-spline. This is a generalisation of Bézier curves, to avoid the one weakness that Bézier curves are still globally dependent on the control points. The B-spline is made to depend only on a number of neighbouring control points, by the fact that it is built with blending functions that are zero outside a prescribed range. With B-splines, we can make the 'interior' blending functions, those sufficiently far from either end, position independent.

The effect is that B-splines are still continuous in the first and second derivatives, but allow local control, and are also good for closed curves. For detailed discussion of Bézier curves and B-splines, and other curves, see [5].

2.2 Beta Splines

A more recent development is the Beta-spline [2]. This takes advantage of the fact that B-splines are 'over-compliant' and provides some extra control parameters to the shape of the curve.

B-splines are continuous at the joins in the first and second derivatives, thus the tangent and the curvature vectors across a join are the same in magnitude and direction. (remember that the curves are parametric, providing the notion of the 'speed' at which the curve is traversed with respect to the parameter.)

Beta-splines differ in that only the unit tangent and curvature vectors need be continuous, this provides two extra parameters, beta-1 and beta-2, termed bias and tension. *Bias* controls the change in velocity over a join, a large bias causes the curve to hug the line from the corresponding control point to the next one. A large *tension* causes the join to move towards the control point.

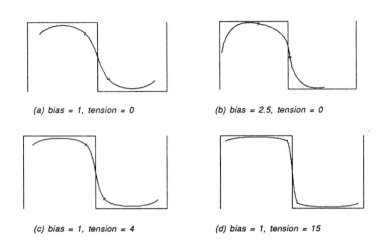

Figure 3. Beta Splines with positive tension.

Clearly, everything that has been said about curves can be extended to surfaces, and much recent work in this area of mathematics has had to do with the generation of smooth surfaces.

Bias and tension are global properties of the beta-spline, but more recent work has been carried out on local control of bias and tension [2]. This is still a very active area of research; at the 1988 SIGGRAPH [1], there were two papers on further refinements to B-splines, one allowing very detailed interactive manipulation of B-spline surfaces: [3,7].

3 Fractals

Now curves and surfaces are easy. All we have to do is look very closely, and we can see that they are nothing more than lots of little straight lines and planes. Our traditional geometry was quite powerful enough to handle them. Our curves and surfaces were adequate for the designer of cars or planes, and given a bit of work on reflection and refraction, and some clever functions to apply texture, we can produce very realistic images of things being designed. We can have tables and chairs, bottles and glasses, but they are just a little bit 'hollow'. Our scenes are

---

[1] Special Interest Group in Graphics of the Association for Computing Machinery (ACM).

Figure 4. Successive approximations to the Cantor Set

rather too smooth and plastic looking, and if we dare try to model anything like a tree, a leaf or a mountain, then we come completely unstuck.

Fortunately, this is where some of our new mathematics comes to our aid. One of the fundamental skills of mathematicians is the ability to spot counter-examples. Without this skill, we may be doomed to a life of trying to prove untrue theorems. Now many counter-examples are contrived in the extreme, simple to define, but complex in their effects. We start by looking at three related but peculiar sets.

The Cantor Set (Fig.4) is the result of taking an interval in the Reals, say [0,1] and successively removing the middle third of each of the bits that remain.

The Koch curve (Fig.5(a)) is a similar creature, with the middles displaced instead of removed, it is a good example of continuous curve that is differentiable nowhere. The Koch curve is also called the snowflake curve, which gives us our first clue that this unnatural object is in some ways very natural.

Sierpeinski's Gasket (Fig.5(b)) is a two-dimensional version of the Cantor set, it is produced from a triangle by successively removing a middle triangle from the initial and then the remaining triangles.

Returning to our trees, leaves and mountains; in the early days of computer graphics, these were often drawn as lollipops, polygons, and triangles. Alternatively, they were drawn by laboriously digitising hundreds or thousands of points but reflected the mathematicians' inability to handle natural complex objects con-

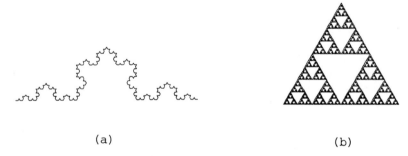

Figure 5. (a) The Koch Curve. (b) Sierpeinski's Gasket

vincingly. With hindsight we can now say that if our use of mathematics can only deal with natural objects by over-simplifying them or over-complicating itself, then it is a poor use of mathematics.

Fortunately we were saved from this poverty when Benoit B. Mandelbrot gave us the word and the concept of the 'fractal'. The richness of this concept and its attendant mathematics are being increasingly appreciated. Reference [3] is essential reading for anyone interested in fractals as seen by the inventor of the term. Reference [5] is an excellently illustrated survey of much recent work, with plenty of mathematics and copious further references.

Fractals give us a handle on things with infinitely fine structure. By definition, a *fractal* is a set with Hausdorff dimension (defined in Equation 3.1) that is fractional, and is therefore greater than its topological dimension which must be integral.

The easiest fractals to construct, and use as examples, are those with the property of 'self-similarity'. Self-similar fractals are those that are constructed from copies of themselves, and can easily be defined recursively. Thus the Cantor set is composed of two third-size Cantor sets with a similar sized gap between them. The fractal dimension of self-similar fractal can be determined from its construction; if it is built of N parts, each scaled by a factor r, then its *Hausdorff dimension* or *fractal dimension* is given by:

$$D = Log\, N/log\, r \qquad (3.1)$$

To confirm (3.1) intuitively, a line can be formed of N parts each scaled down by a factor of N, thus $D = log\, N/log\, N = 1$. A square can be formed of N^2 parts, scaled down by a factor of N, giving $D = log(N^2)/log N = 2$. Similarly a cube has dimension 3. Trying this out on the Cantor Set, it is composed of 2 parts, each scaled down by a factor of 3, so it has fractal dimension $D = log\, 2/log\, 3 \approx .63$. The Koch curve, similarly, is composed of 4 parts, each scaled down by a factor of 3, giving $D = log\, 4/log\, 3 \approx 1.26$. Sierpeinski's Gasket is composed of 3 parts, each scaled down by a factor of 2, giving $D = log\, 3/log\, 2 \approx 1.58$. There are other definitions and estimators available, which give a more rigorous view of dimension.

Figure 6. (a) A regular tree, and (b) a random tree

It is clear how a tree, for example, could be defined in terms of smaller branches that are similar to the whole tree, giving a fractal tree of unlimited detail. Such constructions however are too regular to look real, and the theory and methods must therefore be widened.

Many classes of natural object possess a form of self-similarity in essence, but with random or unpredictable variations. This is termed *Statistical Self Similarity*. It is simple to think of a generated tree with statistical self-similarity, merely by providing for random variation in the scaling or orientation of the branches, and this gives far more satisfying images. Figure 6 shows simple examples of these. Such fractals are called *random fractals*, whereas the regular ones described previously are called *deterministic fractals*.

3.1 Brownian Motion

The simple concept of Brownian motion, or random walk, provides a powerful mechanism for generating fractal images. Brownian motion in one variable is the simplest form of random fractal. In Brownian motion, a particle moves in a random way, tracing a jagged path that is fractal in appearance. The expected distance between two points on the particles path varies with the square root of the time between them, so Brownian motion over time is not strictly self-similar, as the time and distance dimensions have different scaling factors. This property is know as self-affinity.

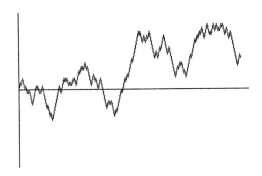

Figure 7. Brownian Motion in one variable

The concept of Brownian motion has been extended to a general case where the mean square variation in the particles position varies with the time difference to the power $2H$.

$$E|V_H(t_2) - V_H(t_1)|^2 \quad \text{varies as} \quad |t_2 - t_1|^{2H} \qquad (3.2)$$

This generalisation is called *fractional Brownian Motion*, or fBM for short. When $H = 1/2$, fBM corresponds to ordinary Brownian Motion. The fact that fBM is not self-similar makes estimating its fractal dimension less obvious, but fractals generally reduce their dimension by one when intersected with a suitable plane or hyperplane. The zeroset of an fBM is defined as the set of points where the trace intersects the t-axis, and the zeroset is found to have dimension $(1-H)$. The fractal dimension of the self-affine fBM is therefore $(2-H)$. Fractional Brownian Motion is one of the most widely used forms of fractal used for computer generation of landscapes, and many techniques have been developed to approximate fBM.

It can be seen that fractals provide a good means of generating realistically complex images. They make a little definition go a long way. It is inevitable therefore that they are under the same pressures as were splines for curve generation. How can we produce a fractal 'to order'. One solution to this is provided by fBM. Given the desired H-value, a set of points can be interpolated with a fractal curve. One feature of the resulting curve is that the large scale shape of the curve does not influence the small, which may or may not be what is desired.

3.2 Iterated Function Systems

Another technique is provided by what are called *iterated function systems* [6]. An iterated function system operates on a d-dimensional space, having a set of affine-mappings M_i and a set of associated probabilities P_i summing to one. The mappings are all contractions, having eigenvalues less than one. The system is iterated by doing a random walk from a chosen start point, choosing an i according to the probabilities, and applying the corresponding M_i to the previous point.

It is found that an iterated function system has a unique *attractor*, to which all random walks are drawn, and which they cannot leave once they have reached it. A fuller treatment of strange attractors and chaos can be found in [1] in this volume. There is an associated theorem that the attractor is the union of its images under each of the affine mappings. The probabilities P_i do not affect the iterated function system other than in the way it is traversed, but they have considerable effect on how it is drawn.

The attractors for the following two examples are the Cantor set and the familiar 'dragon curve':

$$\text{For real } x: \quad f(x) = x/3; \quad g(x) = x/3 + 2/3; \quad P = (1/2, 1/2) \qquad (3.3)$$

$$\text{For complex } z: \quad f(z) = (i+1)z/2 + 1; \quad g(z) = (i+1)z/2 - 1 \qquad (3.4)$$

A practical example of the use of this technique is the generation of the black spleenwort fern [9]. A botanical drawing of the fern was examined to determine a set of affine mappings, including a singular mapping to generate the stems, resulting in a set of four mappings which when used for the random walk produce the image of the fern.

The dimensions of the images generated by the iterated function system can be found as follows: given n affine maps and constant scaling factor, $|eigenvalue| = s$, then

$$D = \log n / \log(1/s) \qquad (3.5)$$

Iterated function systems provide another method for fractal interpolation; the resulting self-affinity of the generated curve meaning that unlike Brownian interpolation, the large scale shape is reflected in the small scale shape.

3.3 Natural Fractals

Many naturally occurring fractal patterns occur in dynamical systems. A *dynamical system* is essentially anything that has a state that changes with time, where the state at time t depends on the state at time $(t - dt)$. The state at a time in the future is therefore the result of possibly very many successive applications of the same function.

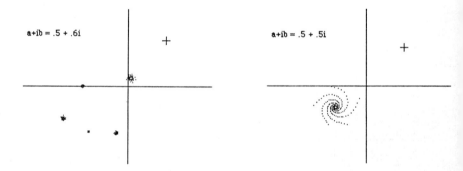

Figure 8. Orbits of z under (3.6)

Sometimes these systems are easy to predict but often they are very unstable; a minute difference in state at time t_1 resulting in a very large difference in state at time t_2 [1].

Around the start of the century, two mathematicians, Gaston Julia and Pierre Fatou, embarked on the study of repeated rational mappings in the complex plane. The study then became dormant for around fifty years, before its was resumed with the impetus and enablement provided by computer graphics.

When a mapping is repeatedly applied to a point in the complex plane, the trace may go off to infinity, it may tend to some finite value, or it may progress around some orbit. The *Julia set* is defined as the set of points whose orbit is unstable; that is where two points, arbitrarily close, have diverging orbits. As a simple example, the mapping from z to z^2 takes points with modulus less than one to zero, and points with modulus greater than one to infinity. Points with modulus equal to one, i.e. those lying on the unit circle remain on the unit circle. The Julia set for this mapping is therefore the unit circle centred at the origin. In this case the Julia set is not fractal, but in general they are of much greater complexity. The very similar mappings:

$$z_{t+1} = z_t^2 + c \tag{3.6}$$

for complex c, have far more detailed Julia sets. The orbits of z under such mappings are intricate and very sensitive to small variations in starting point, as illustrated in Fig. 8.

One way to determine the Julia set is to iterate backwards, where the trace tends to the unstable set. The inverse mappings are:

$$z_t = +\sqrt{z_{t+1} - c}; \quad z_t' = -\sqrt{z_{t+1} - c} \tag{3.7}$$

thus providing an iterated function system (as described earlier) for generating the Julia set. Figure 9 illustrates such an approximation to a Julia set.

Figure 9. Julia set drawn by Inverse Iteration

Related to any Julia set, there is a 'filled-in Julia set', more simply defined as the set of points that do not go off to infinity under the repeated mapping. In some senses the filled-in Julia sets are more convenient, they also look prettier, so for the rest of this paper 'filled-in' will be implicit.

The Julia sets of the mapping z to $z^2 + c$ take a variety of forms, but are of two distinct kinds, those that are connected and those that are (Cantor-like) totally disconnected. The proof of this dichotomy is quite neat, and I will attempt to outline it (see [6] for more details).

For any c, there is a circle centred on the origin outside which all points map towards infinity. Take this circle and apply the iterated function system-like inverse mapping. As the mapping is applied repeatedly, the resulting 'curve' will tend to

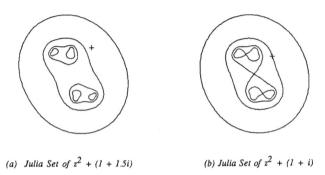

(a) Julia Set of $z^2 + (1 + 1.5i)$ (b) Julia Set of $z^2 + (1 + i)$

Figure 10. Circle mappings for a Cantor-like Julia set

the boundary of the Julia set. It can be proved that the result of an application will be one of:

- a closed circle-like curve
- a figure of eight
- two disjoint closed curves

depending on whether the point c lies inside, on or outside the curve being mapped (repectively). Figure 10 shows examples of these curves. The result is that when c lies in the Julia set, it is connected; when c lies outside the Julia set, it is Cantor-like.

3.4 The Mandebrot Set

We have followed one of the many roads leading to the Mandelbrot set, named after its dicoverer [4]. One of the definitions of the *Mandelbrot set* is that it is the set of all c in the complex plane for which repeated iterations of

$$z_{t+1} = z_t^2 + c \qquad (3.8)$$

starting at zero, do not go off to infinity. From the previous result, we see the Mandelbrot set is also the set of c values for which the related Julia set is connected. The Julia sets take on intricate forms, but the Mandelbrot set is an order of magnitude more complex. One of the more interesting features of its relation to the Julia sets is the fact that if you 'zoom in' very close to the Mandelbrot set at a point c, the image you see is very similar to the Julia set corresponding to that point. This gives a flavour of the immense variety of detail present in the Mandelbrot set.

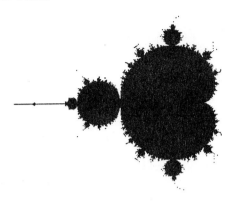

Figure 11. The Mandelbrot Set

4 Conclusion

What I have told is very much a travellers tale, I have not conducted research of any great depth into the matters discussed here, but I have seen some interesting sights, and some interesting mathematics.

Since first encountering fractals, I have been repeatedly struck by analogies in a little book by J.R.R. Tolkien, "Tree and Leaf" [8]. Tolkien's subject is 'Fairy Stories', or literary fantasy, whereas fractals seem to represent a visual fantasy. Tolkien stresses the importance of a story having the 'inner consistency of reality', whereas fractals achieve the 'inner complexity of reality'. Tolkien shows how, having met the fantastic, we can return to the natural with greater appreciation and understanding. Mandelbrot, through his work, and the research sparked off by it, has shown us the fantasy of the Julia and Mandelbrot sets; leaving us better equipped to appreciate, and handle mathematically, the fractal intricacies of the world about us.

So mathematics has turned full-circle. It was initially much inspired by the visual world, but diverged from it to establish its own foundations, and in doing so became increasingly abstract and unimaginable. Now, with the demands of computer graphics, geometrical problems have received a new lease of life. Also through the toolset provided by computer graphics, some of the concepts from the foundations of mathematics have been given, or found to possess, a visual nature. And conversely, some real-world objects that previously defied mathematical representation, have been made respectable.

Therefore, although it has returned to its visual origins, it has not been a fruitless journey. Mathematics is older, wiser, more powerful and more beautiful than when it first began.

Finally, the advance of computer graphics is bringing mathematics more into the public eye. For some time mathematics had been regarded by many people as dry and esoteric, but the association with computer graphics is changing this. The annual ACM SIGGRAPH conference is one of the most popular of the ACM conferences; with many of the papers presented being of significant new mathematical content. Broadcast media, adverts and films are involving geometrically modelled sequences and fractal landscapes, computer art is growing in maturity, and utilising (though not depending on) sophisticated mathematics.

The field of computer graphics, therefore, is having a profound effect on mathematics. It is contributing a new set of problems, a new toolkit, and a new glamorous public image. What more could mathematicians ask of a revolution?

Acknowledgements: I would like to thank Jeff Johnson of the Open University for his suggestions and encouragement during the preparation of this paper, and also Dave Marshall and Andrew Hutt of ICL for their support.

References

[1] Arrowsmith, D. K. 'The Mathematics of Chaos', in *The Mathematical Revolution Inspired by Computing*, J Johnson and M Loomes (eds), I.M.A. & Oxford University Press, Oxford, 1990

[2] Barsky, B.A. and Beatty, J.C. 'Local Control of Bias and Tension in Beta-splines', *ACM Transactions on Graphics*, **Vol.2** No.2. 109-134, 1983

[3] Forsey, D.R. and Bartels, R.H. 'Hierarchical B-spline Refinement', *ACM Computer Graphics*, **Vol.22**, No.4, 205-212, 1988

[4] Mandelbrot, B.B. *The Fractal Geometry of Nature*, W.H. Freeman and Co. New York, 1982.

[5] Mortenson, M.E. *Geometric Modelling*, John Wiley & Sons, New York, 1985

[6] Peitgen, H-O and Saupe, D. (Eds) *The Science of Fractal Images*, Springer Verlag, Berlin, 1988.

[7] Shantz, M. and Chang, S-L. 'Rendering Trimmed NURBS with Adaptive Forward Differencing', *ACM Computer Graphics*, **Vol.22** No.4, 205-212, 1988.

[8] Tolkien, J.R.R. *Tree and Leaf*, George Allen and Unwin Ltd, London. 1964.

[9] Demko, S., Hodges, L. and Naylor, B. 'Construction of Fractal objects with Iterated Function Systems', *ACM Computer Graphics* **Vol.19**, No.3, 1985

The Mathematics of Chaos

D K Arrowsmith
School of Mathematical Sciences
Queen Mary & Westfield College, University of London
Mile End Road, London, E1 4NS

Abstract

The mathematical ideas underlying the theory of chaos for both continuous and discrete dynamical systems are introduced using three key examples. The doubling map is considered to show that dynamical complexity can even arise in maps given by elementary formulae. Its behaviour is discussed by showing that iterations of the map are equivalent to a symbolic dynamic given by a shift on infinite binary sequences. The Logistic and Horseshoe maps are also studied from this symbolic viewpoint. The properties of these maps are used to highlight some of the key concepts in dynamics such as periodicity, recurrence and bifurcation as well as to show their seminal importance in giving explicit illustration of some of the usual ingredients of chaos such as sensitive dependence on initial conditions, homoclinic points, period doubling cascades and strange attractors. The basic ideas elucidated here are fundamental to an understanding of nonlinear dynamics and chaos.

1 Introduction

A *dynamical system* is a collection of states which evolve with time. The time variable t can be either continuous or discrete. Continuous dynamical systems are usually described by an ordinary differential equation

$$\frac{d\mathbf{x}}{dt} = \mathbf{X}(\mathbf{x}). \tag{1.1}$$

Here x represents the *state* of the system and takes values in the *state* (or *phase*) *space* which can be either a Euclidean space or, more generally, a *manifold M*. Manifolds are topological spaces which are locally Euclidean, e.g. a sphere, cylinder or torus. The *vector field* X gives the instantaneous direction of motion at $\mathbf{x} \in M$. For example, the dynamics of a pendulum is described on a cylinder where the circle variable, θ, denotes the angular displacement and the real variable, v, denotes the tangential speed. The system is given by $d\theta/dt = v, dv/dt = -sin(\theta)$ and can be written in the form of (1.1) with x= (θ, v) and $\mathbf{X}(\mathbf{x}) = (v, -sin(\theta))$.

The discrete time model is usually interpreted as the iteration of a function

$$x_{t+1} = f(x_t) \tag{1.2}$$

where $f : M \to M$. An example of such a system is a multi-species model of population which is measured yearly. The solution curves of (1.1) and the sequences of states which satisfy (1.2) are both called *orbits*.

Typically we impose *initial conditions* on such systems and specify a point $x = x_0$ when $t = t_0$. The goal is to predict the state x at an arbitrary time t. The basic ingredient of *chaos* is the *sensitive dependence* of the future evolution of the system on the initial conditions. As we shall see the evolution of some systems depends so precisely on initial conditions that they appear to have random behaviour although mathematically they are deterministic. Thus it is possible to have orbits which follow arbitrarily large prescribed random configurations by choosing the initial point in an increasingly constrained way (see Section 3).

Following the invention of the calculus, there was a strong belief that the solution to a problem consisted of finding functions, no matter how esoteric, which fitted the dynamic formulation and initial conditions of the problem. Thus many mathematicians devoted themselves to finding ever more sophisticated analytical methods for developing solutions to equations. This approach continued unabated for more than a century after Newton and Leibnitz and culminated in the difficulties of Fourier and Laplace in obtaining analytical solutions to various problems in Celestial Mechanics. An alternative approach was hewn out of this impasse. It took the genius of Poincaré to realise that a *qualitative* emphasis was needed if progress was to be made towards an understanding of the complexity of dynamical processes. The new approach forfeited *quantitative* detail and replaced it with an attempt to discover some of the qualitative features of solutions which may give rise to complex dynamical behaviour [27].

For example, the differential equation

$$\frac{dx}{dt} = x + \sin(x) + \tanh(x) \tag{1.3}$$

is *presumably* difficult to solve analytically. However, if we note that

$$x + \sin(x) + \tanh(x) \text{ is } \begin{cases} > 0 & \text{for } x > 0 \\ = 0 & \text{for } x = 0 \\ < 0 & \text{for } x < 0, \end{cases} \tag{1.4}$$

then we can calculate the asymptotics of the solutions with ease: all orbits $x(t)$ with initial condition $x(t_0) > 0 \ (< 0)$ satisfy

$$x \to \infty \quad (-\infty) \tag{1.5}$$

as t increases, and

$$x \to 0^- \quad (0^+) \tag{1.6}$$

as t decreases. Thus, $x(t) \equiv 0$, for all $t \in \mathbb{R}$, is said to be a *fixed point solution* which is *unstable* because of the behaviour of the neighbouring trajectories.

The Mathematics of Chaos

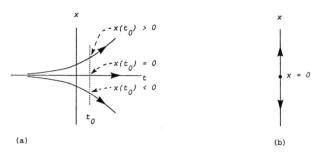

Figure 1. (a) The behaviour of solution curves to the differential equation (1.3), (b) the phase portrait for (1.3) depicting an unstable fixed point at $x = 0$.

We can represent this information by a *phase portrait* for the system described above (see Figure 1). This is a concise way of displaying the asymptotic form of the trajectory behaviour together with the fixed points and does not indicate the explicit dependence of the state x on t. When the asymptotic behaviour of a system is known from a qualitative analysis, we have a stronger framework for discussing quantitative results obtained from computation over long time intervals. Without the benefit of the qualitative analysis, the computer can only give an indication of specially selected trajectories and may indeed sometimes mislead. The latter problem would not arise in the above example where, except for the dichotomy of behaviour in any neighbourhood of the origin, there are no difficulties. If we wish to consider the behaviour of the solution $x(t)$ satisfying $x(t_0) = x_0 \neq 0$, then the same asymptotic behaviour is shared by all nearby initial conditions provided the sign of x_0 remains unchanged. Such insensitivity is not shared by all systems. Note that equation (1.3) is nonlinear and yet the asymptotic behaviour is highly predictable.

Nonlinearity does not inevitably give chaos. The certainty is that *linear systems* on $\mathbb{R}^n$ i.e.,

$$\frac{d\mathbf{x}}{dt} = \mathbf{A}\mathbf{x}, \tag{1.7}$$

$\mathbf{x} \in \mathbb{R}^n$, and $\mathbf{A}$ an $n \times n$ matrix, *never* give chaos. All such systems have a fixed point solution at the origin and stability is determined by the eigenvalues of the matrix $\mathbf{A}$. A particularly important class of linear systems consists of those which are *hyperbolic*, i.e. $\mathbf{A}$ has no eigenvalues with zero real part. For such a system, let $s(u = n - s)$ be the number of eigenvalues which have real part less (greater) than zero. Then the system (1.7) is *topologically equivalent* to the system

$$\frac{d\mathbf{x}_s}{dt} = -\mathbf{x}_s, \qquad \frac{d\mathbf{x}_u}{dt} = \mathbf{x}_u \tag{1.8}$$

where $\mathbf{x}_s \in \mathbb{R}^s, \mathbf{x}_u \in \mathbb{R}^u$. *Topologically equivalent* means that there is a bicontinuous bijective map from $\mathbb{R}^n$ to $\mathbb{R}^n$ which maps orbits of one system onto those of the other. The subspaces $\mathbb{R}^u$ and $\mathbb{R}^s$ of (1.8) are called the *unstable* and *stable manifolds* associated with the fixed point at 0 (see Figure 2). If both $u, s \neq 0$ then the fixed point is of *saddle* type.

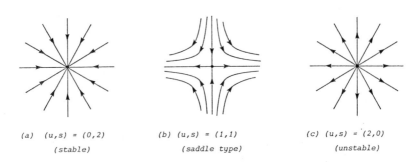

(a) (u,s) = (0,2) (stable)
(b) (u,s) = (1,1) (saddle type)
(c) (u,s) = (2,0) (unstable)

Figure 2. The various types of phase portrait for equation (1.8) when $n = 2$.

For more complex asymptotic behaviour more sophisticated systems are needed and can be obtained in two obvious ways: (i) consider systems on spaces other than $\mathbb{R}^n$; or (ii) introduce nonlinear systems on $\mathbb{R}^n$. *Recurrence* of orbits is a necessary ingredient of complexity and both of the above manoeuvres allow its introduction in less regular ways than within linear systems – the only orbits of linear systems which return arbitrarily closely on themselves are periodic orbits. To help in the discussion of recurrence we first of all note a crucial step in seeing the dynamical equivalences between discrete and continuous systems.

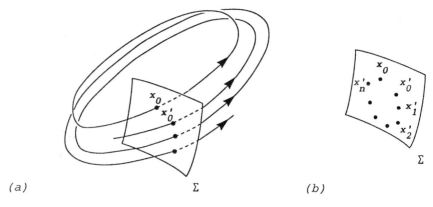

(a)   Σ   (b)

Figure 3. (a) The construction of the Poincaré local section Σ at the point x_0 of a periodic orbit γ and the behaviour of a neighbouring orbit through x'_0. (b) the fixed point orbit at x_0 and the neighbouring orbit through x'_0.

Suppose we have system (1.1) with a solution curve $x(t)$ which is *periodic*, i.e. there exists a minimum positive T such that $x(t+T)=x(t)$, for all $t \in \mathbb{R}$. Then a periodic orbit is topologically a circle in the phase portrait. Let γ denote the periodic orbit so formed and let $x_0 \in \gamma$. Now introduce a *local section* i.e. a (n-

1)-dimensional hyperplane Σ containing x_0 which is transverse to the orbit γ (see Figure 3). Let x_0' be a point of Σ sufficiently close to x_0. Then the orbit through x_0' will return to the section Σ as time increases and will pierce Σ at some point which we will call $f(x_0')$. Thus successive piercing of Σ by orbits of (1.1) gives rise to a map $f : \Sigma \to \Sigma$ defined on a neighbourhood of x_0. Moreover, the periodic orbit γ is replaced by a fixed point x_0 of f and more generally, the system (1.1) will have other periodic orbits close to γ if and only if the map f has a periodic orbit, i.e. $f^n(x) = x$ for some $n \in \mathbb{Z}$, the integers. In this case, the periodic orbit will pierce the section Σ several times before closing up. Thus the dynamics of the vector field X is now interpreted by the discrete dynamics of the map f on Σ and vice-versa. The importance of this construction was observed by Poincaré and was subsequently used extensively by Birkhoff. The map obtained in this way is known as a *Poincaré map*. Such maps are special in the sense that the are *diffeomorphisms*, i.e.

(i) f is smooth;
(ii) f^{-1} exists (just reverse the time t in (1.1));
(iii) f^{-1} is also smooth.

In the next section we will examine this correspondence between discrete and continuous dynamics in more detail.

2 Basic dynamical definitions

To emphasise the increased role of the initial condition for solutions of ordinary differential equations we introduce the *flow* notation ϕ_t to denote time t evolution. More precisely, we define $\phi_t(x_0)$ to be the state point on the solution curve of (1.1) after time t has elapsed given that $x = x_0$ when $t = 0$. Thus a flow satisfies:

(i) $\phi_0(x) = x$, for all $x \in M$;
(ii) $\phi_t(\phi_s(x)) = \phi_{s+t}(x)$; (2.1)
(iii) $\phi_{-t}(x) = \phi_t^{-1}(x)$.

The *orbit* of the flow is the set $\{\phi_t(x) | t \in \mathbb{R}\}$. Sometimes orbits escape to infinity in finite time and so the flow in (2.1) is only *local* and not defined for all t. Nevertheless, the properties in (2.1) still hold whenever they are defined. A *fixed point* of the flow satisfies $\phi_t(x) \equiv x$ for all $t \in \mathbb{R}$. The orbit through x is *periodic* if there exists a positive real T such that $\phi_T(x) = x$ and $\phi_t(x) \neq x$ for all $t \in (0, T)$. The periodic orbit's *stability* is dependent on the asymptotic behaviour of its neighbouring trajectories. Let M be a differentiable manifold and suppose $f : M \to M$ is a diffeomorphism as defined in Section 1. For each $x \in M$, the iteration (1.2) generates a sequence which is the *orbit* or *trajectory* of f. The orbit can also be tracked through negative time by using f^{-1}. Thus the orbit of f is $\{f^m\{x\}|\forall m \in \mathbb{Z}\}$. For $m \in \mathbb{Z}^+$, f^m is the composition of f with itself m-times and $f^{-m} = \{f^{-1}\}^m$. Typically, we might expect a bi-infinite sequence of points ordered by the integers $\mathbb{Z}$. However, important exceptions can occur. A point x^* is a *fixed point* of f if $f(x^*) = x^*$ and x^* is a *periodic point* of period q if q is the minimum positive integer such that $f^q(x^*) = x^*$. Thus fixed points are period-1 points. The periodic orbit is then

$$\{x^*, f(x^*), f^2(x^*), \ldots, f^{q-1}(x^*)\}. \tag{2.2}$$

Fixed and periodic points can be classified according to the behaviour of nearby orbits as is the case with flows. A fixed point x^* is said to be *stable* if, for every neighbourhood $\mathcal{N}$ of x^*, there exists a neighbourhood $\mathcal{N}' \subseteq \mathcal{N}$ of x^* such that the *forward* orbit of every point $x \in \mathcal{N}'$ is contained within $\mathcal{N}$. Otherwise the fixed point is *unstable*. If the fixed point x^* is stable then it is said to be *asymptotically stable* if there exists a neighbourhood $\mathcal{N}$ of x^* such that for all $x \in \mathcal{N}$, $lim_{m\to\infty} f^m(x) = x^*$. There are analogous definitions for flows, and Linearization Theorems for both flows and diffeomorphisms which ensure that properties of the linear derivative $DX(x^*)$ and $Df(x^*)$ at hyperbolic fixed points x^* give the topological types and hence the stability types of X and f respectively at x^*. For example, if $Df(x^*)$ has no eigenvalues with modulus equal to 1, then x^* is said to be a *hyperbolic* fixed point of f at x^*. If all the eigenvalues have modulus less than one, then x^* is asymptotically stable. Conversely, if at least one eigenvalue has modulus greater than one then x^* is not stable (i.e. it is *unstable*).

Equivalence of diffeomorphisms is defined by a conjugacy - the diffeomorphisms f and g on M are said to be *topologically equivalent* if there is a homeomorphism $h : M \to M$ which provides a conjugacy $f(x) = h \circ g \circ h^{-1}(x)$, for all $x \in M$, or

$$f \circ h = h \circ g. \tag{2.3}$$

Again, as with equivalence for flows, the images of orbits of f by h are orbits of g and so the conjugacy preserves orbits (see Section 3). For a more detailed discussion of these basic concepts and theorems, see [3] and [12].

In our discussion of discrete dynamics in Section 3, we will relax the condition that the maps f have an inverse and consider only the *forward* iterations and asymptotics.

The dynamical definitions for maps are slightly different regarding periodic orbits from those for diffeomorphisms. A map f is said to have a *forward periodic orbit* through x_0 if there exists a positive integer N such that $f^m(f^N(x_0)) = f^N(x_0)$ for some $m \in \mathbb{Z}^+$. Note that if f is a diffeomorphism, then N can be chosen arbitrarily and in particular zero. Thus for a diffeomorphism a periodic orbit repeatedly cycles a finite set of points, whereas if f^{-1} does not exist it *eventually* repeatedly cycles a finite set of points.

The qualitative approach, which stems from a discussion of the properties given above has led to the discovery of the key ingredients of dynamical chaos. When allied with the great power of modern digital computers, these features can be displayed easily for systems and maps which have a simple analytical form. First, we consider some simple maps for which the dynamical complexity is relatively easy to uncover.

3 The doubling map and symbolic dynamics

Consider the map f on the interval of real numbers $I' = [0, 1)$ defined by

$$x_1 = 2x_0 \bmod 1. \tag{3.1}$$

It is not difficult to 'solve' this equation and deduce that the n-th iterate $x_n = f^n(x_0)$ of $x_0 \in I'$ is simply given by

$$x_n = 2^n x_0 \bmod 1. \tag{3.2}$$

However, it is soon apparent that the solution is no more illuminating than the equation simply because it is not at all clear how the repeated doubling interacts with the process of reducing mod 1. The temptation at this point is to resort to the computer for help. Unfortunately, we have only a finite number, say m, of binary places stored in the machine for each number we wish to represent in $[0, 1)$ and we loose information at the rate of 1-binary place per iteration since the map effectively shifts the binary expansion by one binary place and deletes the integer part. Thus, if the computer fills out the number to m places for each iteration in a controlled way by adding 0 in the m-th place, we have zero after m-iterations. If the m-th place is filled randomly, then for each iteration we have a random number generator for binary integers. Clearly, the computer is not particularly useful here because of the exponential loss of information. The crucial feature is the way in which orbits move. More precisely, the map offers *sensitive dependence on initial conditions* [8,11]. Given any two distinct points of x, x' of I', their orbits diverge exponentially in the sense that $dist(f^n(x), f^n(x')) = 2^n dist(x, x')$ subject, of course, to the global constraint here that no two points can ever be more than distance one apart. Thus locally, orbits are diverging exponentially. However, the binary expansion offers a clue as to how we can unlock the dynamical secrets held within the formula (3.2).

Any real number $x \in I'$ can be written in the form

$$x = \sum_{n=1}^{\infty} \frac{b_n}{2^n}, \tag{3.3}$$

where $b_n = 0$ or 1. Thus we can represent each point of I' as a sequence $\sigma = \{b_n\}_{n=1}^{\infty}$. Let S denote the set of all such binary sequences. Relative to this new representation, or *coding*, of points in I' the map f takes the form of a *shift* α on binary sequences.

$$\{b_1, b_2, \ldots, b_n, \ldots\} \longmapsto \{b_2, b_3, \ldots, b_{n-1}, \ldots\}. \tag{3.4}$$

The map α is defined precisely by

$$\alpha(\{b_n\}_{n=1}^{\infty}) = \{b_{n+1}\}_{n=1}^{\infty} \tag{3.5}$$

and it is *semi-conjugate* to f by the map

$$\pi(\{b_n\}_{n=1}^{\infty}) = x = \sum_{n=1}^{\infty} \frac{b_n}{2^n} \tag{3.6}$$

The qualification *semi*-conjugacy arises from the fact that π is not a bijection and therefore does not have an inverse. The *dyadic fractions* $\{m/2^n | m, n$ positive integers$\}$ do not have unique binary expansions and thus π is not $1-1$ on the representatives of dyadic fractions in S. It is easy to check that the map π has the semi-conjugacy property

$$f \circ \pi = \pi \circ \alpha. \tag{3.7}$$

(cf. equation (2.3)). This straightforward observation allows us to examine the periodic structure of the map f. We note that the period-m periodic points of α in S are precisely those sequences which *eventually* repeat after m-digits and no fewer. Thus a period-1 point of α is given by the repeating expansions

$$\sigma = \{\overline{0}\ldots\} = \{00000\ldots\}, \tag{3.8}$$

and period-2 points are given by

$$\sigma = \{\overline{01}\ldots\} = \{01010\ldots\}, \qquad \{\overline{10}\ldots\} = \{10101\ldots\}. \tag{3.9}$$

Points which are eventually period one or two points could be obtained by delaying the introduction of the recurrences given in (3.8) and (3.9). We can immediately see that periodic points of all orders can be constructed for α in this way. It is also important to check the implications for our original map f. If $\sigma \in S$ is a period-m point of α, then $\alpha^m(\sigma) = \sigma$. The semi-conjugacy (3.7) implies that for every positive integer m, $f^m \pi = \pi \alpha^m$. It then follows that

$$f^m(\pi(\sigma)) = \pi \alpha^m(\sigma) = \pi(\sigma) \tag{3.10}$$

and so π maps α-orbits to f-orbits. Also, $x = \pi(\sigma) \in I'$ satisfies $f^m(x) = x$. Furthermore, σ is not periodic for any integer smaller than m and this implies that x is m-periodic. Thus we can deduce the periodic behaviour of f by investigating that of α. We can extract further information about f when we observe that π can be seen as a continuous function by taking the usual Euclidean metric on I' and the corresponding metric

$$d(\sigma, \sigma') = \sum_{n=1}^{\infty} \frac{(b_n - b'_n)}{2^n} \tag{3.11}$$

where $\sigma = \{b_n\}_{n=1}^{\infty}$ and $\sigma' = \{b'_n\}_{n=1}^{\infty}$ are elements of S. It is now possible to talk about the *topology* of orbits.

A key feature of chaos is the requirement that the orbits should in some sense be bound together thus making the dynamics indecomposable. This can be achieved by finding an orbit which *densely* fills out the set I'. Such an orbit is given by constructing σ as the sequence obtained by listing all symbol sequences of length 1, then of length 2 and so on for positive integers. Let σ' be any prescribed binary sequence. Given any positive integer n, by construction σ has within it the *symbol block* $\sigma'^{(n)}$ consisting of the first n symbols of σ'. Suppose this block commences at the $(k+1)-th$ decimal place of σ. Then σ' and $\alpha^k(\sigma)$ have the same first n entries and thus $d(\sigma', \alpha^k(\sigma)) < 2^{-n}$. The positive integer n was chosen arbitrarily and so the orbit of σ approaches arbitrarily close to σ'. Thus the orbit of σ is dense on I' and binds the dynamical behaviour together preventing a decomposition into closed subsets with simpler dynamical behaviour.

The dense orbit was, of course, produced to a special recipe. Obviously, the recipe is not unique as other dense orbits can be constructed by simple permutations in the ordering of the blocks which form the sequence. What is astonishing is that almost all orbits are dense. This follows from a classical result of Hardy which when paraphrased says that the set of real numbers in I' which have all possible finite sequences of '0's and '1's in their binary expansions has measure one. Using the codes for describing orbits it becomes quite pleasurable to construct orbits with specific dynamical properties. It is simple to show for example that (a) there exist orbits which move asymptotically towards any given periodic orbit of f, (b) the set of periodic orbits of f is dense in I' and (c) there exist aperiodic orbits.

Aside from the dynamical properties of the orbits the interpretation of the map f as a shift on sequences of the symbols '0' and '1' offers a clash between determinism and randomness. The point $x = \sum_{n=1}^{\infty} \frac{b_n}{2^n}$ is to the left or right side of the interval $[0, 1)$ according as b_1 is 0 or 1 and similarly for $f^n(x)$ according as b_{n+1} is 0 or 1. Thus for any random sequence of "heads and tails", or equivalently "lefts and rights", we can find an orbit of f which follows precisely that same pattern. But the map f is *deterministic* and so we have a possible conflict with *randomness*. The fallacy is settled when we realise that the determinism is assured if we know *exactly* the present (i.e. initial conditions). However, this requires infinite precision and we can only know our initial conditions approximately. In the presence of sensitive dependence on initial conditions uncertainty today implies chaos tomorrow!

The doubling map exhibits the key features of a chaotic map [8]: it possesses *sensitive dependence* on initial conditions, it is *indecomposable* in the sense that it has dense orbits and the periodic points form a *dense set*. A drawback to the acceptability of this map is that it could be argued to be contrived in the use of mod 1 and therefore amongst other properties, it is discontinuous. We will now show how the map, and particularly, its coding, appears in apparently more innocuous maps.

4 Period doubling and chaos in the logistic map

The *logistic map* is essentially a 1-parameter family of smooth maps $F_\rho : I \to I$, $I = [0, 1]$, which depend on the parameter $\rho \in (0, 4]$. The family is defined by

$$F_\rho(x) = \rho x(1-x). \tag{4.1}$$

Thus the graph of the map F_ρ is simply an inverted parabola which is contained within the square $I \times I$. Our aim is to investigate how the dynamical complexity

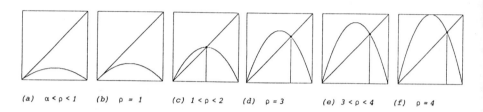

(a) $\alpha < \rho < 1$ (b) $\rho = 1$ (c) $1 < \rho < 2$ (d) $\rho = 3$ (e) $3 < \rho < 4$ (f) $\rho = 4$

Figure 4. The graphs of F_ρ for the values of ρ indicated. Note that: (i) F_ρ has a non-trivial fixed point only when $\rho > 1$; (ii) $-1 < DF_\rho(x^*) < 0$ for $1 < \rho < 3$; (iii) $DF_3(x^*) = -1$; and (iv) $DF_\rho(x^*) < -1$ for $3 < \rho \le 4$.

of the map F_ρ increases as the parameter ρ increases. This is done by investigating the *bifurcations* of (4.1). The *bifurcation values* of (4.1) are those values of ρ for which F_ρ undergoes topological changes in its orbit structure.

We will begin as we did with the doubling map and investigate the occurrence of periodic orbits. To obtain fixed points, solutions of the equation $F_\rho(x) = x$ are required and it is clear that for all $\rho \in (0,1]$, there is exactly one fixed point at the origin (see Figure 4). For $\rho \in (1,4]$ there are precisely two and they are given by $x = 0$ and $x = x^* = (\rho - 1)/\rho$. We can exploit the graph of the logistic map in Figure 4 to investigate the dynamics of other orbits. The following procedure is most easily carried out on a micro-computer. The orbit through $x_0 \in I$ is calculated by finding $x_1 = F_\rho(x_0)$ and then recording x_1 on the x-axis in order to calculate $x_2 = F_\rho(x_1)$. Note that this process of recording the previous x-value for the next iteration can be carried out diagramatically by moving horizontally along $y = x_1$ to the line $y = x$. On the line $x = x_1$ we just move vertically to the graph of F_ρ to obtain x_2. Repeating this construction gives a 'staircase' which records the orbit of F_ρ through x_0 (see Figure 5(a)).

We should also note that the fixed points of F_ρ are given by the x (or y) coordinates of the intersection of the line $y = x$ with $y = F_\rho(x)$. We see in Figure 5(b) that the fixed point at $x = x^*$ is stable. Orbits are apparently attracted towards it with positive iteration and repelled from the fixed point at the origin. At this stage we have no hint of a chaotic map. There is certainly no sensitive dependence on initial conditions for *every* orbit with initial point $x_0 \in (0,1)$ approaches the attracting fixed point at $x = x^*$. However we note that 'overshoot' starts to occur at $x = x^*$ as ρ increases (see Figure 5(c)) and the orbit then oscillates as it moves towards the equilibrium position. The fixed point is becoming *less* attracting. Recall from Section 2 that the fixed point x^* of F_ρ will be asymptotically stable if $|DF_\rho(x^*)| < 1$ and unstable if $|DF_\rho(x^*)| > 1$. The magnitude of the slope of F_ρ at $x = x^*$ decreases as ρ increases and at $\rho = 3$ we have $DF_\rho(x^*) = -1$ which implies that the stability of x^* is transitional. The fixed point x^* becomes an unstable point for F_ρ when $\rho > 3$. We now observe what has been happening to period two

The Mathematics of Chaos

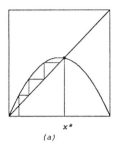
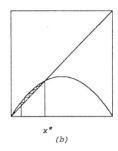
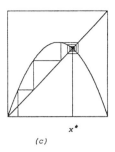

Figure 5. (a) The first few iterations of an orbit of F_ρ through $x = x_0$. Note that we need only record the "staircase" to obtain the dynamics. (b) Orbits moving asymptotically towards the stable fixed point x* for $\rho = 1.5$ and $\rho = 2.5$. Note that the approach to x^* is direct in (b) and oscillatory in (c).

behaviour of F_ρ while the non-trivial fixed point x^* has been undergoing structural changes. For this investigation we need to consider the graph of $F_\rho^2 = F_\rho \circ F_\rho$ and its intersection with the line $y = x$. For $\rho < 3$ the graph of F_ρ^2 intersects $y = x$ at only one non-trivial point (see Figure 6(a)). It is very easy to see that it must be the fixed point $x = x^*$ since any x^* which satisfies $F_\rho(x^*) = x^*$ also satisfies

$$F_\rho^2(x^*) = F_\rho(F_\rho(x^*)) = F_\rho(x^*) = x^*. \tag{4.2}$$

If we explore the graph of F_ρ^2 further we see in Figure 6(b) that there is a tangency with $y = x$ when $\rho = 3$ and three non-trivial intersections for $\rho > 3$. Only one of these points can represent the fixed or period-1 point x^* and therefore the other two points must be period-2 points. Here we have a *flip bifurcation*. The fixed point has changed stability as a consequence of the eigenvalue passing through -1, and a period-2 orbit has been created. In fact, the period-2 orbit has taken on the role of an *attractor* for the map F_ρ vacated by the fixed point becoming unstable.

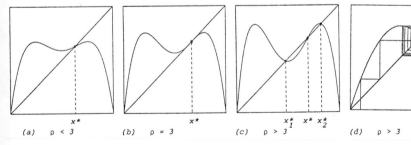

Figure 6. (a)-(c) show the graph of F_ρ^2 as ρ increases through 3. A period-2 orbit $\{x_1^*, x_2^*\}$ is created for $\rho > 3$ in addition to the fixed point x^*. (d) shows an orbit with initial point x_0 which is forwardly asymptotic to the period-2 orbit $\{x_1^*, x_2^*\}$.

On returning to our microcomputer to investigate the changing patterns of orbits we now see orbits moving asymptotically towards the period-2 orbit $\{x_1^*, x_2^*\}$; this behaviour is shown by a square with diagonal vertices at $(x_1^*, x_1^*), (x_2^*, x_2^*)$ in Figure 6(d). The square arises from the oscillatory movement between the period-2 points x_1^* and x_2^* under iteration of F_ρ. A simple computation for the values of ρ currently under investigation will show that at this stage there are no other periodic orbits. However, the scenario changes with increasing rapidity as we continue to increase the parameter ρ. We find that the same behaviour occurs again and the fate of the attracting period-2 orbit is the same as that of the fixed point at $\rho = 3$. The graph F_ρ^2 has slope -1 at the periodic points x_1^*, x_2^* when $\rho = \rho_1^*$ where $\rho_1^* > 3$. Investigation of the graph of F_ρ^4 for this value of ρ will show a tangency with $y = x$ at two points. Increasing the value of ρ beyond ρ^* will produce period-4 orbits as a result of another flip bifurcation. In fact we obtain an infinite *cascade* of flip bifurcations producing periodic points of period-2^n, $n \in \mathbb{Z}$. This is known as the *period-doubling route to chaos*. The phenomenon of period doubling cascades discovered by computer has also given rise to a new universal constant. Let the sequence of flip bifurcations occur at the sequence of parameter values $\{\rho_n\}_{n=1}^\infty$. Then the values ρ_n accumulate at a parameter ρ^* and the sequence

$$\mu_n = \frac{\rho_{n+1} - \rho_n}{\rho_{n+2} - \rho_{n+1}} \qquad (4.3)$$

has a computed limit 4.66920... called the *Feigenbaum number*. The surprising discovery was that this value occurs for a large class of maps which undergo period doubling cascades [10,12,35].

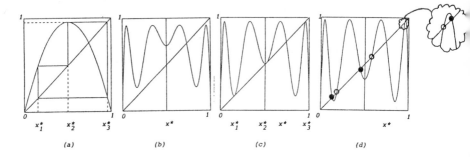

Figure 7. (a) The period-3 orbit of F_{ρ^*}, where $x_2^* = 0.5$. (b) The graphs of F_ρ^3 for: (b) $\rho < \rho^*$; (c) $\rho = \rho^*$ and (d) $\rho > \rho^*$, there are two periodic orbits of period 3, one stable (•) and the other unstable (○). By increasing ρ further, the graph of F_ρ^3 has slope -1 at each point of the stable period-3 orbit. F_ρ^3 undergoes a flip bifurcation and a stable period-6 orbit of F_ρ is created.

Let us now return to the motivation for the discussion of the logistic map. We can show that, in spite of its apparent innocuous form, the logistic map takes on the behaviour of the doubling map when $\rho = 4$.

Consider the map $\Phi : [0,1) \to [0,1]$ defined by $\Phi(\theta) = sin^2(2\pi\theta)$. Then Φ provides a semi-conjugacy for the doubling map f and the special logistic map F_4. Specifically, the map Φ satisfies $f \circ \Phi = \Phi \circ F_4$. Recall that the semi-conjugacy preserves orbit structure and it is straightforward to conclude from the properties of f that F_4 is sensitively dependent on initial conditions, has a set of period points dense in [0,1] and has orbits dense in [0,1]. Thus we have that F_4 is *chaotic*.

The problem now arises as to how the period doubling phenomenon, which was the precursor of chaos, changes into the chaotic behaviour observed for $\rho = 4$? The key to answering this question is in the period-3 behaviour of F_ρ. Observe in Figure 7 that we have a period-3 orbit (with $x_2^* = 0.5$) for $\rho^* = 3.8282\ldots$. As ρ continues to increase, the slope of F_ρ^3 at the stable orbit points decreases and eventually reaches -1. A flip bifurcation then occurs creating stable period-6 orbits. As with the stable 2^k cycles, there follows a cascade of stable 3×2^k cycles. Numerical experiments estimate that these accumulate at $\rho = 3.8495\ldots$ (see [23]).

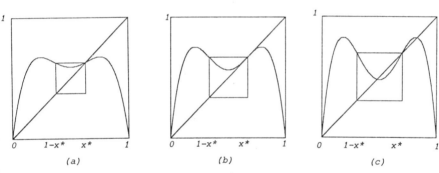

Figure 8. The restriction of the graph of F_ρ^2 to the interval $[1 - x^*, x^*]$
(cf. Figure 4(a), (b), and (c)).

A remarkable result due to Sarkovskii [28] allows us to deduce the underlying complexity of many 1-dimensional maps from the existence of certain periodic orbits.

Theorem 4.1 *Order the positive integers* $\mathbb{Z}^+$ *as follows*

$$3 \triangleleft 5 \triangleleft 7 \triangleleft 9 \triangleleft 11 \triangleleft \ldots \triangleleft 2 \cdot 3 \triangleleft 2 \cdot 5 \triangleleft 2 \cdot 7 \triangleleft \ldots \triangleleft 2^k \cdot 3 \triangleleft 2^k \cdot 5$$

$$\triangleleft 2^k \cdot 7 \triangleleft \ldots 2^k \triangleleft \ldots \triangleleft 2^3 \triangleleft 2^2 \triangleleft 2 \triangleleft 1 \tag{4.4}$$

If $f : \mathbb{R} \to \mathbb{R}$ *is a continuous map which has an orbit of period* n, *then* f *has an orbit of period* m *for every* $m \in \mathbb{Z}^+$ *with* $n \triangleleft m$.

The theorem clearly gives period three a special status. If period three points occur then all other periodicities occur. A famously titled paper "Period three implies chaos", [11], forged the link between the behaviour implied by Sarkovskii and the other ingredients of chaotic motion.

The theorem is applicable to F_ρ even though its domain is restricted here to the interval I. It is straightforward to check that if we extend the domain of F_ρ to $\mathbb{R}$ then all orbits $\{x_n\}$ through points $x_0 \in \mathbb{R}\setminus I$ satisfy $x_n \to -\infty$ as $n \to \infty$. Thus F_ρ has no periodic points in $\mathbb{R}\setminus I$ and any recurrence predicted by Sarkovskii's theorem is restricted to the interval I.

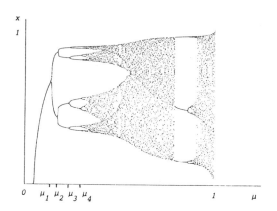

Figure 9. The period doubling cascade for the logistic map. For each $\mu = \mu_0$, the approximate locations of the stable periodic points of F_{μ_0} are depicted. The scaling of the μ-axis is non-linear to emphasise the crucial *period-3 window*.

We see from Sarkovskii's theorem that infinitely many period doubling cascades occur once a period-3 orbit exists. The inevitability of the period doubling cascades in the logistic map arises from a *self-similar structure* when the graphs F_ρ and F_ρ^2 are compared. As ρ is increased the graph of F_ρ goes through the sequence given in Figure 4 and a suitable restriction of the graph of F_ρ^2 goes through the same sequence see Figure 8. This observation can be made precise. For $\rho > 3$ we have the situation illustrated in Figure 8(c) for the graphs of F_ρ and F_ρ^2. If $(1-x^*)$ and x^* satisfy $F_\rho((1-x^*)) = F_\rho(x^*) = x^*$. Then the graph of F_ρ^2 restricted to the interval $[(1-x^*), x^*]$ has, up to a rotation, the appearance of the graph F_ρ on $[0,1]$. There is, in fact, a conjugacy $h : [0,1] \to [(1-x^*), x^*]$ such that

$$h \circ F_\rho = F_\rho^2 \circ h. \tag{4.5}$$

If we introduce an operator T_h defined by

$$T_h(F_\rho) = h \circ F_\rho \circ h^{-1} \tag{4.6}$$

then from (4.5), $T_h(F_\rho) = F_\rho^2$ and T_h is said to be a *renormalization operator*. It is trivial to check that $T_h(F_\rho^n) = F_\rho^{2n}$ and so the operator renormalizes any power of F_ρ to be its square. The effect of the transformation T_h is to say that with a suitable rescaling F_ρ can be described by F_ρ^2. It is easy to check in (4.5) that, via T_h, fixed points of F_ρ and F_ρ^2 correspond. Similarly, period-m points of F_ρ correspond with period-m of F_ρ^2 and hence period-$2m$ points of F_ρ. Thus a flip bifurcation undergone by F_ρ predicts a flip bifurcation for F_ρ^2. In terms of periodicity of F_ρ, a stable period-2 orbit changes stability and throws off a stable period-4 orbit. The renormalization procedure can be applied in exactly the same way to F_ρ^2 and F_ρ^4 and in general to $F_\rho^{2^n}$ and $F_\rho^{2^{n+1}}$ to obtain flip bifurcations from period-2^n to period-2^{n+1} stable orbits. The universality of the Feigenbaum number arises from properties of the operator T_h.

We have confined our discussion here to $\rho \in (0,4]$. For $\rho > 4$, the map F_ρ can also be described using symbolic dynamics. It can be shown that there is an invariant Cantor set Λ (see Section 5) on which F_ρ is described by a shift on binary sequences.

The logistic map has been the most widely studied map where self-similarity leads to the complexity described above. Much more obvious and striking pictures of self-similar structures are obtained in the beautiful computer pictures obtained by zooming in on self-repeating shapes of fractal curves. Some key terms here are *fractal*, *Julia* and *Mandelbrot sets* [4,21,26].

5 The horseshoe diffeomorphism and homoclinic points

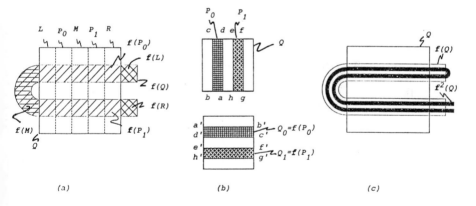

Figure 10. (a) The image of the horseshoe map $f(Q)$ relative to Q in $\mathbb{R}^2$. (b) The horseshoe map restricted to the rectangles P_0, P_1 on which it is affine. The images Q_0, Q_1 are such that $a \mapsto a'$, etc. (c) the image $\mathbf{f}^2(Q) \subset \mathbb{R}^2$

A final and crucial example of the power of coding to understand complex behaviour was described by Smale in [30]. This discovery came well before the more recent interest in one dimensional maps. The *horseshoe* is an example of a diffeomorphism of the plane which displays complex dynamics. Its properties allow it to be described as the Poincaré map of an ordinary differential equation in 3 dimensions [31]. Indeed, the impetus for Smale's example came from studies by Levinson [18] of the Poincaré map of a forced oscillator. We are not able to go into detail here and only give the key ideas (see [3,8,12]).

First of all, we need the concept of a Cantor set. Consider the unit interval of points $I = [0, 1]$ on the real line and the sequence of subsets obtained by splitting intervals into three equal parts and deleting the middle third. Let $T_1 = [0, \frac{1}{3}] \cup [\frac{2}{3}, 1]$, $T_2 = [0, \frac{1}{9}] \cup [\frac{2}{9}, \frac{1}{3}] \cup [\frac{2}{3}, \frac{7}{9}] \cup [\frac{8}{9}, 1]$ etc. and define $T = \cap_{n=1}^{\infty} T_n$. This set is non-empty, has no interior, and yet every point of T is an accumulation point of T. Closed sets with these properties are known as *Cantor sets*.

The horseshoe diffeomorphism $\mathbf{f} : Q \to \mathbb{R}^2$ is defined on a square $Q = \{(x, y) | |x|, |y| \le 1\} \subset \mathbb{R}^2$ and has an invariant set of Cantor type on which the dynamics can be described by a shift on symbol sequences as in the previous two examples. The map $\mathbf{f}$ can be extended to a diffeomorphism of the whole plane.

The image of Q is a horseshoe shaped object (see Figure 10(a)). Essentially, the rectangle Q is stretched out to a fifth of its height and five times its length, then folded and placed across Q in the shape of a horseshoe. We assume that the rectangular segments P_0 and P_1 of Q map linearly onto the rectangles Q_0 and Q_1 respectively as depicted in Figure 10(b). Each iteration of the map results in a doubling of the number of horizontal strips which are images of Q. The map $\mathbf{f}$ folds the square and each iteration corresponds to the process of kneading of dough to make puff pastry (see Figure 10(c)).

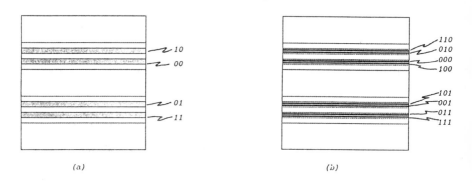

Figure 11. (a) The image by: (a) $\mathbf{f}^2$; (b) $\mathbf{f}^3$, of Q restricted to Q. In (a) strip (1) $\subset \mathbf{f}^2(P_1) \cap \mathbf{f}(P_0)$ has code "10"; strip (2) $\subset \mathbf{f}^2(P_0) \cap \mathbf{f}(P_0)$ and has code "00"; strip (3) has code "01"; strip (4) has code "11". In (b) strip (1) $\subset \mathbf{f}^3(P_1) \cap \mathbf{f}^2(P_1) \cap \mathbf{f}(P_0)$ and has code "110".

The image of f^n on Q consists of 2^n horizontal strips, each of width $2/5^n$. Moreover, the strips can be labelled by sequences of $0's$ and $1's$. Observe $f^n(Q) \cap Q = (f^n(P_0) \cup f^n(P_1)) \cap Q$ and each strip is a subset of either $f^n(P_0)$ or $f^n(P_1)$. Note that the topmost of the four strips in Figure 11(a) is part of the image of $f^2(P_1) \cap f(P_0)$ and thus we label it '10'. Similarly the other strips are labelled '00','01' and '11'. For each interation a further symbol is needed to label the strips on Q that are created. Thus for the eight strips of $f^3(Q) \cap Q$, (see Figure 11(b)), the topmost strip is coded '110', the second strip '010', and the bottom strip '111'. By considering $\cap_{n=1}^{\infty}(f^n(Q) \cap Q)$ we obtain the limiting set of this process which is a Cantor set in the vertical direction and an interval in the x-direction. The vertical coordinates are now coded by an infinite string of symbols.

If we consider the inverse procedure to that described above with f replaced by f^{-1} we find that vertical strips are created in the same way as above (see Figure 11). The set $\cap_{n=0}^{\infty}(f^{-n}(Q) \cap Q)$ is the Cartesian product of a Cantor set in the horizontal direction and a vertical line. Thus the set $\Lambda = \cap_{n=-\infty}^{\infty}(f^n(Q) \cap Q)$ is a cartesian product of two Cantor sets which is also a Cantor set (see Figure 11(b) and (c)). $\Lambda \subseteq Q$ is an invariant set of f by its very construction. Also, $f|_\Lambda$ is hyperbolic in nature with contraction in the vertical direction and expansion in the horizontal direction.

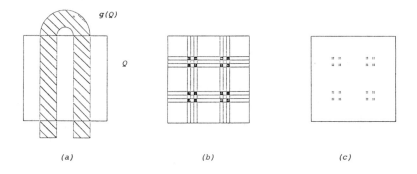

Figure 12. (a) A diffeomorphism $g : Q \to \mathbb{R}^2$ inverse to f, i.e. $g \circ f = f \circ g = \text{id}$, where defined. (b) and (c) The sets $\cap_{n=-1}^{2} f^n(Q)$ and $\cap_{n=-2}^{3} f^n(Q)$ consisting of 16 and 64 squares respectively.

The symbolic dynamics of f on Λ is introduced by coding a point of Λ as a bi-infinite binary sequence. Each point $(x,y) \in \Lambda$ is uniquely obtained from a nested sequence of strips. More precisely, the point (x,y) is either in: Q_0 or Q_1, i.e. $f(P_0)$ or $f(P_1)$; $f^2(P_0)$ or $f^2(P_1)$; and generally $f^n(P_0)$ or $f^n(P_1)$, $n \in \mathbb{Z}^+$. Listing the appropriate P_0 or P_1 at each stage gives a binary sequence. So far we have only locked the *vertical* position y of the point of $(x,y) \in \Lambda$. The abcissa x is used to determine the choices of $f^{-n}(P_0)$ or $f^{-n}(P_1)$ for $n \geq 0$ i.e. by choosing a nested sequence of vertical strips. Thus the point can be written as $(x, y) = \cap_{n=-\infty}^{+\infty} \mathbf{f}^{-n}(P_{\sigma(n)})$ where $\sigma(n)$, $n \in \mathbf{Z}$, is the bi-infinite sequence of 0's and 1's obtained above, i.e. $(x, y) \in \mathbf{f}^{-n}(P_{\sigma(n)})$. Let α be the shift

$$\alpha(\sigma_n) = \sigma_{n+1}. \tag{5.1}$$

and define

$$h(\sigma) = \cap_{n=-\infty}^{+\infty} \mathbf{f}^{-n}(P_{\sigma(n)}) \tag{5.2}$$

It then follows that α and $\mathbf{f}$ are conjugate by h, i.e $\mathbf{f} \circ h = h \circ \alpha$. The symbolic dynamics of α can now be used in the same way as in Section 3 to obtain the complex behaviour of the horseshoe.

The Smale horseshoe first appeared in 1963 as an example of a planar diffeomorphism with infinitely many periodic points. Specifically, the map has an invariant Cantor set Λ on which there is a dense orbit to make it dynamically indecomposable and a dense set of periodic points. It also has rich dynamics arising from its description as a shift on bi-infinite sequences of symbols '0' and '1' (cf. doubling and logistic maps). Crucially, these properties remain intact under small perturbations of the map. We say the map is *structurally stable* and thus its strange properties are not an artifact of the strong assumptions of linearity in its constituent parts. However, the most astonishing feature of the map is its central role in complex behaviour and chaos in systems which we now describe.

Consider a diffeomorphism $\mathbf{f} : \mathbb{R}^2 \to \mathbb{R}^2$ which has a fixed point which is a hyperbolic saddle point $\mathbf{x}^*$. Let W^s and W^u be the stable and unstable manifolds associated with $\mathbf{f}$ at $\mathbf{x}^*$ (see Figure 12). The manifolds W^s and W^u are invariant under both $\mathbf{f}$ and its inverse and are defined by

$$W^s = \{\mathbf{x} \mid lim_{m \to \infty} \mathbf{f}^m(\mathbf{x}) = \mathbf{x}^*\} \tag{5.3}$$

and

$$W^u = \{\mathbf{x} \mid lim_{m \to -\infty} \mathbf{f}^m(\mathbf{x}) = \mathbf{x}^*\}. \tag{5.4}$$

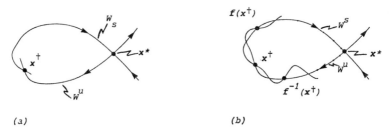

(a) (b)

Figure 13. (a) A homoclinic point $\mathbf{x}^\dagger \in W^s \cap W^u$ of the saddle point $\mathbf{x}^*$. (b) Self-intersections of W^s and W^u to ensure that $\mathbf{f}(x^\dagger)$ and $\mathbf{f}^{-1}(\mathbf{x}^\dagger)$ also lie on both W^s and W^u.

Figure 14. The orbit of $x^\dagger$ lies in $W^s \cap W^u$ and the orbit approaches the fixed point x^* asymptotically for both positive and negative iterations resulting in the folded manifolds. (b) An example of the behaviour depicted in (a) for the map $f(x,y) = (y+x^2, y+x(x-1))$.

The sets defined by (5.3) and (5.4) are the manifolds which correspond to the stable and unstable linear subspaces $\mathbb{R}^s$ and $\mathbb{R}^u$ defined in equation (1.6) (see Figure 2(b)). However, they are not necessarily linear subspaces of $\mathbb{R}^n$ when the system is nonlinear. It is possible for these manifolds to intersect at a point $x^\dagger$ as in Figure 13(a). The invariance of the manifolds W^s and W^u ensures that all iterates of f also lie on both manifolds and this means that they re-intersect as in Figure 13(b). The orbit through $x^\dagger$ approaches the point x^* asymptotically in both forward and reverse iterations of f and so both manifolds continually reintersect and fold as they accumulate at x^* as in Figure 14(a). The point $x^\dagger$ is called a *homoclinic point* to the saddle x^*. If the manifolds W^s and W^u have distinct tangential directions at $x^\dagger$ as in Figures 12-14, then the homoclinic point $x^\dagger$ is said to be *transverse*. For higher dimensions, $n > 2$, the manifolds W^s and W^u need to be in general position. The consequences of the existence of a transverse homoclinic point are given by the following theorem which concludes that some iterate of f behaves like a horseshoe map.

Theorem 5.1 *(The Smale-Birkhoff Theorem). Let $x^\dagger$ be a transverse homoclinic point of the diffeomorphism* $f : M \to M$. *Then there is a Cantor set* $\Lambda \subset M$, $x^\dagger \in \Lambda$, *and positive integer* N *such that* $f^N(\Lambda) = \Lambda$ *and* f^N *is given by a shift automorphism on* Λ.

Pictorially, we can see why the conclusions of the theorem are not unreasonable. As illustrated in Figure 13 the invariance of W^s and W^u under forward and backward iteration of f forces the manifolds to bend back on themselves as $f^n(x^\dagger) \to x^*$ when $n \to \pm\infty$. Now consider a small region R (see Figure 14) containing the point $x^\dagger$ with its sides "parallel" to the manifolds W^s and W^u. The effect of positive iterations of f on R is to contract in the W^s direction and expand in the W^u direction. This results in the image $R_1 = f^{n_1}(R)$. For negative iterations of f, i.e. positive iterations of f^{-1}, we obtain an elongated strip stretched along W^s and contracted along W^u. Consider the strip $R_2 = f^{-n_2}(R)$. It follows that $f^{n_1+n_2}(R_2) = R_1$ and

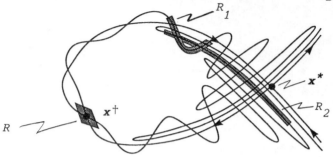

Figure 15. The strips R_1 and R_2 produced from the forward and reverse iteration of the region R resulting in a horseshoe like map from R_2 to R_1.

we conclude that $\mathbf{f}^N$, where $N = n_1 + n_2$, has the appearance of a horse-shoe like map. The Smale-Birkhoff theorem makes this observation precise and the existence of transverse homoclinic points signals the existence of embedded horseshoes in the dynamics of $\mathbf{f}$. However, a severe drawback to the numerical study of the horseshoe diffeomorphism is that the invariant set Λ is virtually impossible to detect by computer. The set of points which asymptotically approach Λ on Q is the set of vertical straight lines passing through Λ. It is a "Cantor set of vertical intervals". The five fold expansion in the horizontal direction ensures sensitive dependence on initial conditions which means that for all practical purposes the set will be missed and the orbit will be flushed outside Q. Even so, it is instructive to model the map on the computer and, unless it is done symbolically, *all* of the complexity is missed. The set Λ is a "generalised saddle" which has non-trivial inset and outset. Ideally, transverse homoclinic behaviour in an attracting environment is required. It is then possible to use the computer successfully to show up the complexity (see Section 6).

6   Strange attracting sets and attractors

Let $\mathbf{f}$ be a diffeomorphism on a manifold M. A closed invariant set $\mathcal{A}$ of $\mathbf{f}$ is said to be an *attracting set* for $\mathbf{f}$ if there exists a closed neighbourhood $\mathcal{N}$ of $\mathcal{A}$ such that $\mathbf{f}(\mathcal{N}) \subset \text{interior}(\mathcal{N})$ and $\mathcal{A} = \cap_{n=0}^{\infty} \mathbf{f}^n(\mathcal{N})$. $\mathcal{N}$ is referred to as a *trapping region* for $\mathcal{A}$. The attracting set is said to be *strange* if $\mathcal{A}$ contains a transverse homoclinic point of $\mathbf{f}$. Furthermore, the strange attracting set $\mathcal{A}$ is a *strange attractor* if $\mathbf{f}|\mathcal{A}$ has a dense orbit and a dense set of periodic orbits. These concepts have their origins in pioneering work of Birkhoff on planar diffeomorphisms with attracting sets other than fixed points or invariant circles [5]. Clearly, there are corresponding definitions for flows. However the Poincaré map construction suggests that similar dynamical behaviour is likely to be found at one dimension lower for diffeomorphisms than for flows.

Many examples of strange attractors appear to have been found but their precise

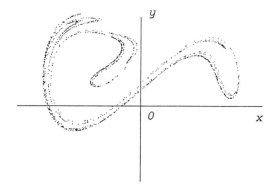

Figure 16. The Duffing attractor of the Poincaré map for the system (6.2) with $\mu = 0.4, \nu = 0.25$ and $\omega = 1.0$.

dynamical properties are often elusive. However, the one common feature seems to be a combination of repeated stretching and folding, this gives rise to attracting sets which usually have non-integer or *fractal* dimension. The stretching is required to obtain sensitive dependence on initial conditions and the folding is necessary if the structure is to remain in a compact region of the phase portrait [12,33]. Eigenvalues give a measure of contraction and expansion at a fixed point or periodic orbit. *Lyapunov exponents* [12] generalise this concept to measure rates of expansion away from periodic orbits and enable checks to be made for sensitivity of neighbouring orbits.

Now, consider the *Duffing equation*:

$$\frac{d^2x}{dt^2} + \nu\frac{dx}{dt} - x + x^3 = \mu cos(\omega t) \quad (6.1)$$

where μ, ν and ω are real parameters. We interpret this system as autonomous by introducing a third variable $\theta = t$. Then (6.1) becomes

$$\frac{dx}{dt} = y, \quad \frac{dy}{dt} = x - x^3 - \nu y + \mu cos(\omega \theta), \quad \frac{d\theta}{dt} = 1 \quad (6.2)$$

and the periodicity in t of equation (6.1) ensures that the trajectories of (6.2) are periodic in θ with period $2\pi/\omega$. Thus we can obtain all the trajectory behaviour by restricting to $\theta \in [0, 2\pi/\omega]$. The Poincaré map associated with system (6.2) is just the $2\pi/\omega$ advance of the flow i.e. $\phi_{2\pi/\omega}: \mathbb{R}^2 \to \mathbb{R}^2$. A picture of the attracting set of the Poincaré map of (6.2) is given in Figure 16. This can be obtained easily on a microcomputer by using an Euler approximation to (6.2) with a suitable step-length and plotting a point of a chosen orbit every $2\pi/\omega$ increase in t. This is time consuming since the computer has to perform many step calculations for each plotted point. The attractor in this case is numerically indistinguishable from the unstable manifold of a saddle periodic orbit. Such oscillators have been a strong

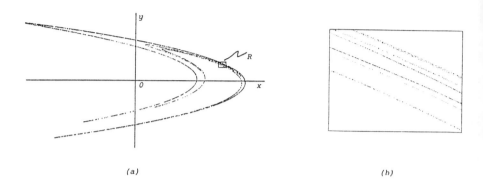

Figure 17. (a) The Hénon attractor of (6.3), (b) magnification of the rectangle R to indicate the braiding.

source of attractors, see [1,12,33], and have been a rich vein for studying chaos in engineering problems.

A much more convenient situation occurs when the attractor arises from an explicit map rather than the implicit ones for nonlinear non-autonomous planar systems. Such an example is the *Hénon* attractor. This arises from a planar diffeomorphism with the simple formula

$$f(x,y) = (y - 1.4x^2 + 1, 0.3x) \qquad (6.3)$$

Starting at any reasonable initial value, the pattern on the computer screen takes up the picture given in Figure 17(a). However, closer investigation reveals a braided structure. A suitable magnification reveals that apparent single curves split into two or more (see Figure 17(b)). Further magnification reveals further splitting. The braiding arises in a similar way to the horseshoe map in the sense that the attractor has an infinitely folded appearance. Unfortunately, in spite of the simplicity of the map and the ease of obtaining pictures of the attractor, there does not appear to be a definitive description of the dynamical properties of the Hénon map. The complexity appears to arise from a tangency between the stable and unstable manifolds of a hyperbolic saddle. A piecewise-linear version of the Hénon map, known as the *Lozi map*, has been shown to have a strange attractor [24]. Many other examples of strange attracting sets can be obtained from simple maps (see [14]) and a major problem is to understand their dynamical properties. Ruelle [28] states that it is difficult to see that the Hénon attractor is not a very long period stable periodic orbit and that proofs of its strangeness have been promised by several authors!

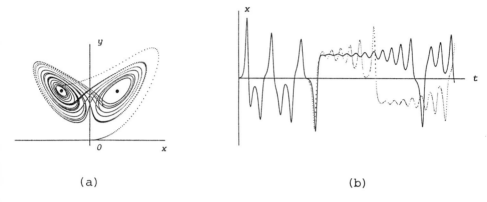

Figure 18. (a) An orbit of the Lorenz equations moving in an irregular way about two fixed points. (b) The x-displacement for two orbits with close initial conditions which show their independent futures.

The Lorenz equation was the first simple system which aroused the interest of nonlinear dynamicists. The system is a first order equation in 3-variables [19]:

$$\tfrac{dx}{dt} = \sigma(y-x),\ \tfrac{dy}{dt} = \rho x - y - xz,\ \tfrac{dz}{dt} = xy - \beta z, \tag{6.4}$$

$(x, y, z) \in \mathbb{R}^3$ and σ, β, ρ positive parameters. The case usually studied is for $\sigma = 10, \beta = 8/3$ and $\rho = 28$, [22,32]. The system possesses three fixed points, one of which is at the origin. They are all unstable of saddle type and therefore each fixed point has non-trivial stable and unstable manifolds. Solutions oscillate about the two fixed points away from the origin in a complex and apparently random manner. This behaviour arises from the relative positions of the stable and unstable manifolds of the three saddle points. It can be easily checked that the way in which orbits move from the vicinity of one fixed point to the other depends critically on the initial condition (see Figure 18). However, all orbits remain within a bounded region of $\mathbb{R}^3$ and this gives rise to the strange attracting set.

7 Applications

Publications in the area of non-linear dynamics are expanding rapidly but many results are of a very isolated nature and do not throw a great deal of light on the fundamental understanding of chaos. For instance, there is no universally accepted definition of strange attractor! An obvious problem is to classify strange attractors, but this is too ambitious if interpreted in its widest sense. However one can obtain special classes of attractors for which the project is not so daunting, for example,

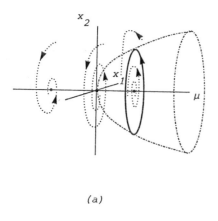

(a)

(b)

Figure 19. (a) The creation of an invariant circle as an attractor of $\mathbf{f}_\mu$ which grows from the origin 0 as μ increases through zero. (b) As μ increases further, the attracting circle can change to a strange attracting set [2].

the well known attractors of the systems of Lorenz, Duffing, Birkhoff-Shaw, Rossler, or the attractors of planar diffeomorphisms such as the Hénon and Leslie attractors [1,12,33].

Another class of attractors which appear to be rich in structure is obtained from the *Hopf bifurcation* for maps. Let $\mathbf{f}_\mu : \mathbb{R}^2 \to \mathbb{R}^2$, $\mu \in \mathbb{R}$, be a one parameter family of planar diffeomorphisms such that $\mathbf{f}_\mu(0) = 0$ for all μ. Consider the linearization of $\mathbf{f}$ given by its derivative $D\mathbf{f}_\mu(0)$ and suppose the eigenvalues are complex conjugate $\lambda(\mu), \overline{\lambda(\mu)}$ with $|\lambda(0)| = 1$ and $\frac{d}{d\mu}(\lambda(\mu))|_{\mu=0} > 0$. Then the fixed point at the origin of $\mathbb{R}^2$ changes stability from stable to unstable as μ increases through $\mu = 0$. Under further non-degeneracy conditions on the family $\mathbf{f}_\mu$ (see [2,3]) we find that not only is there a change of stability but a stable invariant circle is created around the origin which grows with increasing μ. In many examples, the invariant circle undergoes further bifurcation to form more complicated attractors. It seems that these attractors have many characteristics in common and that their classification is a possibility. This is only one aspect of how the map analogy of a well understood vector field bifurcation is an area of current research. Another is the double invariant circle bifurcation. The vector field bifurcation is simply illustrated in Figure 21. Stable and unstable cycles coalesce to form a partially stable cycle which then disappears. The simplicity arises from uniqueness of solution and so when the invariant cycles intersect, they coalesce. This is not the case for planar maps and the behaviour here is very complex and not fully understood although it is believed to involve strange attracting sets of Birkhoff type [5].

Interest has grown substantially of late in the dynamics of maps in the complex plane. This was initated by the pioneering mathematical and computer work of

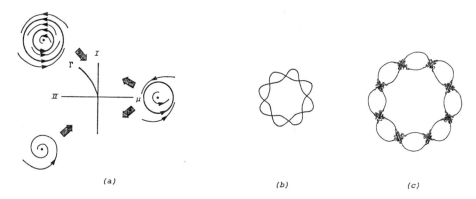

Figure 20. (a) The various phase portraits of the system $\dot{\theta} = 1$, $\dot{r} = \mu r + \nu r^3 - r^5$, $(\mu, \nu) \in \mathbb{R}^2$ show 0, 1, or 2 periodic orbits in the regions indicated. On Γ, the cycles in the region I coalesce to form a single cycle which disappears in region II. (b) The generic behaviour of two intersecting invariant circles for diffeomorphisms analogous to the system in (a) for $(\mu, \nu) \in \Gamma$. (c) The corresponding behaviour of the saddle separatrices.

Mandelbrot based on the theory from earlier results of Fatou and Julia. We have made no attempt to cover this important aspect of modern dynamics (see [8,9]).

The development of applications usually follows leisurely in the wake of the pure mathematics. However, chaos has taken on a popular appeal in general science journals as well as on radio and television. This has resulted in a wide body of scientists looking for chaos in the systems which they study. Often it has been at the novelty stage of trying to find chaotic solutions *per se*. However, this is not without interest particularly when applied to natural systems such as the dynamics of the solar system. In this area of study there have been serious attempts to use the chaotic behaviour to understand hitherto unexplained phenomena. The Kirkwood gaps in the Asteroid belt are believed to be chaotic regions of the Solar system. In particular, chaotic behaviour gives an explanation of how highly eccentric orbits can occur to allow crossings of the Earth's orbit to produce meteors. The chaos has been detected by not only by numerical integration, but also investigations of algebraic maps which model the asteroid behaviour. Hyperion, a moon of Saturn that is shaped like a rugby ball, has been shown to follow a chaotic tumbling motion [36] around the planet. As I write, the new discoveries of Neptune's ring and moon structure by *Voyager II* will undoubtedly be scrutinised from the viewpoint of chaotic behaviour.

At the other end of the scale atomic behaviour is governed by quantum mechanics. Although there is, as yet, little evidence for or against the existence of chaos in quantum systems, much work has been done on corresponding classical systems.

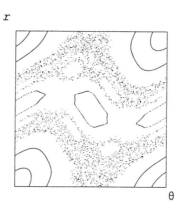

Figure 21. Chaotic regions of the standard map (7.1) when k=1.1.

One of the most widely studied maps is the *standard map* which is a one-parameter family of area-preserving maps of the cylinder [20] given by

$$(\theta', r') = (\theta + r', r - \tfrac{k}{2\pi}sin(2\pi\theta)), \quad (\theta, r) \in S_1 \times \mathbb{R}. \tag{7.1}$$

This is the simplest classical model for the plasma behaviour in a particle accelerator. For $k = 0$, we obtain a continuum of invariant circles. The points on these circles are periodic or not according as r is rational or irrational. As the parameter increases the invariant circles gradually break up. Those which are rational become "island chains" and the fate of the irrational circles depends on how "irrational" is the corresponding value of r. The more irrational the frequency associated with the invariant circle, the greater resistance to break-up. The last circle to break is the "golden circle" with frequency equal to the golden mean $(1+\sqrt{5})/2$ (see Figure 21). The irrational circles are replaced by Cantor sets and the resulting chaotic motion is thus linked to *analytic number theory*. More recently, links have also been made between dynamics and *algebraic number theory* [31,34].

The emergence of chaos has given new direction and impetus to the study of many other areas, for example turbulence, laser physics, theory of vibration, population dynamics, and weather forecasting.

Acknowledgement:
I would like to thank Jeff Johnson and Derek Richards of The Open University for their encouragement and detailed help during the preparation of this paper.

References

[1] Abraham, R. & Shaw, C.D. (1986) *Dynamics: The Geometry of Behaviour*, Part One, *Bifurcation and Chaos in Forced Van der Pol Systems*, Aerial Press, Santa Cruz, CA.

[2] Aronson, D.G., Chory, M.A., Hall, G.R. & McGeehee, R.P. (1982) Bifurcations from an invariant circle for two-parameter families of maps of the plane: A computer assisted study. *Commun. Math. Phys.* **83**, 303-54.

[3] Arrowsmith, D.K. & Place C.M. (1990) *An Introduction to Dynamical Systems*, Cambridge University Press.

[4] Barnsley, M.F., (1988) *Fractals Everywhere*, Academic Press.

[5] Birkhoff, G.D. (1968) Sur quelques curves fermées remarkables, *G.D. Birkhoff Collected Mathematical Papers* **2**, 418-43.

[6] Collet, P. & Eckmann, J.P. (1980) *Iterated Maps of the Interval as Dynamical Systems*, Birkhaüser, Boston.

[7] Cvitanović, P. (ed.) (1984) *Universality in Chaos*, Adam Hilger.

[8] Devaney, R.L. (1986) *An introduction to Chaotic Dynamics*, Benjamin.

[9] Douady, A. & Hubbard, J. (1982) Itération des polynômes quadratiques complexes, *Comptes Rendues de l'Academie des Sciences*, Paris, t**294**, 123-26.

[10] Feigenbaum, M.J. (1983) Universal behaviour in non-linear systems, *Physica* **7D**, 16-39.

[11] Guckenheimer, J. (1979) Sensitive dependence on initial conditions, for one-dimensional maps, *Commun. Math. Phys.* **70**, 133-60.

[12] Guckenheimer, J. & Holmes, P. (1983) *Nonlinear Oscillations, Dynamical Systems and Bifurcations of Vector Fields*, Springer Verlag.

[13] Guckenheimer, J., Oster, G & Ipatchki, A. (1977) The dynamics of density dependent population models, *J.Math.Biol.*, **4**, 101-47.

[14] Gumowski, J. & Mira, C. (1980) *Dynamique chaotique*, Cépadues Editions.

[15] Hénon, M. (1976) A two dimensional mapping with a strange attractor, *Commun. Math. Phys.* **50**, 69-77.

[16] Koçak, H. (1989) *Differential and Difference Equations through computer experiments*, Springer Verlag.

[17] Li, T-Y & Yorke, J. (1975) Period three implies chaos, *Amer. Math. Monthly* **82**, 985-92.

[18] Levinson, N. (1949) A second order differential equation with singular solutions, Ann. of Math. (2) **50**, 127-53.

[19] Lorenz, E.N. (1963) Deterministic non-periodic flow, *J.Atmos. Sci.* **20**, 130-41.

[20] MacKay, R.S. & Meiss, J. (1987) *Hamiltonian Dynamical Systems*, Adam Hilger.

[21] Mandelbrot, B.B. (1977) *Fractals: Form, Chance & Nature*, W.H. Freeman, San Francisco.

[22] Marsden, J.E. & McCracken, M. (1976) *The Hopf Bifurcation and Its Applications*, Springer-Verlag.

[23] May, R.M. Simple mathematical models with very complicated dynamics, *Nature* **261**, 159-67.

[24] Misiurewicz, M. (1980) The Lozi mapping has a strange attractor, in *Nonlinear Dynamics*, ed. R.H.G. Helleman, New York Acad. of Sci., New York, 348-58.

[25] Murray, C.D. (1986) The Structure of the 2 : 1 and 3 : 2 Jovian Resonances, *Icarus* **65**, 70-82.

[26] Peitgen, H.-O. & Richter, P.H. (1986) *The Beauty of Fractals*, Springer Verlag.

[27] Poincaré, H., *The New Methods of Celestial Mechanics*, D. Goroff (ed), Adam Hilger, (1990).

[28] Ruelle, D. (1989) *Elements of Differentiable Dynamics & Bifurcation Theory*, Academic Press.

[29] Sarkovskii, A.N. (1964) Coexistence of cycles of a continuous map of a line into itself, Ukrainian Math. J. **16**, 16-71.

[30] Smale, S. (1963) Diffeomorphisms with many periodic points, *Differential and Combinatorial Topology*, (ed) S.Cairns, 63-80, Princeton University Press.

[31] Smale, S., (1967) Differentiable Dynamical Systems, *Bull.Math.Lond.Soc.* **73**, 747-817.

[32] Sparrow C. (1982) *The Lorenz Equation*, Springer Verlag.

[33] Thompson, J.M.T. & Stewart H.B. (1986) Nonlinear Dynamics & Chaos, Wiley.

[34] Vivaldi, F. (1987) Arithmetical Theory of Anosov Diffeomorphisms, *Proc. Royal Soc. Lond.* A **413**, 97-107.

[35] Whitley, D.C. (1983) Discrete dynamical systems in dimensions one and two, *Bull.Lond.Math.Soc.* **15**, 177-217.

[36] Wisdom, J., Peale, S.J. & Mignard, F. (1984) The Chaotic Rotation of Hyperion, *Icarus* **58**, 137-52.

Computing The Unpredictable: Deterministic Chaos and The Nervous System

Arun V. Holden

Department of Physiology and Centre for Nonlinear Studies
The University, Leeds, LS2 9NQ

Abstract

Spatio-temporal patterned, but irregular, electrical activity in the cells of the nervous system underlies all our behaviour and thought processes. Computer simulation of unpredictable behaviour in a complicated, structured system such as the mammalian brain may appear a thankless task, but some ideas from chaotic dynamics seem helpful. Chaotic behaviour is complicated, unpredictable, irregular behaviour exhibited by relatively simple, nonlinear systems. Irregularity is produced by *sensitivity to initial conditions*, rather than a large number of degrees of freedom, and chaos is found in maps, differential systems and real physical systems. Maps and differential systems that represent the behaviour of elements of the nervous system also exhibit chaos, and signals recorded from nerve cells and neural systems have been identified and quantified as chaotic. Simple nonlinear models can account for some of the irregularity of neural activity: spatio-temporal patterning and irregularity in the nervous system is more of an open problem, but coupled lattices of nonlinear elements (neural networks) provides an approach to the spatio-temporal chaos and coherence of neural activity.

1 Introduction

The nervous system consists of a large number (some 10^{12} in the human brain) of interconnected units, the nerve cells (neurones) and their supporting neuroglial cells. Any one cell has specific connnections from and onto some 10^5 cells. This complexity in connectivity underlies the belief that nervous systems are among the most complicated objects in the natural world. It is relatively straightforward to observe the electrical activity of single cells in the nervous system, and there is a tendency to identify the behaviour of the nervous system with its electrical behaviour. Individual cells show irregular electrical activity; oscillations, fluctuations, and sterotyped pulses or action potentials. Within groups of cells there is a patterning or coherence of this irregularity. Sherrington [27] described the patterns of electrical activity within the waking brain as " the head-mass becomes an enchanted loom where millions of flashing shuttles weave a dissolving pattern, always a meaningful pattern though never an abiding one; a shifting harmony of subpatterns." In my 1976 monograph [14] I commented that stochastic models of

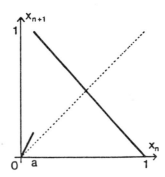

Figure 1. Piece-wise linear interval map for a model neurone, with threshold a.

the nervous system - diffusion processes and their first passage times, networks of logical neurones, field-theoretic approaches to sheets of excitatory and inhibitory neurones - had failed to match the insights of Sherrington's description. However, Bob May's 1976 review article [22] introduced chaotic dynamics to experimental and theoretical biologists (see [7][16]) and ideas from chaotic dynamics now provide simple explanations for some of the complicated behaviours seen in neural systems. Spatio-temporal patterned, irregular activity in the nervous system is not chaotic, neither is fully developed, hydrodynamic turbulence, but chaotic dynamics provides a conceptual framework, and some quantitative approaches to aspects of these complicated, unpredictable behaviours.

2 Interval maps

Studies on the quadratic map, as discussed in the preceeding chapter [2], have led to an awareness that complicated behaviour may be generated by simple nonlinear systems, and raise the notion that complicated behaviour in nerve cells might be generated by simple processes. The problem is to design a map that is a plausible model of a nerve cell. Nerve cells that transmit information over long distances in the nervous system do so by a sequence of action potentials, each of which is a brief (approximately 1 milli-second (ms)) large (about 100 milli-Volt (mV)) change in the electrical potential across the cell membrane. These propagating action potentials (or nonlinear travelling waves) are stereotyped, so can be considered as events; this could lead to a description in terms of stochastic point processes [14]. However, the more fundamental concept is that there is threshold, below which an action potential is not generated, and above which it is. Thus the function that determines the evolution of the membrane potential behaves as if it has a discontinuity, separating a subthreshold range from a suprathreshold range. The simplest possible dynamics in the suprathreshold and subthreshold ranges is linear, and a piece-wise linear

Computing The Unpredictable

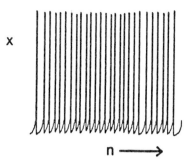

Figure 2. Sample iterations for map (2.1), generating a spike train that resembles a sequence of action potentials recorded from a real neurone.

interval map was introduced by Labos [20] as a model neurone

$$\begin{aligned} x_{n+1} &= mx_n, & x_n \le a \\ x_{n+1} &= ux_n - u, & x_n > a \end{aligned} \quad (2.1)$$

where $m = 1 + (1-a)^2$ and $u = 1/(a-1)$. This recursive funtion is shown in Figure 1 with a threshold $a = 0.1$; this gives the threshold approximately 10% of the peak to peak height of the action potentials. Iteration of this recursive function gives a series of spikes that looks similar to a recorded train of action potentials, and so although this is an extremely abstract representation of a nerve cell, it generates a "biologically plausible" output, as shown in Figure 2.

A bifurcation diagram of values of x_n at different values of the threshold parameter a is not helpful; the behaviours that we are interested in are regular (periodic) and irregular sequences of spikes. For a fixed value of threshold different starting conditions x_0 give rise to different sequences, or spike trains with different interspike intervals, where the interspike interval is the number of iterations (time steps) while the x_n remain on the subthreshold line. If we start at x_0 and remain on the subthreshold line for k steps the iteration will return to the subthreshold line at step $(k+2)$ with a value $x_{k+2} = u^2 m^k x_0 - au^2$. From the equations (2.1) it is easy to obtain an $x_0(k)$, where k is the interspike interval, for a periodic discharge as

$$x_0(k) = au^2/(u^2 m^k - 1).$$

Thus for any number k we can compute an x_0 that will give a periodic solution. However, this computation is unstable, as any error ϵ in x_0 is amplified by $u^2 m^k$ (which is > 1) each cycle. Repeated iteration gives, after several tens of spikes, an irregular discharge: this map acts a chaotic spike train generator.

Thus, for a one-dimensional map model, retaining only the threshold property of a nerve cell gives a chaotic discharge as the normal mode of behaviour. However, this tells us more about about maps, with an average $|f'| > 1$, than neurones. This kind of model is extended to include more realistic biological behaviours, by incorporating more piece-wise linear segments, in [23].

3 Periodically forced neurones

Although there are well studied experimental physical systems whose behaviour has been identified as chaotic (see, for example, [1]), some argue that the current enthusiasm for chaotic dynamics is misplaced - see the recent monograph by Gumowski [11]. What would be convincing would be a system of nonlinear ordinary differential equations that was firmly based on quantitative, detailed experiments, where chaotic behaviour was understood qualitatively, and where quantitative results (bifurcations points, numerical integrations) corresponded closely with experimental observations. It is curious that we are closer to such a system in neurobiology than in mathematical physics, where there are only well studied systems of unreal caricatures, such as the Lorenz equations [21], or detailed experiments on irregular fluid flow phenomena, where rigorous but simple mathematical models are lacking.

Ever since Helmholtz in 1850 not only measured the conduction velocity of frog nerve nerve fibres, but noted the dependence of conduction velocity on temperature, the description of excitation in nerve cells has been in precise and quantitative terms. The development of intracellular recording methods and the voltage clamp technique has provided direct measurements of membrane currents at different membrane potentials. Hodgkin and Huxley [13] summarised the results of an extensive series of experiments on squid giant nerve fibres by a fourth-order system of nonlinear ordinary differential equations that represent the electrical properties of the excitable membrane of a squid nerve fibre, which has three conductance systems, a sodium, potassium and leakage system.

The Hodgkin-Huxley membrane system is a four variable nonlinear differential system; the four variables are membrane potential V, sodium activation m, sodium inactivation h, and potassium activation n. $0 \leq m, n, h \leq 1$, and these variables represent the fraction of a control or gating process that is in the state that allows ionic current to flow through the membrane. For a two variable nonlinear system periodic oscillations are to be expected; for a three variable system, chaos; for a four variable system, hyperchaos, with two positive Lyapunov exponents (see previous chapter for a definition [2]). Systematic exploration of the bifurcation behaviour of the Hodgkin -Huxley system as parameters (ionic conductances, concentrations, applied current, temperature) are varied [18] only shows equilibria and simple periodic solutions, with periodic solutions emerging at sub- or super-critical Hopf bifurcations. This restricted range of behaviour of the Hodgkin-Huxley system means that it is behaving as a two-variable system; although physically the different variables do correspond to independent processes the four variable system may be approximated by a reduced system. Thus to produce chaotic activity in such a simple nonlinear oscillator periodic forcing is necessary.

For some frequencies and amplitudes of applied sinusoidal current density, the solutions of the Hodgkin-Huxley equations are irregular (see Figure 3).

From such numerical integrations attractors can be reconstructed, Lyapunov exponents, fractal dimensions and entropies estimated, and a bifurcation diagram constructed showing the periodic, quasi-periodic and chaotic responses at different

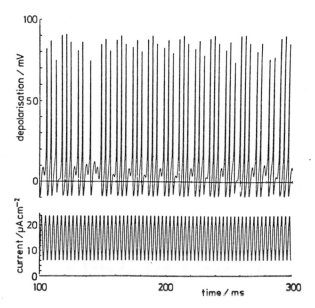

Figure 3. Numerical solutions of the Hodgkin-Huxley membrane equations' maintained response to an applied current density of $15 + 9\cos(2\pi 325 t)$ mA/cm^2. This irregular activity is persistent and is an example of deterministic chaos.

amplitudes and frequencies of sinusoidal forcing [5][7].

These numerical studies on chaotic oscillations in the forced Hodgkin-Huxley membrane system are in close agreement with actual experiments on the membrane of squid giant axons; see the papers by Matsumoto and Aihara in [16]. Although sinusoidal forcing of the Hodgkin-Huxley system provides an example where there is close agreement between experimental results and numerical computations this laboratory system is of little biological significance; what is of more interest is endogenous chaos in nerve cells.

4 Endogenous bursting and chaos in single neurones

Voltage clamp analyses of excitable membranes from different nerve and muscle cells have shown a wide range of conductance processes, all with H-H type kinetics, with α's and β's that are voltage dependent. In addition, there are charge dependent conductances, in that entry of Ca^{2+} charge (current multiplied by time) activates a K^+-selective conductance. Chay [3] has simplified the 5 variable model for pancreatic β cells [5] to give a 3 variable model for molluscan neuronal somatic membrane:

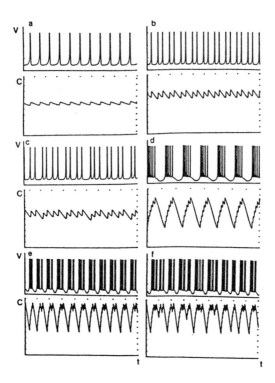

Figure 4. Numerical integrations of (16), with $V(t)$ displayed above $C(t)$. Same voltage scale; different scales for C and time. (a) Simple, (b) period two, (c) period four, (d) six burst and (e) nine burst periodic solutions; (f) irregular (chaotic) solution. These solutions were obtained for different values of g.

$$\begin{aligned} dV/dt &= g_I^* m_\infty^3 h_\infty (V_I - V) + g_{K,V} n^4 (V_K - V) \\ &\quad + g_{K,C}^* C/(1+c).(V_K - V) + g_L^* (V_L - V) + I \\ dC/dt &= \rho \{ m_\infty^3 h_\infty (V_C - V) - k_C C \} \\ dn/dt &= (n_\infty - n)/\tau_n \end{aligned} \qquad (4.1)$$

Time t, measured in msec, is the independent variable. The dependent variables are V, the membrane potential in mV, C, the dimensionless calcium concentration, and n, a probability of activation. The g_I^*, $g_{K,V}$ and g_L are maximal conductances, and the V_I, V_K, and V_L reversal potentials.

Explicit expressions for the steady state values h_∞, m_∞ and n_∞ and their time constants τ_h, τ_m, and τ_n are given. The parameters are I, the applied current in mV, and g_{KC}^*, the maximal conductance of the Ca^+-sensitive K channel divided by the membrane capacitance. The values of the constants are

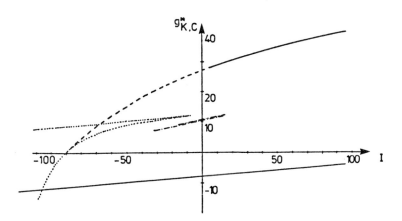

Figure 5. Curve of bifurcation points for the Chay system. Solid line means Hopf bifurcation with stable periodic solution bifurcating, dashed line is Hopf bifurcation with unstable periodic solution bifurcating, dotted line is saddle node bifurcations and dash-dotted lines are period doublings.

$g_I^* = 1800\ s^{-1}$ $g_{K,V}^* = 1700\ s^{-1}$ $g_L^* = 7\ s^{-1}$
$V_I = 100\ mV$ $V_K = -75\ V$ $V_L = -40\ V$ $V_C = 100\ mV$
$k_C = 3.3/18\ mV$ $\rho = 0.27\ mVs^{-1}$

Numerical solutions of (5.1), with appropriate values for the parameters $g_{K,C}^*, I$, show simple, periodic, doublet, triplet, bursting and chaotic discharges - see Figure 4. Hopf, saddle node and period doubling bifurcation curves in the $g_{K,C}^* - I$ plane are shown in Figure 5.

Thus the addition of a charge-dependent conductance has led to an increase in the range of possible behaviours: this may be understood as a result of introducing cross-terms into the kinetics. In the H-H system

$$dm/dt = f_1(m, V)$$
$$dh/dt = f_2(h, V)$$
$$dn/dt = f_3(n, V)$$

the gating variables (m, n, h) which represent chemical quantities have kinetics which depend on voltage and their value only: there are no cross terms in the kinetics. However,

$$dC/dt = f_4(C, m, h, V)$$

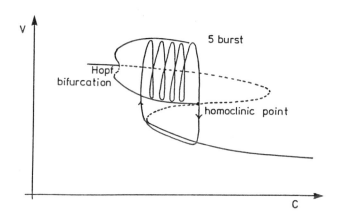

Figure 6. C acts as a slow variable that modulates the dynamics of the V - n subsystem, which has oscillatory solutions arising at a Hopf bifurcation and which terminate by collescing with a saddle point, giving a homoclinic orbit.

and so the addition of g_{KC} introduces cross terms. These cross terms arise because the flow of a chemical species causes a concentration change in a restricted compartment: if the intracellular compartment were semi-infinite, or the Ca^{2+} were buffered, there would be no cross terms. Thus spatial compartmentalization in a chemical system allows chaotic activity.

A more general, dynamic explanation is to consider the (V, n) system for C close to a subcritical Hopf bifurcation at C^* – see Figure 6. At $C < C^*$ there will be large amplitude oscillations, each giving an increment in C, until C^* is reached and the burst terminated. The slow dynamics of C then lead to a reinjection into the oscillatory zone [26].

The Hodgkin-Huxley and Chay systems are biophysically accurate, excitation systems that also suggest a plausible mechanism for excitation. They are obtained from experiments, and so are complicated, often of high order, and are continually being updated by new experimental results. Biophysically relevant problems require quantitatively accurate answers: what happens at what values of what parameter? Thus numerical methods and results are important, and there is a strong tendency for both the model and the method of analysis to be given as computer algorithms: the current model for cardiac membrane dynamics is a computer package (from D. Noble of Oxford University), and the use of path tracking procedures such as AUTO [8] is becoming common. It would be possible to combine model formulation from voltage clamp data, and model analysis by bifurcation and even perhaps singularity theory into an expert system: all the steps can be represented as algorithms.

A different approach is to find the simplest system of ordinary differential equations that can represent the behaviours exhibited by neurones: the Hindmarsh - Rose [12] equations are an example of such a polynomial system. Although these systems are simple enough to analyse, they are of more interest to applied mathematicians than physiologists.

Bifurcation diagrams, such as Figure 5 are becoming common: for a given membrane preparation there are a number of possible models - a complete, biophysically accurate, high order differential system, and a number of simplifications, alternatives, caricatures (o.d.e.'s with polynomial r.h.s.) and even cartoons (piecewise linear o.d.e.'s, or even maps). For each of these a bifurcation diagram can be calculated, and the bifurcation diagram characterises the system, much as a fingerprint characterises an individual. There is a need for objective, algorithmic analysis of these diagrams, to evaluate which provide the best description of the experimental results. A problem is that many features (such as period doubling into chaos) are universal.

Even when a complete, quantitatively accurate bifurcation diagram is available, and is quantitatively consistent with experimental results, there is a need to understand the diagram. This means relating the bifurcation theory results to the underlying singularity theory - see Laboriau in [7].

The results on forced chaos are a good example of close, quantitative agreement between experiment and model: for the assessment of measures of chaos the squid giant axon membrane is an ideal laboratory preparation, since the H-H equations provide an accurate description. However, the mathematical analysis of chaos in periodically forced systems is still rudimentary - see Tomita in [16], and requires a deep understanding of circle maps.

These examples of endogenous chaos show that the irregularity of the activity of single nerve cells, which in some cases may be identified as chaotic (by reconstruction of attractors, Poincaré sections, estimation of Lyapunov exponents (see the previous chapter for definition [2]), can be accurately modelled. The elements of the nervous system are in some cases chaotic; what we now want to consider is the patterned but irregular activity in systems of nerve cells.

5 Chaos in neural systems

So far we have demonstrated that simple models of single nerve cells can exhibit chaotic activity, and chaotic behaviour can be identified in signals recorded from single neurones. The nervous system is a hierarchical organisation of interconnected, interacting cells; and its behaviour is determined both by the properties of its components, and the pattern and nauture of the connectivities between them. Given a knowledge of neuronal properties and connectivities, it should be possible to understand the "meaningful, ..., shifting harmony of subpatterns" of their spatio-temporal activity. A major problem is the lack of detailed experimental observations on the simultaneous activity of large numbers of functionally related neurones. One promising experimental preparation is the skin of some marine molluscs, as the pattern of skin pigmentation provides a simple mapping of the spatio-temporal pattern of activity generated by part of the nervous system [17].

Given this lack of hard data, there is a tendency to use inadequate data (such as recordings of the electro-encephalogram (EEG)) or to speculate on the essential or perhaps pathological role of chaotic activity in neural systems. Chaos has been identified in EEG recordings, and changes in EEG dynamics quantified by reconstruction of attractors, and estimation of their low, and even fractal dimensions ; see Rapp *et al.* in [7]. These intriguing results may be clinically useful, say in the identification of brain death, but the EEG is a far too crude description of brain activity for spatio-temporal descriptions of the EEG to be useful in characterising Sherrington's meaningful, patterns.

6 Towards spatio-temporal chaos

Not only is the nervous system a spatially extensive, complicated structure, but it is an open system, in that information flows into, and out from it. This flow of information across the boundaries of the nervous system represents its actual functioning, since activity in the nervous system can only be biologically meaningful when it gives rise to some change in motor behaviour.

Spatio-temporal patterned activity in such an open system may be explored using as simple as possible a model, that retains the nonlinear, threshold properties of neurones, and the connectivity and interactions seen in neurones. One approach is the theory of neural nets, where each element is a binary element; networks of such formal neurones, with modifiable connections, form cellular automata [10].

However, if the components of the network themselves are chaotic, or generate information, a continuous state space for the component is required. An interval map seems to be the simplest possible model for the component neurones; thus a neural network would be represented by a coupled map lattice.

Coupled map lattices (CML's) are continuous state, discrete time, discrete space dynamical systems. An example in one-dimension may be represented as:

$$x_{n+1}^i = \sum_{j=1}^n \alpha_{ij} f_j(x_n^j), \quad i = 1, ..., n$$

where a simple form might employ the logistic map as a measure of the local dynamics, and appropriate α_{ij} to represent diffusive coupling of some of the n nodes of the system. More realistic CML models of the behaviour of neural systems would use (2.1) and non-local, excitatory and inhibitory connectivity functions α_{ij}. These dynamical systems can *either* directly model specific spatially distributed systems *or* can be viewed as approximations to models based on partial differential equations that form some kind of continuous field representation. In both cases the power of the CMLs is their spatially discrete structure which is suitable for computation, particularly using parallel languages and machines [19].

References

[1] Abraham, N.B., Gollub, J.P and Swinney, H.L., 'Testing nonlinear dynamics', *Physica-D* **11**, 252 -264, 1984

[2] Arrowsmith, D. K., 'The mathematics of chaos', in *The Mathematical Revolution Inspired by Computing*, J. H. Johnson & M. J. Loomes (eds), Oxford University Press, (Oxford) 1990.

[3] Chay, T., 'Chaos in a three variable model of an excitable cell', *Physica-D* **16**, 233 - 242, 1985

[4] Chay, T. and Rinzel, J., 'Bursting, beating and chaos in an excitable membrane model', *Biophys J.* **47**, 357-366, 1985

[5] Crutchfield, J.P. and Kanenko, K. 'Phenomenology of spatio-temporal chaos', In *Directions in Chaos* Hao Bai-Lin (ed), World Scientific, (Singapore), 272-353, 1987

[6] Cvitanovic, P., (ed) *Universality in Chaos*, Adam Hilger, (Bristol) 1984

[7] Degn, H., Holden, A. V. and Olsen, L. F., (eds) *Chaos in Biological Systems*, Plenum Press, (New York) 1988

[8] Doedel, E., 'AUTO86 Users Manual', Department of Computer Science, California Institute of Technology, (Ca 91125, USA), 1986

[9] Feigenbaum, M. J., 'Universal behaviour in nonlinear systems', *Los Alamos Science* **1**, 4-27, 1980 (reprinted in [5])

[10] Fogelman-Soulie, F., Robert, Y and Tchuente, M. (eds) *Automata Networks in Computer Science*, Manchester University Press, (Manchester) 1987

[11] Gumowski, I., *Oscillatory Evolution Processes*, Manchester University Press, (Manchester), 1989

[12] Hindmarsh, J.L., Rose, R.M., 'A model of neuronal bursting using three coupled first order differential equations', *Proc. Roy. Soc. of London* **B221**,87-102, 1984

[13] Hodgkin, A. L., Huxley, A. F., 'A quantitative description of membrane current and its application to conduction and exciation in nerve', *J. Physiol* **117**, 305-355, 1952

[14] Holden, A. V., *Models of the Stochastic Activity of Neurones*, Springer-Verlag, (Berlin) 1976.

[15] Holden, A. V., 'The mathematics of excitation' in *Biomathematics in 1980*, A. C. Scott and L. M. Ricciardi (eds), North-Holland, (Amsterdam), 15-48, 1982

[16] Holden, A. V., (ed) *Chaos*, Manchester and Princeton University Presses, 1986

[17] Holden, A. V. and Matsumoto, G., 'Patterned and irregular activity in excitable media', in *Structure, coherence and chaos in dynamical systems*, P.L. Christiansen and R. D. Parmentier (eds), Manchester University Press (Manchester), 185-198, 1989.

[18] Holden, A.V. and Winlow, W., 'The nerve cell as a differential system', *IEEE Trans. SMC*, **13**, 711-720, 1983

[19] Holden, A.V., Tucker, J and Thompson, B.C., 'The computational structure of neural systems', in *Neurocomputers and Attention*, A.V. Holden and V. I. Kryukov (eds), Manchester University Press, (Manchester), 1990.

[20] Labos, E., 'Periodic and nonperiodic motions in different classes of formal neuronal networks and chaotic spike generators', *Cybernetics and Systems Research* **2** 237-243 Elsevier, (Amsterdam) 1984

[21] Lorenz, E., 'Deterministic nonperiodic flow', *J. Atmos. Sci.* **20**, 130-141, 1963

[22] May, R. M., 'Simple mathematical models with very complicated dynamics' *Nature*, London **261**, 459-67, 1976

[23] Nogradi, E. and Labos, E. 'Pseudorandom interval maps for simulation of normal and exotic neuronal activities', in *Cybernetics and Systems 86*, R. Trappl (eds), D. Reidel Publishing Co., 427-435, 1987

[24] Peitgen, H-O and Saupe, D. (eds) *The Science of Fractal Images*, Springer-Verlag, (Berlin), 1988

[25] Rössler, O. E., 'An equation for continuous chaos', *Phys Letters*, **57A**, 397-398, 1976

[26] Rinzel, J., 'Bursting oscillations in an excitable membrane model', in *Proc. 8th Dundee Conference on the Theory of Ordinary and Partial Differential Equations*, B.D.Sleeman, R.J.Jarvis and D.S.Stein (eds), Springer,-Verlag, (Berlin) 1985

[27] Sherrington, C. S., *Man and his nature*, Cambridge University Press, (Cambridge) 1946

[28] Sparrow, C. T. *The Lorenz Equations: bifurcations, chaos and strange attractors*, Springer-Verlag, (Berlin), 1982

[29] Wolf, A., Swift,J.B., Swinney, H.L., Vastano, J.A., 'Determining Lyapunov exponents from a time series', *Physica* **16 D**, 285-317, 1985

Word Processing Algorithms, Rewrite Rules and Group Theory.

D. B. A. Epstein [1]

Mathematics Institute, University of Warwick, CV4 7AL

Abstract

This article shows how, given an alphabet A, a set of strings arranged in a tree T, and another string w, one can find out in time proportional to $|w|$ whether w is contained in the tree T. If one makes explicit the dependence on T and on A, the time taken is $O(|T|^2|A||w|)$. There are examples showing that one can not do better than the quadratic dependence on $|T|$, but probably this happens only with pathological examples. Since writing the article, I have realised that, at the expense of a certain amount of space, the algorithm can be adjusted to make the estimate $O(|T||A||w|)$, with the hidden constant in the formula also smaller than before. The result is presumably well known to experts on string manipulation.

1 Introduction

During this century we have seen a steady increase in the importance of the constructivist point of view in mathematics. Although most of us are quite happy to accept a non-constructive proof, we usually feel happier about a constructive proof, and such a proof normally feels more substantial, satisfying and aesthetically pleasing than a parallel non-constructive proof.

The distinction is rather clearer to me now and perhaps also to the rest of the mathematical community than it was when I was a graduate student. My first real contact with the constructive approach was an algorithm produced by Wolfgang Haken solving the Knot problem. This problem can be stated as follows. A *knot* is

---

[1]This article was written for an unpublished volume of papers by the members of the University of Warwick Mathematics Department, which was presented to Christopher Zeeman on the occasion of his departure from the University to become Principle of Hertford College, Oxford. The work described here was carried out in part under the auspices of the Geometry Supercomputer Project. The Project receives support from the National Science Foundation and the University of Minnesota and its Supercomputer Institute. The research was also supported by the Science and Engineering Council of the United Kingdom. The author wishes to thank all these bodies for making the research possible. An earlier version of this paper appeared as Report 88/85 of the University of Minnesota Supercomputer Institute.

a topological circle embedded in Euclidean 3-space. In order for it to be possible to examine the knot from an algorithmic point of view, we have to think of it as being given by a finite set of data, for example a *knot diagram* which specifies a finite number of overcrossings and undercrossings with a combinatorial description of how the crossings are joined together. The Knot Problem is to find an algorithm which starts with this finite set of data, and determines whether or not the knot can be deformed to an unknotted circle. Haken's contribution was to give such an algorithm, together with a proof that it worked. This was the first result in a development which has now become a major industry, and is, in my opinion, one of the most important strands of modern mathematics, namely the attempt to produce algorithmic processes which can answer all the important questions about compact 3-dimensional manifolds. It is this development which has been a major impetus for my own research over the last few years.

Haken's result, presented to the International Congress of Mathematicians in Amsterdam in 1954 should have startled the mathematical world. Instead, it was met with the same indifference shown by the native inhabitants of Tierra del Fuego to the arrival of H.M.S. Beagle with Charles Darwin on Board — the evidence of the senses was simply disbelieved. The incredulous response of mathematicians to Haken's claims was conditioned by the remarkable results just obtained previously by Novikov and Boone, whose work was also presented in Amsterdam. They had recently produced explicit examples of groups given by finite collections of generators and relators for which one can formally prove that no algorithms exist which can determine, for any given word in the generators of the group, whether it represents the identity or not. The question of the existence of such an algorithm is called the *Word Problem*. Associated with this was their proof that there is no algorithm, which could take any finite set of generators and relators defining a group, and say whether or not the group is trivial. In the same way, given any fixed group H, it is not possible to produce an algorithm which has as its input any finite set of generators and relators and as its output "Yes" or "No", according to whether the group is isomorphic to H. The way this affected beliefs about the Knot Problem was as follows. At that time it was generally believed (and later proved by Papakyriakopoulos) that one could determine whether or not a knot was knotted by knowing its *knot group*, that is the fundamental group of its complement in 3-space. One can easily read off generators and relators for the fundamental group from a knot diagram. Further, it was felt that knot groups were pretty hard to handle, and were good examples of general groups. It followed, so many believed, that if the Word Problem were insoluble, then the Knot Problem also would be insoluble. Haken's work, contradicting these generally held beliefs, was finally accepted after about eight years (when Schubert gave a lucid account [4]).

If we know how to construct some mathematical object, another question arises - how many steps are required in the construction? The importance of the complexity of an algorithm was not fully appreciated by most mathematicians at the time Haken produced his algorithm. Either it was possible to give an algorithm, or it was not possible. In any case, only the very simplest calculations could be done in practice, for they would have to be done by hand. Computers existed, but they were awkward to use, and not generally available to mathematicians. So the algorithm would in, general, have theoretical interest only. Nowadays, the situation is very different. The development of computers has made the speed of algorithms a matter of intense interest to mathematicians who would like to know the answers to particular questions. Any mathematician who produces an algorithm would like to see it work in practice. It is, I believe, a source of frustration to several mathematicians that the only knot, which has been successfully analysed using Haken's methods, is the trefoil knot, although many hours of supercomputer time have been expended.

Since Haken's pioneering work, researchers in 3-manifold theory have become more ambitious. In particular one of our objectives is to produce practical algorithms which could take a 3-dimensional manifold presented in some combinatorial way, and answer various practical questions about it quickly (*e.g.* is it homeomorphic to some other manifold?). By *practical*, we mean that we want the answer within not more than a couple of hours, working with current hardware. Programs developed in Princeton by Bill Thurston, Daryl Cooper and Jeff Weeks have been notably successful in this respect, finding the unique complete hyperbolic structure on a knot complement in a few minutes on a small computer, provided the knot is sufficiently simple to be drawn by hand. So this is constructive mathematics with a vengeance. Algorithms which would give an answer a thousand million years from now are accepted by constructivists, but might be regarded a little disdainfully by the modern topologist.

Although the word problem is not solvable in general, it is solvable for certain groups, for example a finite group, or a nilpotent group, or the fundamental group of a compact hyperbolic manifold without boundary. Since most 3-dimensional manifolds are hyperbolic, this last result is of great interest. A practical solution to the word problem for such a group (the theoretical solution is already known) would help us in our aim of providing practical algorithms to answer practical questions about 3-manifolds. In fact, not only is the word problem solvable for the fundamental group of most 3-manifolds, but it is solvable by a fast algorithm. A huge amount of theoretical work remains to be done in working out such algorithms, but a great deal of progress has already been made in joint work by the author, Jim Cannon, Derek Holt, Mike Paterson and Bill Thurston [1]. At the same time, Sarah Rees, Derek Holt and the author are writing computer code to give effect to the theory. One of the objects of this paper is to describe some of the ideas embodied in our code. The aim is to provide a computer program which takes as its input generators and relators for a group and has as its output an algorithm for solving the word problem in that group. In the case of a general group, the program will often terminate saying "Sorry, can't do it". But for many groups and, in particular for the fundamental group of 3-manifolds which might arise in practice, we hope that the program will be successful on the examples we are interested in.

2 String Matching

The only contact many mathematicians have with computers is through word-processors. Most of them are unaware of the clever algorithms, which are used in good word-processing programs and which make the programs work amazingly quickly. The type of process we are particularly concerned with is that of searching for a particular string inside a rather long text. For example, if you have consistently misspelled (or misspelt) a word, then you can set the computer to find and correct each instance. How long does it take to find each instance? A brute force procedure, given a text of length N and a word of length k, uses $O(Nk)$ steps in a worst case, and $O(N)$ steps for typical English words and text. (See below for a description of what is meant by this $O(n)$ notation.) This is unnecessarily long. In the Boyer-Moore algorithm, the remarkably short expected time taken for English text is $O(N/k)$, which means that, although one looks at only a small fraction of the characters in the text, one can still find any match with certainty. However, if the word one is trying to match is too long to store in memory, there are substantial problems in having to continually back up the text, which is necessary in the Boyer-Moore algorithm. A typical application where this is a problem is where we are asking whether the contents of one file are repeated as a consecutive piece of another file. Below we will describe in detail another algorithm for such matching, due to Knuth, Morris and Pratt, with worst case time estimate $O(N+k)$, and where no backing up is required.

Suppose you have a finite alphabet A. (An *alphabet* is just another name for a set, though the name normally carries the implication that the set is finite. The elements of the set are called the *letters*.) A *string* over A is a finite (possibly empty) sequence of letter of A. We denote the length of a string s by $|s|$. By a *substring* of s, we mean any consecutive subsequence of letters of A. Let $s_1, \ldots, s_f$ be a finite set of strings over A, which we can think of as short strings. Suppose we are also given a very long string t, which we call the *text*. We would like to find all substrings of t which are equal to some s_i. One possible application would occur if one had the works of Shakespeare recorded in a computer readable form (for example on a magnetic disk), and one wished to find the quotation "tomorrow and tomorrow and tomorrow". (This would actually be rather harder than you might think, even if you knew quite a bit of computer science, because Shakespeare spelled it "to-morrow".) In this case t would be the entire works of Shakespeare, and f would be 1. Another example might occur if one had a partial knowledge of an extinct language, and one wished to locate all occurrences of the few words whose meaning one knew.

We now explain an algorithm, due to Knuth, Morris and Pratt, which finds all substrings of t which are equal to some s_i in time $O(|s_1| + \cdots + |s_f| + |t|)$. A brute force approach will take time $O(|t|(|s_1| + \cdots + |s_f|))$.

Before proceeding, we explain the meaning of the $O(n)$ notation. We say that a procedure takes time $O(n)$ to compute, if there is a constant c, independent of the data for the particular situation, such that the procedure takes a time bounded above by cn where n is some linear function of the input (for example its length when expressed in binary form).

For this to be made more than just an intuitive notion, we need to have in mind a particular mathematical model of the computing machine on which the computation is to be carried out. One model often used is the Turing machine, where we have a bi-infinite tape, divided up into a bi-infinite sequence of boxes. In each box, it is possible to write a 0 or a 1, and, at any one time, all except a finite number of boxes are 0. The machine has a finite number of possible states. At each step in the computation, it scans (*i.e.* reads) the symbol in one box and deletes it. Then, according to its state and the symbol it has just seen, it writes either a 0 or a 1 in that box, and moves the tape either to the right or left. The unit of time is the time taken to perform one of these operations.

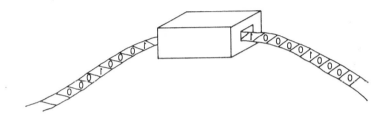

Figure 1. A Turing machine

Any computation which can be performed on any machine can be performed on a Turing machine, and the Turing machine model is often used when proving lower bound results for the length or size of a computation. Unfortunately, the Turing machine does not give a realistic time estimate fo, many realistic problems. The reason for this is that the time taken to retrieve a piece of information from the memory of a modern electronic computer is constant, independent of where the information is located. In a Turing machine, the memory resides on a part of the tape which might be a long distance away from the part where the computation is being carried out, and it takes time for the tape to be moved over to the site of the memory, and then back to the site of the computation. For this reason, computer scientists often use other mathematical models of computing machines, which give a better idea of the time an algorithm would take in practice. We will not describe such machines, but refer the reader instead to [3] for a discussion of several different examples.

Time estimates for algorithms are usually presented as functions of the size of the input data, for example, counting the number of bits needed, and using the $O(n)$ notation. (An indication of the amount of output data required may be included in the input data. For example, in an algorithm to compute the decimal digits of π, the only input would be the number of digits required.) This means that two algorithms whose time estimates are related to each other by a multiplicative constant are regarded as equally good by the theoretical computer scientist. In particular, only the asymptotic behaviour of the algorithm, as the size of the problem grows, is regarded as significant. Of course, in practice, the size of the constant is often important, and for some problems, it is best to use an algorithm which is not the best one asymptotically. There are algorithms where the superiority only displays itself for data sets that are so large that no-one has ever thought it worthwhile to implement them in code. From a theoretical point of view, it is most convenient to have the multiplicative constant available, because this makes the results easier to prove, and because the constant is likely to change, depending on what machine model one uses, even if one restricts oneself to machines with constant time memory retrieval.

One standard technique used to estimate timings is to write the program in some sort of pidgin high level programming language, so that it takes a time which is independent of the input data for each line to be executed. A line may be executed many times, due to looping constructs in the program. One then estimates the number of times each line will be executed, and adds.

3 Data Structures and Knuth-Morris-Pratt

The possibility of producing an efficient algorithm often depends on the way the data are organised. We are all familiar with this principle. For example, suppose one has 5000 names of university students which need to be sorted into alphabetical order. One could try writing the names out on large sheets of paper, leaving gaps between successive names. One could then go through the names systematically, crossing them out and rewriting them into the gaps. When there is no more room in some gap, where one wants to write another name, one rewrites the whole sheet of paper, possibly onto several new sheets, with the gaps between successive names restored. Another procedure is to write each name on a card, and then physically sort the cards. The first procedure is worse than $O(n^2)$, and the second is $O(n \log n)$. (Incidently, the log term can be shown to be inevitable. In the normal process we might use in practice, of looking at each card in turn, and inserting it into the right place, the log term comes from the time taken to find the right place using a binary search.)

We will not given a formal definition of a data structure. It will be good enough to think of a data structure as a blank card, with boxes marked on it, noting which kind of information should be entered in that box (name, telephone number, age, *etc.*). To improve the analogy with a computer, or with a more formal mathematical model, we must not think of the card as being placed in a box of cards able to slide forward and back easily as one inserts or removes cards. Rather, we must think of each card being in a fixed holder at a fixed position. (In the case of a computer, these holders would be particular addresses in memory.) The physical analogy is a little misleading. If we have a huge warehouse full of cards in cardholders, the time taken to retrieve two records which are distant from each other is large, whereas if the records are near each other it is small. In the type of machine model we are using, the time taken to retrieve any record is constant, independent of the relative position of the record examined just previously.

Now we describe a standard data structure for representing a *labelled graph*. Mathematically, this is a connected finite graph, with basepoint. The edges are directed, and each edge is labelled with a single letter from a fixed finite alphabet A. We assume that we can get from the basepoint to any point of the labelled graph by proceeding along directed edges, in the correct direction. Given a vertex v and a letter x of A, we assume that there is at most one edge labelled x issuing from v, and we denote the other end (if the edge exists) by vx.

We represent each vertex by a card, and each card is divided into $|A|$ boxes. If we have an edge labelled x, going from vertex u to vertex v, then the card corresponding to u has the address of v in position x. We get access to this labelled graph via the card representing the basepoint. Thus a representation of a labelled graph in a computer is likely to be scattered all over the memory of the machine. A *labelled tree* is a labelled graph which is acyclic. In a tree with basepoint, we talk of the *parent* w of a vertex v. This means the vertex w comes immediately before v on the unique path from the basepoint to v. We also say that v is a *child* of w. The basepoint itself has no parent, and is the only orphan vertex.

Given k strings $s_1, ..., s_k$ for which we wish to search in a text t, we first form the strings into a labelled tree. By following directed edges, in the correct direction, starting from the basepoint and continuing until a leaf is reached, we find each string s_i exactly once. (If one string is a substring of another, we omit the longer string.) The process of building this tree takes time $O(|s_1| + \cdots + |s_k|)$, because adding in the effect of a single letter amounts either to following from one card to the address written in one of its boxes, or to asking the system to provide the address of a blank card, plus the time to write this new address on the original card. (The fact that the system can provide a blank card in constant time really needs some discussion; the discussion depends on the assumption that the memory is infinite and that data can be retrieved from memory in a time which is independent of where the memory is located.)

By a *prefix* of a string, we mean an initial substring, and by a *suffix* a final substring. Given a vertex v, there is a unique string $s(v)$, such that if we start at the basepoint, and follow the route dictated by $s(v)$, we arrive at v. This gives a bijection between the set of all prefixes of the s_i's and the set of vertices of the tree. A string is said to be *acceptable* if and only if it is equal to some $s(v)$. The next step in the algorithm is to find, for each vertex v, the vertex w, with the property that $s(w)$ is a proper suffix of $s(v)$, and that w is the vertex for which such an $s(w)$ is as long as possible. Then $s(w)$ is the maximal acceptable suffix of $s(v)$. This gives a function f from the set of vertices to itself, called the *failure function* for this tree of strings, defined by $f(v) = w$. If v is the basepoint, then $s(v)$ is the empty string, and so $f(v)$ is not defined by the preceding definition. In that case, we define $f(v)$ to be v. Notice that the sequence of strings $s(f^i(v))$ is arranged in decreasing order of length, and that the length decreases strictly until we reach the empty string,

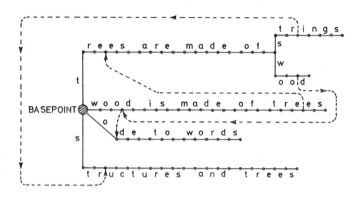

Figure 2. A tree and failure function

after which it remains constant. Moreover, the sequence ranges over all acceptable suffixes of $s(v)$.

The tree and the definition of f are illustrated in Figure 2. Here we have $k = 5$, and the strings we are planning to search for in t are:

"trees are made of wood"
"wood is made by trees"
"trees are made of strings"
"structures and trees"
"ode to words".

Note that in this example our alphabet has a blank as one of the letters. We have drawn long dotted arrows on the diagram, going from v to $f(v)$ for four different examples of v.

Since we are going to use the failure function, we have to find a way of storing it in the computer. What we are doing is to arrange for each of our cards to have one or more box. In that box, on the card corresponding to v, we write the address of the vertex $f(v)$. We can compute the function f in time proportional to the number of vertices. The procedure is to do a *breadth first search* of the graph. That is to say, we look at the vertices, in ascending order according to the distance from the basepoint. (This takes linear time, a not completely obvious fact, though we will not include a justification of the statement.) So, given a vertex v, and having computed $f(w)$ for all vertices w which are nearer to the basepoint than v, we need to compute $f(v)$. Notice that $f(w)$ must be strictly nearer to the basepoint than w. Let w be the parent of v, and suppose the edge from w to v is labelled x. Then we first look at $f(w)$ and see whether x is defined on it. If not, we look at $f^2(w)$, and see if x is defined on that. If not, we look at $f^2(w)$, and see if x is defined on that. Continuing in this way, we are eventually successful at $f^k(w)$ for some value of k, in which case we set $f(v)$ equal to the result of applying x to $f^k(w)$, or else we are never successful, in which case we define $f(w)$ to be the basepoint.

Here is a more formal description of the algorithm, where the indentation indicates how to parse:

1. if (there is another vertex v produced by the breadth-first search)
2.   if (v is the basepoint)
3.   define $f(v)$ to be v and go to line 1
4.   set w equal to the parent of v
5.   set x equal to the unique element of the alphabet with $wx = v$
6.   if (w is equal to the basepoint)
7.   define $f(v)$ to be the basepoint
8.   else
9.   set w equal to $f(w)$
10.   if (wx is defined)
11.   define $f(v)$ to be wx
12.   else go to line 6
13. go to line 1

Lemma 3.1. *Let $s_1, ..., s_k$ be k strings over a finite alphabet A. Suppose we form a tree from these strings in the manner described above. Then the procedure described does indeed define the failure function correctly (that is, $s(f(v))$ is the maximal acceptable suffix of $s(v)$). Moreover, the time estimate for the procedure is $O(|s_1| + \cdots + |s_k|)$.*

Proof. We have already remarked that, for any vertex w, the sequence of strings of the form $s(f^i(w))$ ranges over all suffixes of $s(w)$ which appear in the tree, arranged in order of decreasing length. The correctness of the algorithm follows from this, since if $v = wx$, the maximal accepted suffix of $s(v)$ is equal to some suffix followed by x.

Now, we prove the time estimate. Line 6 is executed at least as often as any other line, so we only need to estimate how often this is done. Note that in line 4, we need to know the parent of v. This is an extra piece of information required in the data structure, which can be inserted in constant time for each vertex, while the tree is being built up. (Most mathematicians who are unused to computer science arguments find subtleties, like the point just made, easy to miss. Since it is clear that the parent of v exists, it is all too easy to ignore the time needed to find that parent.)

Let $d(w)$ be the distance of w from the basepoint, for each vertex in the tree. We attach to each vertex w the label $\lambda(w) = d(w) - d(f(w))$, which is a strictly positive integer, except for the case of the basepoint, when the label is 0. Now consider a directed edge in the tree from w to v. We clearly have $\lambda(v) \geq \lambda(w)$. Moreover in line 9 w is definitely taken nearer the basepoint. It is easy to see that, with a fixed

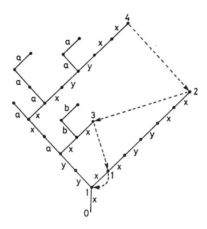

Figure 3. The tree for Lemma 3.1

v, the number of loops for lines 6 to 12 is bounded by $1 + \lambda(v) - \lambda(w)$ (see Figure 3). Hence the total number of steps, as v varies, is bounded by

$$O(|s_1| + \cdots + |s_k|) + \sum_E (\lambda(v) - \lambda(w))$$

where the sum is over the set E of edges of the tree. Each letter in each of the given strings $s_1, ..., s_k$ maps to a particular edge of the graph, and is therefore assigned the number $\lambda(v) - \lambda(w)$, where v and w are vertices at the ends of the edge. Therefore the expression we have just obtained is increased if, instead of summing over the set of edges, we sum over the letters in the s_i's. Since λ applied to the basepoint gives zero, the contribution of a fixed s_i to the sum is bounded by $\lambda(u)$, where u is the vertex (of the tree) corresponding to the end of the path labelled s_i. Now $\lambda(u) \leq d(u) = |s_i|$. This completes the proof of the lemma. □

Now we describe how to use the tree, equipped with its failure function, to find whether one of the strings s_i is a substring of a given text t. At each step in the procedure, we can describe the situation by means of a *state* $p(v)$, where p is a position in the linear text t and v is a vertex of the tree. The initial state is with p at the beginning of t, and with v equal to the basepoint of the tree. We suppose that we are in the middle of the procedure, and describe the next step. Let x be the letter immediately after p in the text. Let i be the smallest non-negative integer such that $f^i(v)$ is either equal to the basepoint or has x defined on it. If x is defined on $f^i(v)$, we set the new value of v equal to $f^i(v)x$, and is x is not defined on $f^i(v)$, we set the new value of v equal to the basepoint. In either case p moves one position to the right. What this means is that at each stage of the procedure, $s(v)$ is the maximal acceptable string which is a suffix of the text up to p. If t contains some s_i, then we will reach a state (p, v), where $s(v) = s(j)$.

Lemma 3.2. *The above procedure takes time $O(|t|)$.*

Proof. We imagine a point P moving on the positive real line, as the procedure continues. When the procedure is in state (p,v), P is at $d(v)$, the distance of v from the basepoint. At each step of the procedure, P moves one unit to the right, or else it jumps i times to the left ($i > 0$), where each jump moves at least one unit, followed by a possible move one unit to the right. Since addition of integers is commutative, P will land up at some point on the real line if all rightward moves by one unit are made first, and then all leftward jumps. The total number of rightward moves is bounded by $|t|$. Therefore the number of leftward jumps is also bounded by $|t|$. Therefore the total number of moves of both types is bounded by $2|t|$. This completes the proof of the lemma. □

The preceding two lemmas are standard results in computer science, due to Knuth, Morris, and Pratt. Together they imply that the while process of finding whether one of the strings s_j is contained in t takes time $O(|t| + |s_1| + \cdots + |s_k|)$.

4 Groups and Knuth-Morris-Pratt

We now show how to use the string matching algorithm of Knuth, Morris and Pratt, to help with the processing of the words of a group.

Let G be a group, with generators $(x_1, ..., x_k)$. It is convenient to assume that if x is in our set of generators, then so is x^{-1}. Let A be the set of generators. There is a map from the set of strings over A onto the group. The word problem is the problem of finding an algorithm which determines for any particular string, whether it maps to the trivial element of G or not.

There is a procedure called the *Knuth-Bendix-Procedure*, which is well-known and often used in various parts of mathematics and computer science. This also goes under the general name of *Rewriting Rules*. R.H. Gilman [2] and others have developed this technique for use when trying to solve the word processing problem for a particular group with given generators and relators. We will only talk about the Knuth-Bendix procedure in the context of such groups, although it has applications in very diverse areas of mathematics and computer science.

The idea is to try to write the group as a semigroup given by generators and relations, rather than as a group given by generators and relators. Formally speaking, we start with the free semigroup over the generators. A *relation* over these generates is simply a pair of words (u, v), written $u = v$. Given a set of relations of the form $u = v$, we form the equivalence relation on the free semigroup, generated by pairs of the form $aub = avb$, as a and b vary over all possible words in the free semigroup. Clearly, we get an induced semigroup structure on the sets of equivalence classes. In this way, we can define a semigroup if we are given a set of generators and relations.

Given generators and relations for a group, we can regard it as a semigroup, by adjoining the inverse of each generator to the set of generators and adjoining relations of the form $xX = 1$ and $Xx = 1$, where X denotes the generator inverse to x, and x varies over all generators. Clearly a semigroup satisfying these additions relations is a group.

Now suppose we have a well-ordering on the set of words of the free semigroup, such that, if $u > v$, then $aub > avb$. At the moment, the only well-ordering we are looking at in our programs is the ordering first by word length, and then, for words of the same length, dictionary ordering. This ordering is clearly a well-ordering and satisfies the condition of compatibility with multiplication. By interchanging u and v, we may assume that for each defining relation $u > v$. We write $u \to v$ and talk of u as a *left hand side* and v as a *right hand side*. We can *reduce* a word, replacing any subword which is a left hand side by the corresponding right hand side. Since the ordering on the words is a well-ordering, and since each reduction makes the word smaller in the ordering, the process must come to an end after a finite number of steps. The only problem is that two different reductions of a word might lead to two different answers. We define a multiplication law on the set of reduced words by juxtaposition and reduction. This is well-defined, and we recover the original semigroup, if and only if any two complete reductions of the same word are equal as strings.

There is only one kind of obstruction to showing that any two complete reductions are equal. Suppose $\alpha\beta\gamma$ is a word, where β is a non-empty word, and $\alpha\beta \to v_1$ and $\beta\gamma \to v_2$. We say that the two left hand sides $\alpha\beta$ and $\beta\alpha$ *overlap*. Then $\alpha\beta\gamma$ can be reduced to $v_1\gamma$ and also to αv_2. Reducing each of these completely, the two words may be identically equal strings or they may be distinct.

We say that a set of rewrite rules is *closed* if the following two conditions are satisfied:

1. Whenever there is an overlap of two left hand sides, $\alpha\beta$ and $\beta\alpha$, there is some complete reduction of $\alpha\beta\gamma$ which starts by reducing $\alpha\beta$ and some complete reduction of $\alpha\beta\gamma$ which starts by reducing $\beta\gamma$, with the property that the two completely reduced words are identical.

2. No left hand side is a substring of another left hand side.

Lemma 4.1. *Suppose we have a set of rewrite rules $u \to v$ which is closed. Then any two reductions of a given word result in identical words.*

Proof. Suppose we have a word and two complete reductions leading to two distinct reduced words. Consider a smallest such example. The first step of a complete reduction involves replacing a left hand side by a right hand side. The two complete reductions start with two left hand sides.

One possibility is that the two left hand sides are disjoint subwords of the given word. Then we can construct a new complete reduction, in which the first two steps are to reduce these two left hand sides, and it does not matter in which order these first reductions are performed.

A second possibility is that the two left hand sides $\alpha\beta$ and $\beta\alpha$ occuring in the first step of the two complete reductions overlap along β. Then we can construct a new complete reduction, which starts by completely reducing $\alpha\beta\gamma$.

The third possibility, that one left hand side entirely contains another, is ruled out by our hypotheses.

Comparing the new method of reduction with either of the given two methods, we see that we have constructed a smaller example, for which we already know that all possible complete reductions lead to the same answer. This completes the proof of the lemma. □

The Knuth-Bendix procedure goes as follows. We start with a finite set of generators and relators for a group. To do this we adjoin all relators of the form xX and Xx, where X is the generator which is the inverse to x in the group. For each relator r, we set up the rewrite rule $r \to 1$. We then look for overlaps between left hand sides $\alpha\beta$ and $\beta\alpha$, and we reduce $\alpha\beta\gamma$ in two different ways. If the two reductions lead to the same word, then we do nothing, but if they lead to different results u and v, then we may assume that $u > v$, and add $u \to v$ to our set of rewrite rules.

We also look for a left hand side u_1 which is entirely contained in another left hand side u_2. Let $u_1 \to u_2$ and $u_2 \to v_2$ be the two rewrite rules. Then, on the one hand u_2 can be completely reduced using the first of these rewrite rules, applied to the substring u_1. On the other hand, we can completely reduce v_2. If the two complete reductions lead to distinct words, we add to our set of rewrite rules. In any case, we throw away the rule $u_2 \to v_2$.

If we are lucky, this process terminates after a finite number of steps and in that case, the rewrite rules give a fast solution to the word problem. But often it will not terminate. Exactly how one proceeds when the process does not terminate is a long story, which need to be taken up elsewhere. The main point, as far as this paper is concerned, is to show why we want to find overlaps and substrings quickly.

5 The Knuth-Morris-Pratt approach to the Knuth-Bendix process.

Suppose we have a number of strings $u_1, ..., u_k$, and we are given a string u. We want to find all overlaps of each u_i with u, where a non-trivial prefix of u_i is equal to a suffix of u. We also wish to find if some $u_i \neq u$ is a substring of u. (There is nothing amiss with the asymmetry here, because we will take each of the left hand sides to be u in turn.) We form the u_i's into a tree, and then treat u as a text string, feeding u into the tree, as in the description above of the Knuth-Morris-Pratt algorithm.

If we want to come to the end of a u_i while feeding u into the tree, then we have found a substring u_i of u. We then throw away the rewrite rule involving u, and possibly insert another rewrite rule, as explained above.

If on the other hand we reach the end of u first, then we have found a maximal suffix of u which is equal to a prefix of some u_i. At this point we will have reached a vertex v in the tree, with $s(v)$ equal to the overlap. We then have to find all the u_i's which will have exactly this overlap with u. This is done by searching further along the tree, finding all the leaves of the tree, whose route to the basepoint is through v. By applying the failure function, we find the next longest suffix of u which is equal to a prefix of some u_i. Once again, we have to find all u_i which possess exactly this overlap. We continue to apply the failure function, each time finding one more overlap, until we reach the basepoint, when consideration of this particular left hand side u is done. The process continues indefinitely, with u cycling endlessly through the current set of left hand sides.

The time estimate for a single pass with u fixed is $O(|u| + |u_1| + \cdots + |u_k|)$. The only aspect of this estimate which we have not yet covered is the time taken to find all the leaves of the tree, whose route to the basepoint is through a given vertex. But this is easily seen to be bounded by the estimate just given. On the other hand, the straightforward method of doing the job without any fancy data structures is $O(|u|^2(|u_1| + \cdots + |u_k|))$. Moreover, in the author's opinion, the speedup is underestimated by this analysis. But the final verdict on this opinion awaits the completion of our programs, when the use of a stopwatch will provide a more objective means of assessment.

In our programs, the actual data structures used are quite a lot more complicated. We have simplified here, because there has only been space to study one aspect of the programming. It should be remarked that the structure actually used amounts to using the Knuth-Bendix process to direct the building of the Cayley graph in particular directions. An interesting aspect of our work is that it would seem to provide the possibility of building a Cayley graph using a parallel algorithm, since all left hand sides can be looked at simultaneously. With the development of parallel architectures and languages, this promises to give a tremendous speed up in the process.

References

[1] Cannon, J., Epstein, D. B. A., Holt, D. F., Paterson, M. S., and Thurston, W. P., 'Word Processing and Group Theory', (to appear).

[2] Gilman, R. H., 'Presentations of Groups and Monoids' , *Journal of Algebra* **57**, 544-554, 1979.

[3] Aho, A. V., Hopcroft, J. E., Ullman, J. D., *Design and Analysis of Computer Algorithms*, Addison-Wesley, 1974

[4] Schubert, H., 'Bestimmung der Primfaktorzerlegung von Verkettungen', *Math. Zeit.* **76**, 116-148, 1961.

Computer Assisted Proof for Mathematics: an Introduction Using the LEGO Proof System

Rod Burstall,
Laboratory for Foundations of Computer Science,
King's Building, Edinburgh University, Edinburgh, EH9 3JZ

Abstract

We give brief account of the use of computers to help us develop mathematical proofs, acting as a clerical assistant with knowledge of logical rules. The paper then focusses on one such system, Pollack's LEGO, based on the Calculus of Constructions, and it shows how this may be used to define mathematical concepts and express proofs. We aim at a gentle introduction, rather than a technical exposition.

1 Automatic Proof and Proof Checking

Attempts to use computers to produce mathematical proofs have followed two main paradigms. In the first, "automatic theorem proving", the machine searches for a proof of a theorem given some premises [4]. Although a considerable amount of work has been done on formulating proof systems which reduce the search space as far as possible, mainly using Robinson's resolution method [13], it is still hard to do interesting proofs. Steady progress has been made but the limitations of machine search are quite severe, and the technology has not been adopted by mathematicians. In the second approach the proof is invented by the human user and the machine is used to check that it has no incorrect or missing steps. This greatly reduces the demand on the machine, but transfers it to the user who has to provide a proof in a notation sufficiently formal to be understood by the machine and sufficiently detailed for each step to be recognised as correct. This is very much more burdensome than the level of proof required for normal communication between mathematicians.

Consider the analogy between formal proofs and programs. A major difference is that we can mechanically recognise the correctness of a proof, but for a program we can only recognise syntactic and type correctness and there is no algorithm for checking that it fulfils its intended purpose. There was a time in the sixties when the term "Automatic Programming" was in vogue, but it proved chimerical and the term has more or less dropped out of the technical vocabulary. However, non-automatic programming is very much with us, a major industry, even though programs are formal objects which people find difficult to produce. People do not write correct programs, but they get rapid feedback from the machine when they make mistakes. In a slow process over the last thirty years the level of languages used in programming has been raised, and the speed of machine response has been reduced from hours to seconds. It seems that there is a similar development for proofs; the logical languages in which we can express proofs to the machine have become more pithy and expressive, and the machine can notify us of mistakes very fast, so that without any change in principle the process of developing a machine checked proof has speeded up by a couple of orders of magnitude. A traditional proof is a series of statements starting with the premises and finishing with the theorem, each statement justified as obtained by the application of some logical rule to some previous ones. For example, if we use "....." to represent various mathematical formulas, then

```
Premise 1   : .....
Premise 2   : .....
Step 1      : .....    by ruleA from premise 1
Step 2      : .....    by ruleB from premise 2
Step 3      : .....    by ruleC from step 1 and step 2
Theorem     : .....    by rule C from premise 1 and step 3
```

When typing a proof into the machine the most tedious part is typing in the formulas represented above by "......". Of course we must put in the premises, but the step formulas and the theorem can then be computed from these if we say which rules are to be use on which arguments. Given the two premises we can write the remainder of the proof as follows

```
step 1   = A(premise 1)
step 2   = B(premise 2)
step 3   = C(step 1, step 2)
theorem = C(premise 1, step 3)
```

In fact all we really need to write down is the following expression which gives the proof of the theorem from the premises

C(premise 1, C(A(premise 1), B(premise 2)))

We can think of such expressions as being proofs; a proof expression is either the name of a premise or a proof rule applied to some proofs. From the proof expression and the premise formulas we can calculate the formula which we have proved (the theorem) by applying the inference rules in the manner indicated by the proof expression. Of course proof expressions are rather hard to write down if one does not write out the intermediate formulas, but this is a task which is mechanical and easily accomplished by the computer. So the technique is to interact with the machine as follows, where everything is typed in by the user except the underlined formulas which are typed out by the machine:

```
premise 1  = ......
premise 2  = ......
step 1     = A(premise 1)
             .....
step 2     = B(premise 2)
             .....
step 3     = C(step 1, step 2)
             .....
theorem    = C(premise 1, step 3)
             .....
```

With a graphical interface one can do a little better: the machine displays a menu of proof rules and the user simply points to the rule she wants to use and the formulas it is to be applied to; the names "step 1", "step 2" etc. can be dispensed with. In fact, once the premises have been typed in the user does not have to type anything else, just point. This is altogether a more relaxed way of doing business with the computer.

In practice it is helpful to work back from the theorem as well as forward from the premises, a method often called *Proof by Refinement*. It is also helpful to use the *Natural Deduction* style of proof system in which trying to prove a fact "if P then Q" we can add P temporarily to our list of premises and just try to prove Q. The proof expressions in a natural deduction system have a somewhat more elaborate structure than those indicated in the example above. They involve expressions for functions, and they use local variables standing for the (as yet unknown) proofs of the assumptions; in fact these expressions are what logicians call "lambda expressions".

It turns out that the problem solved by the machine above is "What formula does the following proof expression prove?", that is filling in the underlined dots above, is a well known one. In fact it is just the same as finding the *type* of an expression in programming languages, provided that we adopt a sufficiently rich notion of type. This idea goes back to Curry and Howard, and it is often referred to as *Propositions as Types*. It forms the basis for the proof systems currently popular in Computer Science, known as *Type Theory* [10] or *Calculus of Constructions* [5][12]. In this paper we will give an overview of one implementation of the Calculus of Constructions, namely the "LEGO" system developed at Edinburgh by Pollack [7,9]; for the pioneering implementation of Type Theory see [6]. A recent tutorial on Type Theory is given in [1], and a more extensive treatment is given in [11].

Having distinguished the automatic proof approach from the proof checking approach, my own preference is for the latter, rather following the story of Achilles and the tortoise. Automatic proof can accomplish some proofs very quickly, but when the search fails to find a proof, the user is left at the point of failure in a machine oriented world and may find it difficult to know what the difficulty was and how to proceed further. In the proof checking paradigm we can work with a more human-understandable representation, and although progress may be slow one is less likely to come up against major barriers. In practice a combination of these two approaches is used. The automatic proof systems are steered through a sequence of theorems chosen by the user, using earlier theorems as stepping stones to later ones. The proof checkers perform some symbolic computation and often use search strategies programmed by the user to fill in automatically some of the more routine steps.

2 Why Work on Computer Proof?

The main motivation for work on computer assisted proof has been the desire to prove that programs do what is intended. This means giving some formal specification of the task to be accomplished, using some logical language adequate to express the mathematical concepts involved, then proving that the result computed by the program is in accord with this specification, *for any input*. The usual method of convincing ourselves that a program is correct, debugging it by trying it on a selection of inputs, can only show this for the finite collection of inputs . Extrapolation to other inputs is the act of faith on which the software industry is built. The mediaeval scholastics might have been pleased to see this demonstration of the inadequacy of reason without faith.

In practice debugging works surprisingly well, but for programs where the cost of error in money or human lives is large one would like to do better. Also in concurrent programs where parts of the program run on different processors proceeding at somewhat unpredictable speed it may not be possible to re-run inputs which gave rise to an error and check that the bug has been corrected, timing differences may give different results even for the same inputs; experiments are not repeatable so empirical testing fails.

However I would like to put forward somewhat speculatively another motivation for working on computer proof systems. The number of mathematicians has not greatly increased since 1950, whilst the number of programmers has increased explosively, many of them largely self taught. To learn to develop mathematical proofs one has to acquire some feeling for what a proof is and when the level of rigour is adequate, otherwise one may be just handwaving. Basically you have to show it to someone who is mathematically skilled and they have to spend an appreciable amount of time looking at it critically. In programming on the other hand the machine, through syntax analysis and debugging, is able to give us a lot of feedback. We still need models to imitate and some advice from our betters, but the computer can point out when we have made a mistake, and provide us with some assurance that we have got something right. This feedback loop is crucial in learning to program. Suppose we had to teach people to program without access to computers. How many programmers would there be in the world, even if we had a real need for programs? Programming would be an arcane speciality.

So it seems to me that it is worth trying to get the same kind of feedback for the activity of producing mathematical proofs. You may say that mathematics is a highly intellectual activity, requiring sophisticated intuitions. But it would be a mistake to dismiss programming as a low level symbol manipulation; it too requires a high level of skill and intuition. Because a program can be executed mechanically we should not be misled into thinking that there is anything mechanical about the process of *inventing* programs. My contention is that many people, not at the genius level, can exhibit a high level of intuitive skill because the concrete feedback from the computer gives them criticism when they are wrong and confidence when they are right. Our intuition always rests on experience. Maybe if the same level of feedback could be given for mathematical proofs the number of people capable of producing proofs would increase markedly. This might not contribute directly to the discovery of more advanced pieces of mathematics, but if many more people could climb the foothills of mathematics then surely some of them might scale the peaks. If I am right about the possibility of a major scaling up in the number of people able to construct mathematical proofs, then this would surely have an impact on our ability to tackle the program correctness problem in practice.

3 A Formal Language for Constructive Mathematics

We now describe the Calculus of Constructions, a logical language with proof rules suitable for formulating constructive mathematics. Intuitionist (constructive) logic, disallowing the law of excluded middle P or *not-P*, seems well adapted to formal proof development on computers, and it has been shown that a considerable portion of mathematics can be expressed constructively [2][3][14]. In particular the parts of mathematics used in computing may be expected to be largely constructive. We will use for our notation that of the **LEGO** implementation of the calculus.

3.1 Type construction operations

The Calculus of Constructions uses the *type construction operators* $\{\_\}\_$ and $<\_>\_$ which are generalisations of the familiar $\to$ and $\times$. Let us explain them as generalisations of corresponding operations on sets. Suppose S and T are sets.

$$f \in S \to T \quad \text{means} \quad fx \in T \text{ for all } x \in S$$

$$(x, y) \in S \times T \quad \text{means} \quad x \in S \text{ and } y \in T.$$

We write fx for the more usual $f(x)$, meaning the value of f for argument x. We write (x, y) for an ordered pair. Now instead of the set T consider a family of sets over S, say F, that is for each $x \in S$, Fx is a set (note that we write Fx rather than F_x).

$$f \in \{x : S\}(Fx) \quad \text{means} \quad fx \in Fx \text{ for all } x \in S$$

$$(x, y) \in <x : S > (Fx) \quad \text{means} \quad x \in S \text{ and } y \in Fx.$$

In other words, $\{x : S\}(Fx) = \{f \mid fx \in Fx \text{ for all } x \in S\}$, and $<x : S>(Fx) = \{(x,y) \mid \forall x \in S, \forall y \in Fx\}$. Notice that if Fx is a constant set T then these reduce to $S \to T$ and $S \times T$ respectively. $\{x : S\}(Fx)$ is often written $\prod_{x:S} Fx$, and $<x : S>(Fx)$ is often written $\sum_{x:S} Fx$, and they are called the *product* and *sum* respectively.

Example for $\{\_\}\_$

Let $\mathbb{N}$ be the natural numbers with zero, $\mathbb{N} = \{0, 1, \ldots\}$, and $\mathbb{R}^n$ be the n-dimensional space of reals. Then a member of $\{n : \mathbb{N}\}(\mathbb{R}^n \to \mathbb{R})$ gives a function from $\mathbb{R}^n$ to $\mathbb{R}$ for each n, that is it is a function which given a number n produces a particular function from $\mathbb{R}^n$ to $\mathbb{R}$. For example *modulus* m is a member of this set, where $mi(x_1, \ldots, x_i) = (\sqrt{x_1^2 + \ldots + x_i^2})$ (such functions are often called *polymorphic* in computer language theory because they work for a whole family of types, here $\mathbb{R}^n$ for any n).

Example for $<\_>\_$

Let *String n* be the set of strings of n characters. Then $<n : \mathbb{N}> (String\ n)$ is the set of pairs (n, s) such that n is a number and s is a string of length n.

Other examples:

A matrix transpose function for m by n matrices, any m, n:

$Transpose \in \{m : \mathbb{N}\}\{n : \mathbb{N}\}(Matrix(m, n) \to Matrix(n, m))$.

A finite state automaton, a tuple of three sets and two functions over them:

$Automaton \in <In : Set><Out : Set><State : Set>$
$\qquad ((In \times State \to State) \times (State \to Out) \times State)$

(a transition function, an output function and a start state)

Now the Calculus of Constructions is a formal calculus in which these two operators are used to build types, reading S and T as types and F as a type valued function. In examples like Automaton we can replace *Set* by *Type*, the type of all types. This of course produces difficulties over paradoxes, so we introduce a hierarchy as follows, $Type_0, Type_1, \ldots$. For example, $\mathbb{N}$ and $\mathbb{N} \to String$ and $\mathbb{N} \times \mathbb{N}$ are in $Type_0$. (In fact we can *declare* basic types like $\mathbb{N}$ and *String* in $Type_0$ with some axioms (see below); then combining them with $\to$ or $\times$ keeps us in $Type_0$.) $Type_0$ and $Type_0 \times Type_0 \to Type_0$, for example, are in $Type_1$ and so on. This enables the calculus to be given a model theoretic semantics. In practice it is rather irritating to have to write the subscripts on the types; **LEGO** is able to deduce the subscripts, thus relieving the user of the obligation to write them.

In the **LEGO** notation we use $->$ for $\to$ and # for $\times$, since these characters are not available in the standard computer character set. We will also use *nat* for $\mathbb{N}$, for the same reason.

3.2 Expressions and Functions

Expressions are formed from constants, variables and the application of functions to arguments, as we have seen in the examples above. We need some notation for functions and adopt the λ notation of Church (see Hyland's paper in this volume [8]) with a different syntax. If E is an expression involving x we write

$$[x : S]E$$

for "E as a function of x, where x is a variable of type S" (Church wrote $\lambda x : S.E$).

Example

$[n : nat](plus(times\ n\ n)\ one)$ — the function f defined by $f(n) = n^2 + 1$ for all n in nat

(The **LEGO** computer input notation currently does not allow infixes like $+$, an infelicity which will be rectified sometime - so I am being honest in the examples in case you want to use the **LEGO** program). We can then regard a definition like

$$f[n : nat] = plus(times\ n\ n)\ one$$

as a syntactic alternative to the basic form

$$f = [n : nat](plus(times\ n\ n)\ one)$$

There are two ways of defining two argument functions. We can define the function to take a pair as argument (pairs are written using a comma and parentheses), or we can define it to take an argument and produce as a result a function which can be applied to the second argument. We use these as follows, assuming $x : s$ and $y : t$,

$f(x, y)$ a function applied to a pair, $f : s\#t \ -> \ u$
$f\ x\ y$ f applied successively to x and y, $f : s \ -> \ t \ -> \ u$

The second is syntactically simpler and we tend to prefer it. (We used it implicitly for plus and times above). To define such a function we write

$$f[m : nat][n : nat] = \ldots$$

or more briefly

$$f[m, n : nat] = \ldots$$

We can define functions to take a type argument (polymorphic functions)

$$id[t : Type][x : t] = x \qquad \text{the polymorphic identity function } \lambda x.x$$

It is often convenient to omit this type argument when using the function, and there is a special notation for this:

$$id[t|Type][x : t] = x \qquad \text{the same function but you do not have to give the type argument explicitly (| replaces :)}$$

With the former definition we would have to write *id nat* 3, with the latter we may write simply *id* 3

3.3 Definitions and Declarations

We have seen how to define functions using the equality sign, $=$. We can define constants similarly. But we can also introduce names of any type by declaration without giving a definition; for this we use $[\ldots : \ldots]$. For example to declare n as a natural number and f as a function from natural numbers to natural numbers

$$[n : nat];$$

$$[f : nat \rightarrow nat];$$

Notice that we use a semicolon after a declaration or definition.

3.4 Propositions and proofs

We wish to develop some logic in our language, in fact a higher order intuitionist logic; for this we will need propositions. We have a type *Prop* of propositions. In fact just as

$$Type_0 : Type_1$$

we have

$$Prop : Type_0$$

Suppose we declare A and B to be propositions, that is values of type *Prop*

$$[A, B : Prop];$$

then we can define further propositions using **and** and **or**, written $\wedge$ and $\vee$ respectively, thus

$$A \vee (B \wedge A)$$

We use $->$ for implication

$$A -> (B \wedge A)$$

and **absurd** for the proposition which has no proof, that is false. Negation is defined by

$$not[A : Prop] = A -> absurd$$

As a matter of fact **and**, **or** and **absurd** are not primitive notions in our language but can themselves be defined, similarly for an equality Q [9]. This shows the power of the basic formalism, but is of less importance to the user who wishes to deal with portions of mathematics, so we omit the definitions here. The use of $->$ for implication is not a notational trick; it really is the same operation as the one used for forming function types. It turns out that the universal quantifier, "for all", is expressed by the dependent function type operation. Thus if we declare predicates P and R over natural numbers and pairs of natural numbers respectively

$$[P : nat -> Prop]; [R : nat \# nat -> Prop];$$

we can form quantified formulas

$\{x : nat\}(Px)$ that is $\forall x : nat.P(x)$

$\{x, y : nat\}(Px -> R(fx, y))$ that is $\forall x, y : nat.Px \supset R(fx, y)$

The existential quantifier, *exists*, can be defined. It takes a function as argument, e.g.

$exists \; ([x : nat](R(x, x))$ that is $\exists x : nat.R(x, x)$

Here the argument of *exists* is $R(x, x)$ as a function of x. Propositions are themselves types and have proofs as their elements.

We can use declarations to introduce assumptions, just as we used them to introduce constants. In fact an assumption is just a declaration of a constant which denotes a proof of the fact being assumed

$$[axiom \; 1 : \{x, y : nat\}(Px -> R(fx, y))]$$

It is possible in **LEGO** to write expressions for proofs, indeed they are formed in just the same way as expressions for other values, but the normal way to create proofs in **LEGO** is by "refinement" (described below), so we will not discuss proof expressions here. It is an important property of the Calculus of Constructions that proofs are values and can be manipulated just like other values, for example they can be arguments and results of functions and elements of pairs.

3.5 Discharging assumptions and declarations

We can discharge an assumption A. In this case any theorem B that we had proved in the context containing the assumption becomes $A \rightarrow B$ in a new context without the assumption A. In discharging a declaration of a variable x any previous function definitions take x as an extra parameter. (These are really particular cases of the same operation.)

4 An Example: Complete Partial Orders

To see how this logical language can be used to describe mathematical systems we consider the definition of a complete partial order. This is an ordering for which every ascending sequence has a least upper bound. For ease of reading I have omitted some brackets and parentheses which LEGO requires.

$t \vert Type;$	assume t is a type (an inferred type since we use vertical bar)
$le : t \rightarrow t \rightarrow Prop;$	assume le is a function from t and t to $Prop$, a binary relation - less than or equal
$Refl = \{x:t\}(le\ x\ x);$	naming a fact about le - reflexivity
$Trans = \{x,y,z:t\}(le\ x\ y \rightarrow le\ y\ z \rightarrow le\ x\ z);$	transitivity
$Antisym = \{x,y:t\}(le\ x\ y \rightarrow le\ y\ x \rightarrow Q\ x\ y);$	antisymmetry
$POrd = Refl \wedge Trans \wedge Antisym;$	Define Partial Order
$seq = nat \rightarrow t$	new type sequence - function from nat to
$Chain[s:seq] = \{n:nat\}(le(s\ n)(s\ (succ\ n)));$	s is a chain iff $s_n \leq s_{n+1}$ for all n
$Ub[s:seq][x:t] = \{n:nat\}(le(s\ n)\ x);$	x is upper bound of sequence s iff $s_n \leq$ for all n
$Lub[s:seq][x:t] = Ub\ s\ x \wedge \{x':t\}(Ub\ s\ x' \rightarrow le\ x\ x');$	x is least upper bound of sequence s iff it is an upper bound and less than or equal to all other upper bounds

Computer Assisted Proof for Mathematics

We have a predicate $Chain$, but we would like to have a type $chain$. This is the type of all sequences satisfying the predicate $Chain$. We define the type $chain$ to be the type of ordered pairs consisting of a sequence and a proof that the sequence satisfies the predicate $Chain$. We are making use of the fact that proofs are values and can be components of pairs.

$chain = <s : seq > Chain\ s;$	Type chain is pairs consisting of (i) a sequence s (ii) a proof that the sequence s is a Chain.
$lub : chain \rightarrow t;$	Least upper bound function.
$Complete = \{c : chain\}(Lub\ c.1(lub\ c));$	Complete means the function lub gives the least upper bound (Lub) for each chain. $c.1$ is c considered as a sequence (first component of c).
$CPOrd = POrd \wedge Complete;$	Complete Partial Order - a property of relation le.
$Discharge\ t;$	Discharge all assumptions back as far as t. If a definition depends on a discharged variable make it a parameter e.g. t and le are parameters of Lub (t is inferred).
$cpord[t : Type] = <le : t \rightarrow t \rightarrow Prop >< lub : chain\ t \rightarrow t > (CPOrd\ lt\ lub);$	
	Complete Partial Order s over t - a new type, whose elements are triples (i) a less than or equal function (ii) a least upper bound function (iii) a proof that these form a partial order.
$s, t	Type;$	
$continuous[C : cpord\ s][D : cpord\ t] = \ldots;$	The type of continuous functions from C to D. These are defined as pairs (i) function from s to t, and (ii) a proof that it is continuous We omit the details.

In this example we have made some simplifying assumptions, for example in using a standard equality, Q, instead of an arbitrary one, but it illustrates the general approach. A convenient representation of set theory in the Calculus of Constructions is still a topic of current research.

5 A Sample Proof

We will illustrate a simple refinement proof in the **LEGO** system (again omitting some brackets and parentheses). We make two assumptions, that the relation R is symmetric and transitive, then show that Rxy implies Rxx. Proof commands typed by the user are in **bold**. The text on the right is commentary. $?n$ is a subgoal derived by machine in response to a proof command. We will explain briefly the two proof commands needed for our example. There are several others not described here.

The **Intros** command can be used in two ways. Applied to a goal with a universal quantifier "for all $x : t$", written $\{x : t\}$ it strips off the quantifier to produce a simpler subgoal, and it adds $x : t$ to the context in which later steps are carried out, as a temporary premise. Applied to an implication, $P \rightarrow Q$, it produces a simpler subgoal Q and adds P as a temporary premise.

The **Refine** command uses an assumption to simplify a goal. If the goal is Q and the assumption is $P \rightarrow Q$ then it produces the subgoal P. If the goal is $Q\ a$ and the assumption is $\{x : t\}(Px \rightarrow Qx)$ then it produces the subgoal $P\ a$. If the goal is $?n$ then **Refine** a replaces $?n$ by the term a. **Refine** also deals with more general cases involving several premises and several variables, but we need not discuss the general case here. (Technically speaking the **Refine** command performs first order unification.)

These commands work on the current top subgoal.

```
t : Type;                                          - assumption
Sym : {x, y : t}(R x y -> R y x);                  - assumption
Trans : {x, y, z : t}(R x y -> R y z -> R x z);    - assumption


Goal {x, y : t}(R x y -> R x x);                   - to be proved
          Intros x y;                              - strip quantifier off goal


x : t
y : t
      ?1 : R x y -> R x x                          - new (sub)goal
          Intros h1;                               - strip premise off goal?1


h1 : R x y
      ?2 : R x x
          Refine Trans;                            - use fact Trans for goal?2
```

```
?3 : t
?4 : R x ?3
?5 : R ?3 x
    Refine y;                    -instantiate ?3 to y

?6 : R x y
?7 : R y x
    Refine h1;                   - use assumption h1 for goal ?6
    Refine Sym;                  - use fact Sym for goal ?7


?8 : R x y
    Refine h1;                   - use assumption h1 for goal ?8

QED                              - LEGO says that the proof is
                                   complete.
```

It only remains to save the resulting proof under an appropriate name for use in proving later theorems.

Acknowledgements
I am very grateful to Randy Pollack for the opportunity of using his **LEGO** proof system and to various other users of the system who have helped me to understand it, notably to Paul Taylor and to Claire Jones. The **LEGO** users manual, largely written by Zhaohui Luo, has been very helpful. I am grateful to the Science and Engineering Research Council and the European Community BRA for supporting our research in this area. I would like to thank the referees and editors for their helpful comments.

References

[1] Backhouse R., Chisholm P., Grant M. and Saaman E., 'Do-it-Yourself Type Theory', *Formal Aspects of Computing*, 1,1, Springer International ,1989.

[2] Beeson, M. J., *Foundations of Constructive Mathematics*, Springer, 1989.

[3] Bishop, E., *Foundations of Constructive Analysis*, McGraw Hill, 1967.

[4] Bundy, A., *The Computer Modelling of Mathematical Reasoning*, Academic Press, 1983.

[5] Coquand Th. and Huet G., 'The Calculus of Constructions.' *Information and Control*, **76**, 1988.

[6] Constable R.L. et al. *Implementing Mathematics in the NuPrl Proof System*, Prentice Hall, 1986.

[7] Harper R. and Pollack R., 'Typechecking, Universe Polymorphism and Typical Ambiguity in the Calculus of Constructions', *Proc TAPSOFT Conference*, Barcelona, Springer LNCS 352, 1989.

[8] Hyland, J. M. E., 'Computing and Foundations', in *The Mathematical Revolution Inspired by Computing*, J H Johnson & M J Loomes (eds), Oxford University Press, 1991.

[9] Luo Z., Pollack R. and Taylor P., 'How to use **LEGO**' (A preliminary Users' Manual), unpublished draft, LFCS, Dept of Computer Science, Edinburgh University, 1989.

[10] Martin-Lof P., *Intuitionistic Type Theory*, Bibliopolis, Naples, 1984.

[11] Nordstrom B., Peterson K., and Smith J., *Programming in Martin-Lof's Type Theory*, Clarendon Press (Oxford), 1990.

[12] Projet Formel, 'The Calculus of Constructions: Documentation and User's Guide', INRIA, Rocquencourt BP105, France, Technical Report 110, 1989.

[13] Robinson, J. A., 'A machine oriented language based on the Resolution Principle', *Journal of the ACM*, 23-41, 1965.

[14] Troelstra, A. S., van Dalen, D., *Constructivism in Mathematics: An Introduction (Vols 1 & 2)*, North Holland, 1988.

A New Method of Automated Theorem Proving

YANG Lu

Institute of Mathematical Sciences
Chengdu Branch, Academia Sinica
610015 Chengdu, Sichuan, China

Abstract

Can a mathmatical theorem be proved by carrying out a series of experiments as people used to do for physical laws? By 'mathematical experiment' we mean the numerical verification of one certain instance. This becomes possible in practice since a new algorithm for automated theorem proving due to Zhang Jingzhong and Yang Lu. This is the *Parallel Numerical Algorithm* which replaces most of the symbolic algebra used in existing methods with numerical computation. In this way, for example, we can prove a theorem concerning general triangles by verifying it numerically for a finite number of triangles with given side-lengths. A lot of numerical computing, sometimes with high complexity, is often needed for our purpose. This is why such a method had not been suggested before the computer age. This paper is an exposition of Principles of the Zhang-Yang algorithm to prove equality-type theorems whose hypotheses and conclusions both can be expressed as polynomial equalities . It is shown that the algorithm is effective in not only proving known non-trivial theorems, but also in discovering new ones. Some unexpected and interesting results of non-Euclidean geometry have been proved in this way.

1 Introduction

In recent years Chinese mathematicians and computer scientists have developed several different methods for automated theorem proving. As a pioneer, Wu Wen-tsün introduced in 1977 an algebraic method [4], using successive pseudo division, which was extremely successful in proving mechanically non-trivial theorems in some fields including elementary geometry. Hong Jiawei's method [2] [3] published in 1986 aroused wide theoretical interest. Further research into the computational complexity and the practical feasibility of Hong's method is under way. Zhang Jingzhong and Yang Lu suggested a method [11] [12], suitable for parallel computing, which replaces most of the symbolic algebra used in the usual algorithms with numerical computation.

Both Hong's method and the Zhang-Yang method were inspired by Wu's outstanding work. ShangChing Chou recently published a monograph [1] in which 512 geometry theorems were proved by using Wu's algorithm.

The Principles of Wu's method will be sketched in Section 2, but first, we need recall two simple algorithms concerning polynomial division. One is the well known *pseudo division* which is available in some symbolic algebra software systems such as MAPLE.

For any polynomial Φ we denote the degree of Φ in x by $\deg(\Phi, x)$. Pseudo division is an algorithm for given polynomials f and g and a given variable x to find polynomials P, Q and R such that

$$Pf + Qg \equiv R \qquad (1.1)$$
$$\deg(R, x) < \deg(g, x), \quad \deg(P, x) = 0 \qquad (1.2)$$

where R is called *pseudo remainder* and is denoted by $\text{prem}(f, g, x)$.

Mutual division is another algorithm which finds, for given polynomials f and g and a given variable x, the Polynomials $\hat{P}$, $\hat{Q}$ and $\hat{R}$ such that

$$\hat{P}f + \hat{Q}g \equiv \hat{R} \qquad (1.3)$$
$$\deg(\hat{P}, x) \leq \deg(g, x), \quad \deg(\hat{Q}, x) \leq \deg(f, x) \qquad (1.4)$$
$$\deg(\hat{R}, x) = 0 \qquad (1.5)$$

where $\hat{R}$ is called *mutual remainder* and denoted by $\text{Eprem}(f, g, x)$. Mutual division, which is described in greater detail in [11, Lemma 2] where it is called the *division algorithm*, is just equivalent to a series of pseudo-divisions, so it is also supported by current software systems.

2 Wu's Algorithm: A Sketch

A theorem which can be reduced to an equivalent algebra theorem in which all the hypotheses and conclusion can be expressed by polynomial equalities, is called an *Equality-Type Theorem*. They have the form:

hypotheses:

$$\begin{aligned} f_1(y_1, y_2, ..., y_{n+s}) &= 0 \\ f_2(y_1, y_2, ..., y_{n+s}) &= 0 \\ &\cdots \\ f_s(y_1, y_2, ..., y_{n+s}) &= 0 \end{aligned} \qquad (2.1)$$

conclusion:

$$G(y_1, y_2, ..., y_{n+s}) = 0. \qquad (2.2)$$

Since there are s hypothesis equations in $n + s$ variables, we may regard n of the variables as independent, denoted by $u_1, u_2, ..., u_n$, and the rest as dependent, denoted by $x_1, x_2, ..., x_s$. Thus we can rewrite the theorem as follows:

hypotheses:

$$f_1(u_1, u_2, \ldots, u_n, x_1, x_2, \ldots, x_s) = 0$$
$$f_2(u_1, u_2, \ldots, u_n, x_1, x_2, \ldots, x_s) = 0 \qquad (2.3)$$
$$\ldots$$
$$f_s(u_1, u_2, \ldots, u_n, x_1, x_2, \ldots, x_s) = 0$$

conclusion:

$$G(u_1, u_2, \ldots, u_n, x_1, x_2, \ldots, x_s) = 0 \qquad (2.4)$$

The first step of Wu's algorithm reduces a geometry theorem to this form. The second step is known as the *Well Ordering Principle* which means that the hypotheses should be transformed into *triangular form*:

$$F_1(u_1, u_2, \ldots, u_n, x_1) = 0$$
$$F_2(u_1, u_2, \ldots, u_n, x_1, x_2) = 0$$
$$F_3(u_1, u_2, \ldots, u_n, x_1, x_2, x_3) = 0 \qquad (2.5)$$
$$\ldots$$
$$F_s(u_1, u_2, \ldots, u_n, x_1, x_2, \ldots, x_s) = 0$$

i.e. for each hypothesis polynomial F_k, only one new dependent variable (namely, x_k) is introduced. The conclusion remains the equation

$$G(u_1, u_2, \ldots, u_n, x_1, x_2, \ldots x_s) = 0$$

It was proved that any equality-type theorem, as defined above, can always be transformed into this "triangular form". Wu remarked that for most planar geometry problems starting from some given points, new points can be successively adjoined as the intersection of line and line, line and circle, or circle and circle in a constructive way. Thus the new dependent variables, as the coordinates, are to be added at most two by two. So it is very easy to transform the problem into triangular form. Of course, Wu implemented a computer program for the triangulation procedure.

The third step is a key step, to reduce the degrees in the dependent variables, and is known as *successive pseudo division* which is done as follows:

$$\begin{aligned}
R_{s-1} &:= \text{prem}(G, F_s, x_s) \\
R_{s-2} &:= \text{prem}(R_{s-1}, F_{s-1}, x_{s-1}) \\
R_{s-3} &:= \text{prem}(R_{s-2}, F_{s-2}, x_{s-2}) \\
&\ldots \\
R_1 &:= \text{prem}(R_2, F_2, x_2) \\
R_0 &:= \text{prem}(R_1, F_1, x_1)
\end{aligned} \qquad (2.6)$$

where R_0 is called the *final pseudo remainder*. *Wu's identity* follows easily from the definition of R_0:

$$Q_1 Q_2 \ldots Q_s G \equiv P_1 F_1 + P_2 F_2 + \ldots + P_s F_s + R_0 \qquad (2.7)$$

where $Q_1, Q_2, \ldots, Q_s, P_1, P_2, \ldots, P_s$ are polynomials, and R_0 the final pseudo remainder obtained by successive pseudo division. According to this identity, Wu established his fundamental theorem:

Theorem 1 $R_0(u_1, u_2, \ldots, u_n, x_1, x_2, \ldots, x_s) \equiv 0$ is a sufficient condition for the theorem to be true under the non-degenerate condition

$$Q_1 Q_2 \ldots Q_s \neq 0 \qquad (2.8)$$

in other words, the theorem holds 'generically' if $R_0 \equiv 0$.

So the fourth step is to verify whether R_0, the final pseudo remainder, is identical to zero or not. Wu pointed out that $R_0 \equiv 0$ is also a necessary condition in the 'irreducible cases' and gave an algorithm to reduce the general cases to the irreducible.

The non-degenerate condition means the geometric figure is in a normal position or a non-degenerate position. The conclusion of the geometry theorem my be invalid if some Q_k becomes zero.

In Wu's algorithm, some symbolic algebra is necessary for successive division and that is used as well for verification of $R_0 \equiv 0$. Wu edited programs and implemented them highly efficiently on computers. Shang-Ching Chou used Wu's algorithm on a SYMBOLICS 3600 computer and proved many non-trivial geometry theorems [1]. For more details of Wu's method, please refer to [1] or [4-10].

3 Basic Theorems

From now on and throughout this paper, we suppose the theorems we deal with have been reduced to irreducible form. This is always possible by using the *decomposition* algorithm but sometimes the computational complexity would be rather high.

By replacing each occurance of pseudo division with mutual division (the latter in [11] was also known as the *division algorithm for polynomials*), one can get a modified Wu's algorithm whose key step we call *successive mutual division*. This is done as follows:

$$\begin{aligned}
\hat{R}_{s-1} &:= \text{Eprem}(G, F_s, x_s), \\
\hat{R}_{s-2} &:= \text{Eprem}(\hat{R}_{s-1}, F_{s-1}, x_{s-1}), \\
\hat{R}_{s-3} &:= \text{Eprem}(\hat{R}_{s-2}, F_{s-2}, x_{s-2}), \\
&\ldots \\
\hat{R}_1 &:= \text{Eprem}(\hat{R}_2, F_2, x_2), \\
\hat{R}_0 &:= \text{Eprem}(\hat{R}_1, F_1, x_1)
\end{aligned} \qquad (3.1)$$

where $\hat{R}_0$ is called the *final mutual remainder*. From the definition of $\hat{R}_0$ we get a modification of (2.7), that is

$$\hat{Q}_1\hat{Q}_2\ldots\hat{Q}_sG \equiv \hat{P}_1F_1 + \hat{P}_2F_2 + \ldots + \hat{P}_sF_s + \hat{R}_0 \qquad (3.2)$$

where $\hat{Q}_1\hat{Q}_2\ldots\hat{Q}_sG, \hat{P}_1, \hat{P}_2, \ldots, \hat{P}_s$ are polynomials, generally speaking, which are different from $Q_1, Q_2, \ldots, Q_s, P_1, P_2, \ldots, P_s$ appeared in (2.7), and $\hat{R}_0$ the final mutual remainder obtained by successive mutual division.

It follows from (1.5) that for every dependent variable x_k, the $\hat{R}_k$ is independent of x_k, so all the dependent variables are elimilated one by one, and $\hat{R}_0$, the final mutual remainder obtained by the successive mutual division, is just in the independent variables and not in any dependent variable.

Corresponding to Theorem 1, Wu's fundamental theorem, we have the following

Theorem 2 $\hat{R}(u_1, u_2, \ldots, u_n) \equiv 0$ is a sufficient condition for the theorem to be true under the the non-degenerate condition

$$\hat{Q}_1\hat{Q}_2\ldots\hat{Q}_s \neq 0 \qquad (3.3)$$

in other words, the theorem hold "generically" if $\hat{R}_0 \equiv 0$.

Our Parallel Numerical Algorithm is based on numerical verification for algebraic identity. Let us begin with the concept of 'lattice array' and following Theorem 3.

Definition   Let $S_1, S_2, \ldots, S_n$ be n subsets of the set K. Let each S_j have t_j distinct elements, for $j = 1, 2, \ldots, n$. We call the Cartesian product of the above n subsets

$$S = S_1 \times S_2 \times \ldots \times S_n$$

an *n-dimensional lattice array* on K, with *size* $(t_1, t_2, \ldots, t_n)$.

Theorem 3   Let $\Phi(u_1, u_2, \ldots, u_n)$ be a polynomial in variables $u_1, u_2, \ldots, u_n$ with coefficients in the field K. Let the degree of Φ in u_j be not more than l_j, for $j = 1, 2, \ldots, n$. If there exists an n-dimensional lattice array S with size $(l_1 + 1, l_2 + 1, \ldots, l_n + 1)$, such that Φ vanishes over S, i.e. for any $(\bar{u}_1, \bar{u}_2, \ldots, \bar{u}_n) \in S$,

$$\Phi(\bar{u}_1, \bar{u}_2, \ldots, \bar{u}_n) = 0 \qquad (3.4)$$

then Φ identical to zero. This is easy to prove to n by induction.

How can Theorem 3 be applied to the final mutual remainder $\hat{R}_0$? If the exact expression of $\hat{R}_0$ is obtained, we can numerically verify whether or not $\hat{R}_0 = 0$ for every element of a sufficiently large lattice array. That means, however, we have to do successive mutual division which we would like to avoid. Theorem 4 allows us to estimate the size of the lattice array to be examined for verifying whether or not the final mutual remainder is identical to zero. This makes the successive mutual division calculations unnecessary in practice.

Theorem 4 Let f and g be polynomials in variables $u_1, u_2, \ldots, u_n$ and x over the field K; $\deg(f, x) = m$, $\deg(g, x) = h$; $h \geq m \geq 1$. Also let $\hat{P}$, $\hat{Q}$, and $\hat{R}$ be polynomials defined by mutual division so that (1.3)-(1.5) hold. Let

$$\begin{aligned} A_0 &= \deg(f, u_j), \\ A_1 &= \deg(g, u_j) + (h - m + 1)A_0, \\ A_{k+1} &= 2A_k + A_{k-1} \quad \text{for } k = 1, 2, \ldots \end{aligned} \quad (3.5)$$

Then

$$\deg(\hat{R}, u_j) \leq A_m.$$

For the proof, please refer to [11, Lemma 2]. By using Theorem 4 successively we can estimate the upper bound of the degree of $\hat{R}_0(u_1, u_2, \ldots, u_n)$ in every u_j; for $j = 1, 2, \ldots, n$. It clearly follows

Corollary In Theorem 4, if $m = 1$ or $m = 2$, the inequality $\deg(\hat{R}, u_j) \leq A_m$ can be expressed simply as

$$\deg(\hat{R}, u_j) \leq m \deg(g, u_j) + (mh - m + 1)\deg(f, u_j). \quad (3.6)$$

For applications to most problems in elementary geometry, this corollary suffices perfectly.

4 Description of the Algorithm

Now, let us turn to the description of our algorithm.

STEP 1 Choose the independent and dependent variables, write down the set of polynomial equations which express the hypotheses and conclusion of the theorem to be proved, and transform the hypotheses into the corresponding triangular form. (This step is the same as in Wu's algorithm.)

STEP 2 Estimate the upper bound for the degree of $\hat{R}_0(u_1, u_2, \ldots, u_n)$, the final mutual remainder obtained by using successive mutual division, in every independent variable u_j to determine the size of the lattice array to be examined. Of course we need not get the detailed expression for $\hat{R}_0$.

Theorem 4 gives a simple algorithm for this step, and successive mutual division is not necessary. It is very easy to write a program for this, and in elementary geometry the procedure is so simple that it can be done by hand.

STEP 3   The size of the lattice array was established in Step 2. Let S be a lattice array with this size. The elements of S may be chosen arbitrarily, only the size is important. Examine each element of S, say, $(\bar{u}_1, \bar{u}_2, \ldots, \bar{u}_n)$, as follows. Substitute it in the triangular form of the hypotheses, which gives

$$\begin{aligned}
F_1(\bar{u}_1, \bar{u}_2, \ldots, \bar{u}_n, x_1) &= 0 \\
F_2(\bar{u}_1, \bar{u}_2, \ldots, \bar{u}_n, x_1, x_2) &= 0 \\
F_3(\bar{u}_1, \bar{u}_2, \ldots, \bar{u}_n, x_1, x_2, x_3) &= 0 \\
&\vdots \\
F_s(\bar{u}_1, \bar{u}_2, \ldots, \bar{u}_n, x_1, x_2, \ldots, x_s) &= 0
\end{aligned} \quad (4.1)$$

Since $\bar{u}_1, \bar{u}_2, \ldots, \bar{u}_n$ are constants, the system of equations is easily solved by computer. Let the solution be

$$x_1 = \bar{x}_1, \quad x_2 = \bar{x}_2, \quad \ldots, \quad x_s = \bar{x}_s \quad (4.2)$$

Then substitute $\bar{u}_1, \ldots, \bar{u}_n, \bar{x}_1, \ldots, \bar{x}_s$ in the desired conclusion, and verify whether or not $G(\bar{u}_1, \ldots, \bar{u}_n, \bar{x}_1, \ldots, \bar{x}_s) = 0$ is true. In this way, examine all elements of the lattice array one by one. If for every element we examine, the equality $G = 0$ is always true, then the theorem is proved. Otherwise, the theorem is disproved. Usually Step 3 is done by computer.

This algorithm has the advantage that the program is simpler than the usual ones which use symbolic algebra only, and suitable for parallel computing. Chou proved a difficult theorem on a SYMBOLICS 3600 computer with CPU time 44 hours. If one uses the parallel algorithm on a parallel computer, then every processor could examine a few elements of the lattice array and the running time would be reduced accordingly. Also the numerical computation occupies less memory than symbolic algebra, so it can be used to prove non-trivial theorems on a microcomputer or even on a pocket computer. In practice, we implemented the algorithm efficiently on a PB 700 pocket computer and not only proved a lot of well-known geometry theorems such as Morley Theorem, Feuerbach Theorem and the Butterfly Theorem, etc., but also discovered new ones including some unexpected interesting results in non-Euclidean geometries, see [11].

5   How To Get Exact Zero

Naturally a question arises concerning the computer representation of numbers. If it is shown by computer that $G(\bar{u}_1, \ldots, \bar{u}_n, \bar{x}_i, \ldots, \bar{x}_s) = 0$, does G equal exactly zero or only a number with small absolute value ? The latter does not suffice for theorem proving. This presents no problem for theorem proving in elementary geometry since in this case $\deg(F_k, x_k) \leq 2$ for $k = 1, 2, \ldots, s$, so that we can get each x_k by solving linear or quadratic equations with one unknown. This can he done without any calculation error using *exact arithmetic* which is available in computer algebra packages such as *MAPLE, MACSYMA, REDUCE* or *MATHEMATICA*.

In general, we can deal with theorems involving equations with higher degrees, for which floating point computation is necessary, as follows. Suppose $F_1, F_2, \ldots, F_s$ and G are polynomials with *integral coefficients*. Then $\hat{R}_0$ also has integral coefficients. Let $\bar{u}_1, \ldots, \bar{u}_n$ be integers. Then $\hat{R}_0(\bar{u}_1, \ldots, \bar{u}_n)$ is also integer. If $\hat{R}_0(\bar{u}_1, \ldots, \bar{u}_n) \neq 0$, its absolute value is greater than or equal to unity:

$$|\hat{R}_0(\bar{u}_1, \ldots, \bar{u}_n)| \geq 1 \tag{5.1}$$

On the other hand, we can infer

$$\hat{Q}_1 \hat{Q}_2 \ldots \hat{Q}_s G \equiv \hat{R}_0 \tag{5.2}$$

from (3.2), the modified Wu's identity obtained by successive mutual division,

$$\hat{Q}_1 \hat{Q}_2 \ldots \hat{Q}_s G \equiv \hat{P}_1 F_1 + \hat{P}_2 F_2 + \ldots \hat{P}_s F_s + \hat{R}_0$$

with $F_k = 0$ for $k = 1, 2, \ldots, s$. We then estimate the upper bound M of $|\hat{Q}_1 \hat{Q}_2 \ldots \hat{Q}_s|$ for $\bar{u}_1, \ldots, \bar{u}_n, \bar{x}_1, \ldots, \bar{x}_s$. If the non-degenerate condition holds, then $G \neq 0$ implies $|\hat{R}_0| \geq 1$, so we have

$$|G| = |\hat{R}_0|/|\hat{Q}_1 \hat{Q}_2 \ldots \hat{Q}_s| \geq 1/M \tag{5.3}$$

That is, under the non-degenerate condition, either

$$G(\bar{u}_1, \ldots, \bar{u}_n, \bar{x}_1, \ldots, \bar{x}_s) = 0$$

exactly, or

$$|G(\bar{u}_1, \ldots, \bar{u}_n, \bar{x}_1, \ldots, \bar{x}_s)| \geq 1/M$$

So we need only estimate G accurately such that the error is less than $1/2M$. In this case, $G = 0$ exactly if and only if $|G| < 1/2M$. This answers our question.

6 Examples: The Verification of New Theorems

Now let us see an example, a new theorem of spherical geometry which we discovered by means of our algorithm. The first proof was given by a PB 700 pocket computer with running time 150 seconds, then by an AST 286 and a VAX 11/785 with time about 5 seconds and one second, respectively.

Theorem 5 If the area of a spherical triangle equals one quarter of the area of this sphere, then the midpoints of the three sides form an equilateral triangle with side-length $\pi r/2$, where r is the radius of the sphere.

It seems very strange that the original triangle is non-equilateral but its midpoint triangle is equilateral! It is most surprising. There is no corresponding version of Euclidean or Lobacheviskian geometry.

There are different ways to reduce the theorem to an equivalent algebra theorem. For instance, we may set

$$\begin{aligned} u_1 &= \exp(iA), \\ u_2 &= \exp(iB), \\ x_1 &= (u_2^2-1)(u_1^2u_2^2-1)\cos(a/r), \\ x_2 &= (u_1^2-1)(u_1^2u_2^2-1)\cos(b/r), \end{aligned} \qquad (6.1)$$

where A and B are two interior angles of the triangle, a and b the opposite sides, respectively. By the spherical law of cosines, we get the following program, written in BASIC, which suffices for verifying the above theorem

```
100 for u₁ = -5 to 5
110   for u₂ = -5 to 5
120     x₁ = 2u₂²(u₁²+1) + (u₂²+1)(u₁²u₂²+1)
130     x₂ = 2u₁²(u₂²+1) + (u₁²+1)(u₁²u₂²+1)
140     x₃ = (u₁²u₂²+1)²((u₂²-1)(u₁²u₂²-1) - x₁)((u₁²-1)(u₁²u₂²-1) - x₂)
150     x₄ = 4u₁²u₂²((u₂²-1)(u₁²u₂²-1) + x₁)((u₁²-1)(u₁²u₂²-1) + x₂)
160     if x₃ - x₄ ≠ then print "the theorem is false"
170   next u₂
180 next u₁
190 print "the theorem is true"
200 end
```

We set the lattice array to be examined as

$$S = \{-5,-4,-3,-2,-1,0,1,2,3,4,5\} \times \{-5,-4,-3,-2,-1,0,1,2,3,4,5\}$$

because these numbers have smallest possible absolute values which can be an advantage for the computation. However, they can be replaced by any set of distinct numbers in a lattice array of the same or greater size.[1]

Trivial implies non-trivial. How can one imagine such a simple program would imply a surprising theorem outside our geometric intuition? By private communications, H.S.M. Coxeter and D. Pedoe recognised Theorem 5 and later Theorem 6 as new theorems and Coxeter pointed out Theorem 5 is easy to prove using ordinary methods. G. Cairns and J.B. Strantzen have shown me their very nice proofs for the two theorems by conventional means.

In general, of course, the running time would be longer for more difficult problems. Let us see another example, a new theorem of 3-dimensional geometry which we discovered by means of our algorithm:

---

[1] Readers trying this program should note that it assumes exact integer arithmetic, and that although any set of distinct numbers can be used in the lattice array, large numbers may cause overflow on BASIC's with two or four byte integers (see Section 5).

Theorem 6   Let a *width* of a tetrahedron be the common perpendicular of a pair of opposite edges, and a *height* be the distance of a vertex from its opposite face. Each tetrahedron has three widths and four heights. By h_1, h_2, h_3, h_4 and w_1, w_2, w_3 denote the heights and widths respectively. It holds for any tetrahedron that

$$h_1^{-2} + h_2^{-2} + h_3^{-2} + h_4^{-2} = w_1^{-2} + w_2^{-2} + w_3^{-2}. \tag{6.2}$$

Without loss of generality, let the Cartesian coordinates of vertices of the tetrahedron be $(0,0,0)$, $(0,0,1)$, $(0, y_3, z_3)$ and (x_4, y_4, z_4), that is to say, the problem has five independent variables.

After reducing it to an equivalent algebra theorem, we estimate the needed size of the lattice array (y_3, z_3, x, y_4, z_4) where $x = x_4^2$. According to the Corollary of Theorem 4, it suffices to examine any lattice array with size $(19, 19, 11, 21, 21)$. We may choose, for instance, $y_3 = -9, \ldots, 9$; $z_3 = -9, \ldots, 9$; $x = -5, \ldots, 5$; $y_3 = -10, \ldots, 10$; $z_4 = -10, \ldots, 10$. In fact, considering the symmetry and the non-degenerate condition, we need only examine for $y_3 = 1, \ldots, 9$; $z_3 = -9, \ldots, 10$; $x = -5, \ldots, -1, 1, \ldots, 5$; $y_4 = -10, \ldots, 10$; $z_4 = 0, \ldots, 11$ and get the following (exact integer arithmetic) program written in ALDES.

(1) BEGIN2
(2) FOR Y3=1,...,9 DO (FOR Z3 = -9,...,10 DO(FOR X= -5,...,5 DO(
    IF IABSF(X) NE 0 THEN FOR Y4 = -10,...,10 DO(FOR Z4 = 0,...,11 DO(
    A2 = ISUM(IPROD(Y3,Y3),IPROD(Z3,Z3)).
    A3 = ISUM(X,ISUM(IPROD(Y4,Y4),IPROD(Z4,Z4))).
    B1 = IDIF(Z4,Z3).
    F1 = IDIF(1,Z4).
    B2 = IDIF(IPROD(Y3,Y4),IPROD(Z3,F1)).
    B3 = IDIF(IPROD(Y4,Y3),IPROD(Z4,IDIF(1,Z3))).
    C1 = ISUM(X,ISUM(IPROD(IDIF(Y3,Y4),IDIF(Y3,Y4)),IPROD(B1,B1))).
    C2 = ISUM(IPROD(F1,F1),ISUM(X,IPROD(Y4,Y4))).
    C3 = ISUM(IPROD(Y3,Y3),IPROD(IDIF(1,Z3),IDIF(1,Z3))).
    F2 = IDIF(IPROD(Y3,IDIF(Y3,Y4)),IPROD(Z3,B1)).
    F3 = IDIF(X,IDIF(IPROD(Y4,IDIF(Y3,Y4)),IPROD(Z4,B1))).
    G1 = IDIF(ISUM(IPROD(Y4,IDIF(Y3,Y4)),IPROD(B1,F1)),X).
    G3 = IDIF(IPROD(IDIF(Z3,1),B1),IPROD(Y3,IDIF(Y3,Y4))).
    D1 = IDIF(C1,IPROD(B1,B1)).
    D4 = IDIF(IPROD(F1,C1),IPROD(G1,B1)).
    D7 = IDIF(IPROD(F1,B1),G1).
    D2 = IDIF(IPROD(A2,C2),IPROD(B2,B2)).
    D5 = IDIF(IPROD(F2,C2),IPROD(G1,B2)).
    D8 = IDIF(IPROD(F2,B2),IPROD(G1,A2)).
    D3 = IDIF(IPROD(A3,C3),IPROD(B3,B3)).
    D6 = IDIF(IPROD(F3,C3),IPROD(G3,B3)).
    D9 = IDIF(IPROD(F3,B3),IPROD(G3,A3)).
    T1 = IPROD(X,IPROD(IDIF(D7,D1),IDIF(D7,D1))).
    T4 = IDIF(IPROD(D7,IDIF(Y4,Y3)),IPROD(Y4,D1)).
    T7 = ISUM(IDIF(IPROD(D7,B1),D4),IPROD(D1,F1)).
    T2 = IPROD(X,IPROD(IDIF(D8,D2),IDIF(D8,D2))).

```
T5 = ISUM(IDIF(IPROD(Y4,D8),IPROD(Y3,D5)),IPROD(D2,IDIF(Y3,Y4))).
T8 = IDIF(IDIF(IPROD(D8,IDIF(Z4,1)),IPROD(Z3,D5)),IPROD(D2,B1)).
T3 = IPROD(X,IPROD(IDIF(D3,D6),IDIF(D3,D6))).
T6 = ISUM(IDIF(IPROD(Y3,D9),IPROD(Y4,D6)),IPROD(D3,IDIF(Y4,Y3))).
T9 = ISUM(IDIF(IPROD(D9,IDIF(Z3,1)),IPROD(Z4,D6)),IPROD(D3,B1)).
S1 = ISUM(ISUM(T1,IPROD(T4,T4)),IPROD(T7,T7)).
S2 = ISUM(ISUM(T2,IPROD(T5,T5)),IPROD(T8,T8)).
S3 = ISUM(ISUM(T3,IPROD(T6,T6)),IPROD(T9,T9)).
S  = ISUM(ISUM(IPROD(IPROD(D1,D1),IPROD(S2,S3)),IPROD(IPROD(D2,D2),
     IPROD(S1,S3))),IPROD(IPROD(D3,D3),IPROD(S1,S2))).
L  = IDIF(IPROD(Y3,Z4),IPROD(Y4,Z3)).
U  = ISUM(ISUM(ISUM(X,IPROD(Y4,Y4)),ISUM(IPROD(Y3,Y3),IPROD(2,
     IPROD(X,IPROD(Y3,Y3))))),ISUM(ISUM(IPROD(X,IPROD(Z3,Z3)),
     IPROD(X,IPROD(IDIF(1,Z3),IDIF(1,Z3)))),ISUM(IPROD(L,L),
     IPROD(ISUM(L,IDIF(Y4;Y3)),ISUM(L,IDIF(Y4,Y3)))))).
P  = IPROD(S,IPROD(X,IPROD(Y3,Y3))).
Q  = IPROD(IPROD(U,S1),IPROD(S2,S3)).
IF ICOMP(P,Q) NE 0 THEN PRINT $(THE THEOREM IS FALSE)))))).
(3) PRINT $(THE THEOREM IS TRUE). END1. STOP..
```

Using the SAC-2 Arithmetic System we ran this program successfully on a terminal of a VAX 11/785 with running time 25589.47 seconds; i.e. more than 7 hours. Maybe it is better to process such a problem by the parallel computer. Recently I learnt that Liu Li succeeded in proving Theorem 6 mechanically using the same algorithm on a VAX 11/785 with CPU time 1 hour and 43 minutes only, and his program was written in BASIC! His work is to appear. More examples can be found in [11].

Acknowledgements.
I am grateful to Dr. Scott McCallum and Dr. Garry Newsam for their help concerning computer languages and programs.

References

[1] S.C. Chou, *Mechanical Geometry Theorem Proving*, D. Reidel Publishing Company, (Amsterdam) 1988.

[2] Hong Jiawei, 'Can the single example prove a geometry theorem?', *Scientia Sinica (Chinese ed.)*, 29, 234-242, (1986).

[3] Hong Jiawei, 'The growth of significant digit length in approximate calculation is faster than that of geometrical series', *Scientia Sinica (Chinese ed.)*, 29, 225-233, (1986).

[4] Wu Wen-tsün, 'On the decision problem and the mechanization of theorem proving in elementary geometry, *Scientia Sincia* 21, 157-179, (1978).

[5] Wu Wen-tsün, 'Basic principles of mechanical theorem proving in geometries', *J. Sys. Sci. and Math. Sci.* 4 (3), 1984, 207-235, republished in *J. Automated Reasoning* 2 (4), 221-252, (1986).

[6] Wu Wen-tsün, 'Some recent advances in mechanical theorem proving of geometries', in *Automated Theorem Proving: After 25 Years, AMS Contemporary Mathematics* 29, 235-242, (1984).

[7] Wu Wen-tsün, 'Basic Principles of Mechanical Theorem Proving in Geometries', (in Chinese), Beijing, (1984).

[8] Wu Wen-tsün, 'Toward mechanization of geometry-some comments on Hilbert's *Grundlagen der Geometrie*', *Acta Mathematica Scientia*, 2, 125-138, (1982).

[9] Wu Wen-tsün, 'Mechanical theorem proving in elementary geometry and differential geometry', *Proc. 1980 Beijing Symposium on Differential Geometry and Differential Equations* 2 (1982), 125-138, Science Press, (1980).

[10] Wu Wen-tsün, 'On zeros of algebraic equations-an application of Ritt principle', *Kexue Tongbao*, 31(1), 1-5, (1986).

[11] Zhang Jingzhong, Yang Lu and Deng Mike, 'The parallel numerical method of mechanical theorem proving', Research Report IMS-30, Oct. 1988, Institute of Mathematical Sciences, Chengdu, China. *Theoretical Computer Science*, 74, 253-271 (1990).

[12] Zhang Jingzhong and Yang Lu, 'Principles of parallel numerical method and the single-instance method of mechanical theorem proving', *Math. in Practice and Theory (in Chinese)*, 1, 34-43 (1989).

Making Discrete Mathematics Executable on a Computer

R. D. Knott
Department of Mathematics, University of Surrey
Guildford, Surrey, GU2 5XH.

Abstract

We show how ordinary mathematical statements about sets and sequences, using ordinary logic, can be viewed as a programming notation which may be run directly on a computer. Making mathematics executable in this way involves little effort at learning new concepts for the mathematician. The notation presented is a subset of the programming language Prolog but using the ordinary mathematician's style of writing. A short tutorial style introduction to Prolog and the style recommended is included. Advantages of the notation include unification of similar ideas in a way which has not been the norm in conventional mathematical texts. The four basic combinatorial methods of selection are presented to illustrate this. Most of the work presented is in the area of Graph Theory although the approach presented is applicable to many areas of discrete mathematics. Definitions written in this style allow the generation of examples and the direct statement of theorems which can be verified across the examples. Further, the examples can be used to find conjectures which the mathematician can then prove. This approach is acceptable in computing speed.

1 Introduction

At Surrey University we have developed a software library that allows the user to write ordinary mathematical statements about sets and sequences, using ordinary logic, which is able to be run directly on a computer. This work has grown out of two Alvey Projects at Surrey University [12],[13],[14] using the Z notation [10], [11], [16]. One objective of the two Projects has been to provide a system which could 'animate' the mathematical descriptions (of software systems), by making the mathematical notation executable on a computer. A spin-off has been that the library of software can stand alone and be a useful aid to the working mathematician.

There have been earlier successful attempts to computerise Graph Theory and combinatorics but using conventional languages (for example [1],[2],[5],[8], [15],[18]). However, the mathematician has had to learn a whole new notation which, when executed, is quite dissimilar in its approach to that of the mathematical discipline.

We hope to show in this paper that a different approach is available and makes the actual mathematical definitions executable (that is, we can 'animate' them on a computer). The technique we use employs the logical and mathematical style which is second nature to the mathematician. Our notation is, in fact, just a subset of Prolog (a declarative computer programming language).

We have applied the method to aspects of Number Theory (e.g. modular arithmetic, g.c.d., partitions and compositions, random numbers), Logic (e.g. 'for all' and 'there exists'), sets and sequences (e.g. set operations, permutations, powerset), algebra (e.g. operations on generating functions). The main effort has been in the area of Graph Theory where we have been able to generate graphs, find their properties, verify that known theorems hold across all of the examples (thus increasing our confidence in the correctness of the rules), and even to generate conjectures from the examples. Our standard example set is the complete collection of all 208 graphs which have between one and six nodes (no edges repeated). The execution times for generating these graphs with their properties are measured in minutes, with times of seconds for generating conjectures that are inferred from these examples. Thus the aim of clarity in mathematical expression has not meant that we have to wait unacceptably long for answers.

The library has been written using a subset of Prolog ([3],[4],[7],[17]) with three main aims. First it should be *declarative* in that we allow the problem to be stated as opposed to a solution method which is the common practice with conventional programming languages. Second, the mathematical nature of the rules should allow programs to be *transformed into more efficient forms*, if necessary, in just the same way that algebraic equations are manipulated. Third, it should be possible to have our programs *proved correct* (and this is often difficult with conventional programming languages).

This paper will illustrate the mathematical style mentioned above. We hope to convince the reader that this style of Prolog is both easily learned and used by the working mathematician. Hence, the first section is a quick tutorial introduction to expressing mathematics in a form which a computer can obey directly, with examples to show the benefits of such a notation for mathematics. The mathematics we illustrate will be in the area called Discrete Mathematics, which we take to include logic, sets, number theory and combinatorics together with graph theory, number theory and group theory, although this selection is not meant to be exhaustive.

2 Mathematics and Prolog

In this section, we illustrate some of the advantages of writing mathematics in Prolog and show that mathematical notation can be simply translated into Prolog. We look at logic (strictly, boolean algebra here, but the library in use at Surrey contains other familiar logical quantifiers and operators) and the fundamental principle of recursion which is a powerful tool with which to capture mathematical definitions. First an overview of Prolog is given for the reader unfamiliar with the language, in order to save interupting the following sections with explanations of the notation.

Discrete Mathematics on a Computer 129

2.1 An overview of Prolog notation.

Our programs are definitions about relationships and other facts. In order to simplify the various notations of mathematics (infix, prefix and postfix operators with superscripts and subscripts and Greek and Roman symbols intermixed), we will restrict ourselves to a prefix form with letters, numbers, and a few mathematical operators (all available on standard keyboards).

We will need to distingush between constants and variables. In our notation variables will begin with upper case letters, whereas constants and symbols (e.g. member, apple) will begin with lower case letters. Thus $3 \in S$ may be respresented in prefix form using no greek letters as

```
member(3,S).
```

A Prolog program consists of collections of such rules, which may have conditions attached. We will issue queries to ask Prolog to find instances (*instantiations*) for the variables which conform to the definitions we have given (that is, which are *unifiable* with the definitions).

Definitions may be qualified. A conditional rule is written with the conditions following the 'if' sign (written as :-) and all definitions in Prolog end with a fullstop or period. We can combine conditions with the Boolean operators of 'and', 'or' and 'not' which we will write as ampersand (&), semicolon(;) and 'not' respectively. We will assume the conventional symbols of arithmetic including numbers. Comparison operations will be written as $>=$ to mean $\geq$ and $=<$ to mean $\leq$. One or two rules will be assumed as outlined here. (For further notes on these rules, see [17], although it must be pointed out that different authors and different Prolog systems do not use the mathematical definition of set_of that we shall assume. Our guiding principle will be to implement rules which are equivalent to the mathematical definitions.)

```
set_of(X,R(X),S)
```

means S is the set of values of X satisfying $R(X)$, that is, $S = \{X|R(X)\}$. Other examples will be introduced as needed.

For example, to state in Prolog the definition that a parent P of person X is a father or mother of X, we write:

```
isparentof(P,X) :- isfatherof(P,X); ismotherof(P,X).
```

which we read as "P isparentof X if P isfatherof X or P ismotherof X." A grandparent is a parent of a parent, and we may write this using the conjunction (&):

```
isgrandparentof(G,X):- isparentof(G,P) & isparentof(P,X).
```

We are now ready to examine how to express mathematics in this notation.

2.2 Expressing Mathematical notation in Prolog

Consider the mathematical definition of the *union* U of two sets S and T:

$$S \cup T = \{x|\ x \in S \vee x \in T\}.$$

To define a rule for union, we have three parameters, the two sets S and T and their union U. Thus, our mathematical definition becomes the Prolog rule as follows:

```
union(S,T,U) :-
    set_of( X, (member(X,S); member(X,T)), U ) .
```

The definition of set_of ensures that elements only appear once in U, and allows for S, T or U being the empty set. We may read the Prolog rule as "The union of S and T is U holds if the set of X, where X is either in S or in T, is U". A similar rule holds for the intersection of two sets

$$S \cap T = \{x|x \in S \wedge x \in T\}$$

which, in Prolog becomes

```
intersection(S,T,I) :-
    set_of( X , (member(X,S) & member(X,T)), I ) .
```

To illustrate the use of logical negation, we have the difference of two sets defined by

$$S \setminus T = \{x|x \in S \wedge x \notin T\}$$

which becomes

```
setdifference(S,T,D):-
    set_of( X , (member(X,S) & not member(X,T) ), D).
```

2.3 Finding answers to questions in Prolog

To use our Prolog rules to find the union and intersection of two sets, we precede the clause with a question mark (?) so that any variables in the clause are now taken to be the answer that Prolog will report. Sets will not be written in curly brackets (which are used to enclose comments in Prolog) but in square brackets. "What is the intersection of $\{1, 2, 3\}$ and $\{4, 3, 2\}$?" becomes

```
? intersection( [1,2,3] , [4,3,2] , What ).
What = [2,3]
```

We must be careful when using lists (enclosed in square brackets) to represent sets, since lists are sequences and the order of presenting elements becomes important. Thus we have the query

```
? intersection( [o,r,a,n,g,e],[1,e,m,o,n],[o,n,e]).
```

which produces the reply

Yes

since the order of [o,n,e] matches that of [o,r,a,n,g,e], but the similar query

?intersection([o,r,a,n,g,e],[l,e,m,o,n],[e,o,n]).

produces the contradictory answer

No.

One solution is to decide in which ways a rule will be used - for instance, that intersection acts like a function on two given lists - and to define it suitably. The two given lists can have their elements presented in any order. To check that the intersection of two lists is a particular set, we call on an 'equalsets' predicate. It is also possible to write Prolog rules which do this automatically, checking which form of definition is appropriate to each use. The interested reader is referred to [7] and [17] for further discussion on this.

2.4 Recursion and Multiple Solutions

Although Prolog definitions are often as simple as those above, we need both a way of writing recursive rules which mimic mathematically recursive definitions, and also rules that allow for relations which have several "results" rather than just a function's unique "result". As an example, consider the relation that an integer X lies in a given range between two given values Lo and Hi. Mathematically, this is written $Lo \leq X \leq Hi$. We will define this from first principles in Prolog as an illustration of a common definitional form familiar to the mathematician - that of the recursive definition.

As a Prolog definition, there are two cases: X could equal Lo or it could exceed Lo. Both cases only apply if there is a genuine non-empty range (i.e., $Lo \leq Hi$) and the second only holds if Lo is strictly less than Hi. Thus we have two conditional rules in the definition of 'range'. The first case is:

range(Lo,Hi,X):- Lo =< Hi & Lo=X .

The second case codes to:

range(Lo,Hi,X):- Lo < Hi & NextLo is Lo+1
       & range(NextLo, Hi, X).

Prolog treats all expressions symbolically, so that $1 + 1 = 2$ will fail since the expression $1 + 1$ is not the same symbolically as 2. The built-in operation is evaluates its right hand side first and is what we want in this case. Note that both rules apply if $Lo < Hi$ and so Prolog may report several solutions:

```
? range( 2,4,A).
A=2
A=3
A=4
No more solutions.
```

To produce the set of integers in a given range, we could issue the query

```
? set_of(I, range(2,4,I), Ints).
Ints = [2,3,4]
```

Finally, as an illustration of a more complex query, we seek integers in the range 0 to 9 whose squares are in the range 30 to 50:

```
? range(0,9,X)    &    XX is X*X    &    range(30,50,XX).
X=6    XX=36
X=7    XX=49
```

Note here that we have used **range** to *generate* solutions (for X) in its first use and used it to *check* a condition (on XX) in its second call. Also note how we have merely specified what the problem is and the Prolog system finds the answer. This example shows that we need to pay some attention to the order of events when Prolog queries are being answered, but that the Prolog 'program' is much closer to the mathematical statement of the question than it would have been in conventional programming languages.

2.5 Obeying rules 'backwards'

In this section we will show how the relational aspects of Prolog allow us to do something that mathematicians do in practice, but which is not normal for programmers. This is to find which arguments to a function will produce a given result. In Prolog, this is the ability to drive our rules 'backwards' and is a consequence of describing relations rather than functions (in the mathematical sense).

2.5.1 Joining sequences

As an illustration, we will describe one rule called **append**, which joins two sequences (lists and sequences are synonymous terms). This rule will then be used in the next section to produce some elegant descriptions for combinatorial functions which have a pleasing similarity - a feature which is not expressed in the usual mathematical descriptions.

Considered as an operation on two sequences, **append** joins them together into one, preserving the order within each. If items are repeated in the sequences to be joined then they will appear repeated in the join. Hence **append** is not the same as union.

To indicate that an item X is at the front of a sequence L, the bar (|) operator is supplied in Prolog. Thus the list L has an element X joined at the front in the expression [X|L]. Also, we can use the bar operation to investigate which element is first in a list and discover the remainder of the list:

```
? [ First | Rest ] = [1,2,3,4].
First = 1        Rest = [2,3,4]
? [ Head | Tail ] = [ singleton ]
Head = singleton     Tail = []
```

where the empty list is denoted []. The following definition for append is *not* what we want:

```
append(L, M, [L|M]).
```

since the answer to the query

```
?append([1,2,3],[4,5],X).
```

is

```
X=[[1,2,3],4,5]
```

i.e. a list of 3 elements, the first being another list, and not the single list that we require:-

```
X=[1,2,3,4,5].
```

The correct rule for stating that L joined to M is A is "If L is empty the join A is the second list M. If L is not empty, the join A has the head of L as its first item, the remainder of A being the join of the rest of L to M.", or, in Prolog:

```
append(L,M,A ):- L = [] & A = M.
append(L,M,A ):- L = [H|T] & append(T,M,R) & A = [H|R].
```

or, in a more succinct but equivalent form as follows

```
append([],M,M).
append([H|T],M,[H|R] ):- append(T,M,R).
```

Thus when we ask

```
?append([1,2,3],[4,5],Z).
```

we have the answer we required:-

```
Z=[1,2,3,4,5]}
```

A surprise is in store for those used to conventional programming languages or mathematical packages. Our definition is of a relation rather than a function. Hence we can ask "What list M can follow $[1,2]$ if the join is to be $[1,2,3,4,5]$?"

```
? append( [1,2], M, [1,2,3,4,5]).
M=[3,4,5]
```

or even, "What lists A and B can be appended to produce [1,2,3,4,5]?" and get several solutions:

```
?append( A, B, [1,2,3,4,5]).
A=[]         B=[1,2,3,4,5 ]
A=[1]        B=[2,3,4,5 ]
A=[1,2]      B=[3,4,5 ]
A=[1,2,3]    B=[4,5 ]
A=[1,2,3,4]  B=[5 ]
A=[1,2,3,4,5]   B=[]
```

Unfortunately, not all our Prolog rules will have this ability to run backwards, or, given the result of a 'function', find which arguments could have produced it. This feature of relations is an added bonus when it does occur and contributes much to the expressive power of Prolog which we will exploit in the definitions of our four combinatoric laws in Section 3.

2.5.2 Membership of a sequence or set

In Section 2.2, we assumed that the member rule was supplied. Finally, in this section we show how the definition of append can be used to provide a definition of member. The last query of the previous section gave six answers. Note how each element of the combined list occurred in turn as the first item in list B. Thus if X is a member of list L, it is possible to find lists P and Q which can be appended to give L, with X as the head of Q (i.e. Q is $[X|R]$, say). That is,

```
member(X,L) :- append(P,Q,L) & Q=[X|R].
```

Since we only mention P and R once, we can replace them by the *anonymous variable* (_) meaning "anything will do here", and replace Q with $[X|$_$]$:

```
member(X,L) :- append(_, [X|_], L).
```

3 Executable Combinatorics

One advantage of using such a mathematically flavoured Prolog style, is that sometimes we can spot patterns in definitions that are not clear in the mathematical notations. Such an example is found in the four basic operations of combinatorics examined in this section. We will show how they can be written in a common form which shows their common structure. Our Prolog notation has become a tool which not only allows us to get an impartial machine to produce or verify solutions, but also becomes a notation that reveals the structure of related problems.

3.1 Choosing sets or sequences with or without repetition

Consider the operation of choosing items from a given collection of objects. We may be interested in the sets of objects we can choose (that is, the collection is viewed as a whole and the order in which we pick the items is irrelevant) or in *sequences* (when the order in which the objects can be chosen does matter). Also, we may wish to choose an object many times in our collection (*replacing* it in the collection for future choices after we have noted which one we chose) or not (that is, *removing* any item chosen so that it is not available again in our selection). We will use the terms sets and sequences with replacement or removal (where a set has unique elements, but a collection in which order does not matter and there may be repeated elements is called a *bag* or *multiset*). The four kinds of choice method will be denoted as follows:

> setrem - sets with removal,
> seqrem - sequences with removal,
> bagrep - bags (multisets) with replacement,
> seqrep - sequences with replacement.

All four methods are defined recursively, with two cases each. The key to all our definitions is that they all use the rule append defined in Section 2.5.1.

3.2 Choosing sets with removal

Now we are ready to define the first combinatorial rule setrem. The first parameter is the list of objects chosen from amongst those of the second parameter - another list of objects. There are two cases - the choices list can be empty or not. The simplest case - the base case for the recursion - is when the list of chosen objects is empty. We can always chose this option no matter what list of objects we are choosing from.

> setrem([] , _).

If our list of chosen objects is not empty, any member of the collection will do as our first choice. Since we are interested in sets then we only want each set of choices defined once. That is, we can use the order of elements in the collection (second parameter) as the preferred order for elements in our choice set. Thus, the remaining choices must be made from those in the collection occurring *after* our first selection :

```
setrem( [ H|T] , Objs ):-append(_ , [H|RestOfObjs], Objs)
                      & setrem(T,RestOfObjs).
```

Thus, like append, there are two alternative rules in our definition of **setrem**.

3.3 Choosing sequences with removal

If we had wanted our choices to be sequences, not sets, but still with no repetition (i.e. we are defining seqrem) then the items both before or after our first choice from the collection from which remaining choices are made:

```
seqrem([],_).
seqrem([H|T],Objs):-append(Before,[H|After],Objs)
        & append(Before,After,OtherObjs)
        & seqrem(T,OtherObjs).
```

It can be seen that all the collections (with removal) which are sets are amongst those collections which are sequences (with removal), since OtherObjs includes all the objects in After.

3.4 Choosing sequences with repetition

We can adapt seqrem to provide a definition for sequences where choices may be repeated - seqrep. Having selected our first item, the collection for the remaining choices is just that of the first choice, but now including the item just chosen. Thus for seqrep the OtherObjs includes H also; which means that the remainder of our choices (T) are chosen from the original collection Objs:

```
seqrep([],_).
seqrep([H|T],Objs):-append(_,[H|_],Objs)
                        & seqrep(T,Objs).
```

3.5 Choosing (multi)sets with repetition

For the sets with items repeatable (or rather, the *bags* with repetition), we take the definition for setrem but include H in the objects for the remaining choices:

```
bagrep([],_).
bagrep([H|T],Objs):-append(_,[H|RestOfObjs], Objs)
                        & bagrep(T, [H|RestOfObjs]).
```

Again, observation shows that, for a given collection *Objs*, the $X's$ produced as solutions to ?bagrep(X,Objs) are amongst the $X's$ produced by ?seqrep(X,Objs); that seqrem(X,Objs) are included in the seqrep(X,Objs) and, finally, that the $X's$ of setrem(X,Objs) are contained in the $X's$ of bagrep(X,Objs).

Also, though we do not present the details here, it is very straight-forward to use a variation of proof by induction to *prove* that the observations of the last paragraph are true in general and to prove that the usual formulae hold for the

	samples with removal	samples with replacement

s
e
q
s
	seqrem([],_).
	seqrem([H|T],Objs):-
	    append(Before,[H|After],Objs)
	    & append(Before,After,Others)
	    & seqrem(T, Others).
		seqrep([],_).
		seqrep([H|T],Objs):-
		    append(Before,[H|After],Objs)
		    & append(Before,[H|After],Others)
		    & seqrep(T, Others).

s
e
t
s
	setrem([],_).
	setrem([H|T] , Objs):-
	    append(Before ,[H|After],Objs)
	    & setrem(T, After).
		bagrep([],_).
		bagrep([H|T],Objs):-
		    append(Before,[H|After], Objs)
		    & bagrep(T, [H|After]).

Table 1. The symmetry in the Four Combinatorial Sampling Methods (Rewritten to show that samples with removal are included in samples with repetition and that sets are included among the sequences).

number of choices of a given size from a given size of collection for each of the four rules.

Finally, we present the complete Prolog code for the four combinatorial methods, in Table 1, with slight variations to show the symmetry that was promised at the start of Section 3.

3.6 Permutations, subsets and powersets for free

Again, as an added bonus, our declarative relations give us some extra definitions for free. If the list of choices is as long as the list of objects from which we are chosing, and, if we are chosing sequences with removal (i.e. seqrem), we have a permutation of the collection:

```
samelength([][]).
samelength([_|T1], [_|T2]):- samelength(T1,T2).
permutation(Perm,Seq):- samelength(Seq,Perm)
                      & seqrem(Perm,Seq).
```

If we consider choices of arbitrary size, we can find all the subsets of a given set of objects:

```
hassubset(Set,Subset):-setrem(Subset,Set).
```

and the collection of all subsets is the powerset:

```
powerset(Set,Powerset):-
          set_of(S, hassubset(Set,S), Powerset).
```

Although the details are not included here, we can further extend the use of append by defining partitions of a set, compositions of a number and the use of sequences and cycles to produce the Stirling numbers [9] of both kinds.

4 Verifying Theorems and Generating Conjectures

All the above rules have been incorporated in a growing library of just under 100 Discrete Mathematics rules at Surrey. It has an extension in a library of Graph Theoretic rules together with a large collection of example graphs together with some of their properties. As a basis for definitions and range of material covered, we have chosen to provide rules to cover the topics of Wilson's excellent little book on introductory graph theory [19]. We have then taken several theorems proved by Wilson and verified that they hold across the large collection of graphs generated by the rules. We have also found that the library can be used to generate conjectures from the examples. Some of the conjectures turn out to be obviously true whereas others would need to be formally proved by the mathematician if indeed they were true of all (simple) graphs and not just for the particular example set we have used. This section discusses the approach to theorem verification and conjecture generation. Although we use the Graph Theory library as our example, the techniques and methods are quite general.

4.1 Graph Theory

At present, we have about 150 rules for Graph Theoretic concepts and a library of examples incorporating all 208 graphs with up to 6 vertices which we generated in about ten minutes of computing time. The major problem with graph generation is that no algorithm is known that tests for graph isomorphism which will do better than "test all possibilities" in certain cases [6]. Even so, we have been able to determine that all graphs of less than six vertices are distinguishable by the number of vertices, vertex degrees and whether they are bipartite or not (i.e. their vertices may be coloured either black or white such that no edge has the same colour at both ends). These conditions have been added to the definition of isomorphism as a quick test on two graphs before searching for an actual isomorphism mapping.

The properties that we can test for are:

Discrete Mathematics on a Computer 139

- **number** of vertices, number of edges, degrees of vertices (valency), whether all vertices have the same degree (regular), bipartite graphs;

- **isomorphism**;

- **named graphs**: complete graphs $k(N)$, $k(M,N)$, path graphs, circuit graphs, wheels, stars, platonic graphs, kcubes;

- **graph generation**: all distinct (non-isomorphic) graphs on a given number of vertices, generating functions produced by Polya's Enumeration Theorem (for a given number of vertices N, a polynomial in x where the coefficient of x^p is the number of graphs on N vertices which have p edges);

- **trees**: generation or counting of trees which are free, binary, rooted, labelled; spanning trees of a graph, Prufer sequences;

- **derived graphs**: line graphs, graph union, complement, vertex removal, edge removal;

- **connectivity**: the number of connected components in a graph, disconnecting sets of vertices, bridges, edge connectivity;

- **paths and trails**; circuits, girth (length of the shortest circuit in a graph), Eulerian type (whether there is a trail of connected edges which includes each edge exactly once, beginning and ending at the same vertex), Hamiltonian type (whether there is a path of distinct edges which goes through each vertex exactly once ending back at its starting point);

- **colourings**: of vertices, of edges, chromatic polynomial (a generating function in x which, given the number of colours available (x) determines in how many ways the vertices may be coloured with no edge having the same colour at each end).

A graph is represented in a two-parameter rule called **absgraph**. The first parameter is the number of vertices in the graph, the second the abstact graph itself. An abstract graph has N vertices, numbered 1 to N, a list of edges (pairs $V - W$ where V and W are vertex numbers) and a list of vertex degrees ($V = D$ where V is a vertex, D the number of edges meeting at that vertex); e.g.

    absgraph(3 ,   [[1,2,3],   [1-2,1-3],   [1=2,2=1,3=1]]).

4.2 Theorem Verification

Since we have no automatic theorem provers available yet, we have taken the approach that one way the computer can help check that our rules are correct is to check that certain known theorems (proved or mentioned in [19]) hold across the complete set of examples (which includes all graphs of one to six vertices). Of course, such a test will not prove the Prolog rules are correct, but the more theorems we verify, the greater our confidence in our executable definitions. When the theorems were found not to hold, since the rules in the libraries are written in such a mathematically perspicuous way, it was often just a case of a spelling error in the rule that was the mistake. However, we have been able to find a couple of 'misprints' and special cases in Wilson's book [19], which demonstrates one of the benefits of computer checking.

The following two examples of statements from [19] and their expression in Prolog took just 13 seconds to verify on all 208 graphs with Salford Prologix on a Prime computer:

"The line graph of $k(5)$ is isomorphic to the complement of the Petersen graph"

```
? defgraph(petersen,P) & complement(P,PC) &
defgraph(k(5),K5) & linegraph((K5,LK5) & isomorphic(LK5,PC).
```

"A graph and its complement are not both disconnected"

```
? forall ( (absgraph(N,Graph) & complement(Graph,GComplement)),
(not (disconnected(Graph) & disconnected(GComplement)) )  )
```

We have verified 35 theorems from Wilson [19] and found this useful in identifying special cases such as, "Is the graph with no vertices connected or not?".

4.3 Conjecture Generation

Apart from checking that known theorems hold, we are able to generate statements that hold across the collection of examples. First we generate the graphs for the 208 examples, which takes about ten minutes of computer time (for the Prologix system on a Prime computer), then we generate all the properties of each graph (another 15 minutes). Next we see if there are any correlations between the properties and report the results as conjectures. We have found that conjectures are produced quite abundantly (about one every 3 seconds when correlating across 72 properties per graph in a collection of 208 graphs). Certainly, execution time seems no obstacle. A sample of some of the conjectures generated follows, numbered for reference (slightly modified for readability):

1. All 'complete graphs' are 'isconnected'.
2. 'Complete graph' and '2 edged graph' are mutually exclusive properties.
3. All 'complete graphs' are '1-component' graphs.
4. 'Complete bipartite' graphs are 'star' graphs.
5. 'Complete bipartite' graphs have a chromatic number of 2.
6. Hamiltonian and non-Hamiltonian are mutually exclusive properties.
7. All pathgraphs are semi-Eulerian.
8. Non-Eulerian includes all graphs regular of degree 3.

9. Semi-Eulerian and regular of degree 3 are mutually exclusive properties.

It can be seen that some of the conjectures are trivially true, for instance 1 and 6. Some are consequences of others (e.g. 3 follows from 1 since the property "isconnected" is another name for "has 1 component"). Others are indeed true, such as the fifth conjecture that complete bipartite graphs, $k(M, N)$ can be vertex-2-coloured (have a chromatic number 2). Others, while true of the examples, may or may not be true in general. For example, the ninth conjecture poses the problem "Are there any graphs regular of degree 3 which are also semi-Eulerian (that is, the path need not end back at the starting point)?". These conjectures are now open to mathematicians to label as trivial or prove rigorously.

$P \wedge Q$	$P \wedge \neg Q$	$\neg P \wedge Q$	$\neg P \wedge \neg Q$	Conclusion Reported
√	√	√	√	(no conclusion)
√	√	√	X	P and Q are exhaustive
√	√	√	X	P includes all Q
√	X	√	-	All P are Q
√	X	X	-	P and Q are equivalent
X	√	√	-	P and Q are mutually exclusive
X	√	X	-	(Q has not occurred in the examples)
X	X	√	-	(P has not occurred in the examples)
X	X	X	√	(both P and Q did not occur - " -)
X	X	X	X	(impossible)

Table 2. Deriving Conjectures From Examples: those found with properties P and Q. Here √ means some examples were found for this case, X means no examples were found for this case, and - means it does not matter if some or no examples were found for this case.

The author is not a graph theoretist but has found some new results by this exercise - at least, the results were new to him. While investigating the chromatic polynomials it was noticed that the degree of the polynomial is the number of vertices in the graph; the coefficient of the next highest power (degree-1) was the negative of the number of edges; that the constant coefficient is always zero; that the lowest power in the polynomial is the number of components of the graph and that its coefficient is 11 iff there are no circuits in the graph. Some of these results were later found in Wilson's book [19] and the rest in other works, together with proofs and the property that the signs of the coefficients alternate. The conjectures are produced from the property lists of the example graphs according to Table 2. Given two properties, we count how many examples have either both properties or just the first or just the second or none of the two. From the four numbers found we then report a conclusion consistent with these figures. The conclusion is not the most general but the way we report the results is to avoid too many results concluded from no evidence. Thus, if no graphs had property P and some had property Q, we could conclude that "All P are Q" is consistent with the examples as well as "No P is Q". We have chosen just one message for each valid combination of counts (since all four counts of zero is not possible) summarised in Table 2.

5 Discussion and Conclusion

We have seen that, with a very small amount of effort, some parts of discrete mathematics can be made directly executable on a computer using the logic programming language Prolog, which seems particularly amenable to such an objective. The advantages to mathematicians are that they are already familiar with the notation and style of use.

5.1 Constructive Mathematics

Not all of mathematics can be made executable however, only that part which is called "constructive", which is, fortunately, quite a large part. At present we are not expecting a notation which allows the definition of $\sqrt{2}$ as "that real number which, when squared, gives 2" to be executable directly on a computer. (With the methods described here, this would necessitate generating real numbers in turn until one was found whose square was 2, and the real numbers are not denumerable in this way.)

5.2 Execution Speed

We have demonstrated that the speed of execution of these mathematical definitions is also not a problem in general. Certainly there are properties that are time-consuming to compute (for instance, isomorphisms of some large graphs). The knowledge of a programmer can be drawn upon to produce faster versions of the simpler, but clear, initial code. Indeed, we have a strong belief that systems will soon be available which allow the user to transform such clear, correct, declarative - but slow - definitions into faster but more complex forms in much the same way that mathematicians transform equations. One result that we have found is that "Theorems = Speed". The more theorems that we know about a property, the more information we have available when we write down a rule. Thus, we could check a graph is a tree if it is connected and has no circuits, both properties being slow to execute on large graphs. But we know that a graph which is connected and has one less edge than it has vertices must be a tree (see for example [19] for a proof). The new property "number of edges = number of vertices -1" is, in general, much faster to check than the presence or absence of circuits, and so leads to a faster form of the rule for determining if a graph is a tree. Unlike conventional programming where we try to eliminate redundancy, the selective introduction of extra information leads to an increase in the speed of execution.

5.3 Compactness of notation

A further advantage of a compact executable notation for mathematics was discovered when investigating tournaments (sets of matches between pairs of players where each player meets every other, all N players taking part in each round of $N/2$ games). There are two cases to consider depending on whether the number of players is even or odd. If odd, then on each round, one player has a "bye" and does not take part in a game. When writing the mathematics in Prolog, two cases were duly written down. On closer inspection, it was found that the two cases had identical code. In fact, to arrange a tournament with N players when N is odd is just the same as arranging a tournament for $N + 1$ players with an extra player called "bye". Whoever plays him wins automatically. On reflection, this is obvious but, again, was brought to light merely by the exercise of expressing the definitions in Prolog.

5.4 Packages versus languages

Although this paper deals with combinatorics and graph theory, its principles are general. We have extended our library of rules to cover planarity testing of graphs and some elementary group theory. We have also demonstrated the use of a computer to act as a super-fast research assistant, generating large collections of examples, verifying conjectures across them and producing new conjectures. There are several advantages of using a genuine computer language like Prolog as opposed to a prepared package such as those which only deal with computer algebra. Undoubtedly, we could have manipulated our chromatic polynomials with any of the several computer algebra packages available. What we could not have done with such packages is to generate the graphs and generate their chromatic polynomials. Some simple rules have been included in our library for handling polynomials (as generating functions) but work on complex equations is better done with specialist packages. However, there is no reason why a complete suite of rules for algebraic manipulation could not be written in Prolog and introduced as another, optional, library. The point here is that the working mathematician has a vast experience and range of techniques that can be used already (with Prolog - available on a large range of computers) to investigate the parts of mathematics of interest without the need to look for specialist packages.

Acknowledgments The author wishes to acknowledge the helpful suggestions of the referees and especially to Jeff Johnson for his interest and encouragement.

References

[1] Baase, S. *Computer Algorithms*, Addison-Wesley, (Second Edition), (Reading, Mass.), 1988.

[2] Brassard, G. *Algorithmics*, Prentice-Hall, (London), 1988.

[3] Bratko, I. *Prolog Programming or Artificial Intelligence'* Addison-Wesley, (Wokingham, England), 1986.

[4] Clocksin, Mellish, *Programming in Prolog*, Springer-Verlag (Third Revised and Extended Edition), (Berlin), 1988.

[5] Even, S., *Algorithmic Combinatorics*, Macmillan, (New York), 1973.

[6] Garey, M. R., Johnson D. S., *Computers and Intractability*, W H Freeman and Co, (San Francisco), 1979.

[7] Gibbins, P., *Logic with Prolog*, Clarendon Press, (Oxford), 1988.

[8] Gould, R., *Graph Theory*, Benjamin/Cummings Publishers, (California), 1988.

[9] Graham, R. L., Knuth D. E., Patashnik O., *Concrete Mathematics*, Addison-Wesley, (Reading, Mass.), 1989.

[10] Hayes, I. *Specification Case Studies*, Prentice-Hall, (London), 1988.

[11] Ince, D. C., *An Introduction to Discrete Mathematics and Formal System Specification*, Oxford University Press, (Oxford), 1988.

[12] Knott, R. D., Krause P. J., "An Approach to Animating Z Using Prolog", Alvey Project SE/065 and SE/090 Report A1.1, Department of Mathematics, University of Surrey, GU2 5XH, July 1988.

[13] Knott, R. D., Krause, P. J., "A Book Library System: an example of the rapid prototyping of a Z specification in Prolog", Alvey Project SE/065 and SE/090 Report A1.2, Department of Mathematics, University of Surrey, GU2 5XH, June 1988.

[14] Knott, R. D., Krause, P. J., "On the Derivation of an Effective Animation : Telephone Network Case Study", Alvey Project SE/065 Report A1.3, Department of Mathematics, University of Surrey, GU2 5XH, November 1988 .

[15] Sedgewick, R., *Algorithms*, Addison-Wesley, (Reading, Mass), 1983 .

[16] Spivey, J. M., *The Z Notation*, Prentice-Hall, (London), 1989.

[17] Stirling, L., Shapiro, E., *The Art of Prolog*, M.I.T. Press, (Cambridge, Mass.), 1986.

[18] Wilf, H. S., *Algorithms and Complexity*, Prentice-Hall, (London), 1986.

[19] Wilson, R. J., *Introduction to Graph Theory*, Longman, (Third Edition) (Harlow, England), 1985.

The Mathematical Revolution Inspired by Computing. J H Johnson & M J Loomes (eds)
©1991 The Institute of Mathematics and its Applications. Oxford University Press

The Wider Uses of the Z Specification Language in Mathematical Modelling

Allan Norcliffe

Department of Mathematical Sciences, Sheffield City Polytechnic

Pond Street, Sheffield, S1 1WB

Abstract

Mathematically-based specification languages, such as Z or VDM, now offer the software engineer a real alternative to English as a specification medium. Specifications written in Z or VDM are unambiguous, concise, and more amenable to formal reasoning than their English counterpart. Their use by industry holds out the promise of better software quality and improved productivity. In this paper we investigate what specification languages, such as Z, can offer the applied mathematician, and report on our experiences of introducing mathematicians to Z in modelling classes at Sheffield City Polytechnic.

1 Introduction

One area of mathematics that has certainly undergone a revolution as a result of developments in computing is that of mathematical modelling. We have only to look at the vast and growing array of problems that were once impossible to solve, which are now being tackled by both students and practitioners alike, to realise the magnitude of this revolution. Not only is there the raw computing power of the super computer, to assist with numerical work, and enhanced colour graphics, to see what the computations mean, but there are now sophisticated computer algebra packages to help with much of the analytical work.

With the availability of all this computing power, the ability to formulate a problem properly, and specify precisely what needs to be done, is becoming increasingly more important. Unfortunately, most mathematicians still tend to concentrate on developing methods of solution, once problems have been formulated, instead of developing the vital skills of problem formulation and specification.

Software engineers have realised the central importance of the specification stage in problem solving, and industry-standard specification languages, such as Z [Spivey, 88] and VDM [Jones, 86], now exist to assist in the production of correct coding. At present, however, Z and VDM tend only to be used by software engineers and computer scientists. Few applied mathematicians really know of the existence of these languages, or of their potential for developing good problem formulation skills.

In this paper we aim to show that specification languages, such as Z, can have a valuable role to play in mathematical modelling work, in developing these skills, and in helping to put the model formulation stage on a more rigorous footing. We also wish to report, briefly, on our experience of using Z in modelling classes at Sheffield City Polytechnic, where Z has been used in a wider context than just software engineering. We begin with a very short introduction to the Z specification language for those unfamiliar with the notation.

2   What is Z?

Z is nothing more than a formal notation, based on set theory and predicate logic. It was developed by J-R Abrial at the Programming Research Group (PRG) in Oxford, and is fast becoming an industry-standard specification language. At the heart of Z is the structure known as a schema, in which the mathematics of what is being specified is set out.

The best way of explaining Z is to see it in action. Suppose we wish to develop a simple software system that will be part of a security system for a building. Suppose that one feature of the system is that it must be able to tell us who the valid users of the building are, and who is currently in or out of the building at any given time. Let $PERSON$ be the set of legitimate users of the building, and let $\mathbb{P}(PERSON)$ be its powerset (set of all subsets). A simple schema that specifies the state of the building can be written in Z as follows.

―― $StateOfBuilding$ ――――――――――――――――
$users, in, out: \mathbb{P}(PERSON)$

$in \cap out = \{\ \}$
$in \cup out = users$
――――――――――――――――――――――

All schemas in Z have a name; ours is called $StateOfBuilding$. The section above the middle line is called the signature and introduces the relevant variables and identifiers into the schema. The notation $x : X$ means x is of type X, in other words x is an element of the set X. Thus users, in, and out are introduced as subsets of the set called $PERSON$. These are the state variables in our simple system.

The text below the middle line is known as the *predicate*. Each line in this part is assumed to be conjoined with the rest of the predicate using the logical AND connective. The notation $\{x_1, ..., x_n\}$ means the set containing exactly $x_1, ..., x_n$. Thus $\{\}$ is the empty set, which can also be written $\emptyset$. The predicate specifies how the variables, introduced in the signature, must be related. In our system we require that no person can be both in and out of the building at the same time, and that those in the building, together with those out of the building, make up the valid users of the building. This predicate constitutes what software engineers would call a *data invariant*. Any operation, that might change the current state of the building, must not violate this invariant. For example, when a valid user is checked out of the building, the state of the building changes but this predicate is still satisfied.

Operations can be specified as schemas in Z and a check out operation might therefore be specified as follows.

―― *CheckOut* ―――――――――――――――――――――
Δ *StateOfBuilding*
user? : *PERSON*
―――――――――――――――――――――――――
user? $\in$ *in*
in' = *in* \ {*user?*}
out = *out* $\cup$ {*user?*}
―――――――――――――――――――――――――

The Δ notation that appears is simply a device for introducing all the variables into *Checkout* that are in *StateOfBuilding*, together with *decorated* versions of these which have an apostrophe symbol after their names, e.g. *in'* and *out'*. The undecorated and decorated variables represent the before and after states of the *CheckOut* operation. The input to the *CheckOut* operation is *user?*, as signified by the query mark after its name. Outputs carry a shriek or exclamation mark.

For *Checkout* to work we require that *user?* has previously been checked into the building. This requirement is catered for in the first line of the predicate by writing *user?* $\in$ *in*. When we checkout *user?* then *user?* is taken out of the subset *in* and added to *out*. This requirement is specified using the set difference and set union operations as indicated. As usual the notation $S \setminus T$ stands for the set of S members which are not in T.

Two other operations that change the state are specified below without further comment.

―― *CheckIn* ―――――――――――――――――――――
Δ *StateOfBuilding*
user? : *PERSON*
―――――――――――――――――――――――――
user? $\in$ *out*
in' = *in* $\cup$ {*user?*}
out' = *out* \ {*user?*}
―――――――――――――――――――――――――

```
┌─ AddUser ─────────────────────────────
│ Δ StateOfBuilding
│ newuser? : PERSON
│ ──────────────────────────────────────
│ newuser? ∉ users
│ users' = users ∪ {newuser?}
│ out' = out ∪ {newuser?}
│ in' = in
└───────────────────────────────────────
```

There are obviously many other operations that need to be specified to make the system robust and complete. We will not specify them here, but point out simply that, in principle, the full system can be methodically built up. Before leaving our example we might usefully note two things. First, we should always check that operations we specify do preserve the data invariant and, second, that an *initial state* can always be set up that similarly satisfies the invariant. Readers can check that each of the above operations do, indeed, preserve the data invariant, and that the following operation gives an initial state with the required invariant property.

```
┌─ InitState ───────────────────────────
│ Δ StateOfBuilding
│ ──────────────────────────────────────
│ users' = { } ∧ in' = { } ∧ out' = { }
└───────────────────────────────────────
```

3 What can Z Offer the Mathematician?

If we look at the Z schemas we have written, we see that we have begun to build a precise and clear model of what the software system is required to do. Everything we have written down is amenable to formal reasoning in a systematic way. We can be critical about what we have written and use the Z as a basis to communicate our ideas unambiguously. We have formulated in a rigorous and highly-structured way a mathematical model that can then usefully provide the solution to our problem - which is the coding for the system. We can also use the Z to verify whether the solution we provide matches the specification. This approach to software development is well documented in the literature (see for example [2][4]) and holds out the promise of better quality, and improved productivity in the software industry.

Let us now look at a quote taken from Ron McLone's [3] influential report of the early seventies relating to the training of mathematicians.

"Good at solving problems, not so good at formulating them, the graduate has a reasonable knowledge of literature and technique, he has some ingenuity and is capable of seeking out further knowledge. On the other hand, the graduate is not good at planning his work, nor of making a critical evaluation of it when completed; and in any event he has to keep his work to himself as he has apparently little idea of how to communicate it to others"

Today, on our degree courses, we would argue that these criticisms of mathematicians now receive considerably more attention than, perhaps, they used to. But if we are honest, we would have to admit that we still tend to neglect vital issues of model formulation and specification in our teaching. A wider appreciation of specification languages, such as Z or VDM, we submit, would not go amiss on many mathematics degrees.

In the next section we set out some examples of the use of Z, in a wider context than just software engineering, to show that Z has a role to play in the formulation and specification of systems that more naturally fall into the province of the applied mathematician. We have deliberately kept the examples simple for clarity; specifying more complex systems should, in principle, present few additional problems.

4 Wider Uses of Z

4.1 Particle Dynamics

The state of a particle, of mass m, might be specified in Z as follows.

```
┌─ StateOfParticle ─────────────────────────
│ m : M
│ a : T ↛ LT$^{-2}$
│ v : T ↛ LT$^{-1}$
│ r : T ↛ L
├───────────────────────────────────────────
│ ∀t : T | t ≥ t$_0$ •
│   $\frac{d}{dt}$ v(t) = a(t)
│   $\frac{d}{dt}$ r(t) = v(t)
└───────────────────────────────────────────
```

Although liberties have been taken with the standard notation, Z is a developing language and perhaps a standard scientific version will soon be devised. In our schema M is just the set of masses, and the mass m is being declared to be a member of this set. The particle's acceleration vector, a, its velocity vector, v, and its position vector, r, are all being modelled as partial functions of time. T is the set of possible times, L the set of vector lengths, and so on. In the predicate we are defining a data invariant that must hold, whatever happens, for all times, t, after the particle came into existence at some constant time t_0.

The initial state of the particle can be fixed by the following operation called InitStateOfParticle.

─── InitStateOfParticle ──────────────────────────────
$\Delta StateOfParticle$

$m' = m_0$
$\forall t : T \mid t \geq t_0 \bullet$
$\mathbf{a}'(t) = \mathbf{0} \wedge \mathbf{v}'(t) = \mathbf{0} \wedge \mathbf{r}'(t) = \mathbf{0}$
──

Here we are simply giving the particle a definite constant mass, m_0, and setting its acceleration, velocity, and position vectors arbitrarily to zero.

If a force F? is now applied, at time $t?$, to the particle, then the following *ApplyForce* schema will specify the ensuing motion. As can be seen the state of the particle remains unaltered for times t that are less than $t?$ but, for times greater than or equal to $t?$, Newton's second law modifies the acceleration accordingly.

─── ApplyForce ───────────────────────────────────────
$\Delta StateOfParticle$
$\mathbf{F} : T \rightarrowtail MLT^{-2}$
$t? : T$

$t? > t_0$
$dom(\mathbf{F}?) = \{\tau \mid \tau \in T \wedge \tau \geq t?\}$
$\forall t : T \mid t \geq t_0 \bullet$
$(t < t?) \Rightarrow \mathbf{a}'(t) = \mathbf{a}(t) \wedge m' = m$
$(t \geq t?) \Rightarrow \mathbf{a}'(t) = \mathbf{F}?(t)/m' \wedge m' = m$
──

Various other operations could be defined, such as a collision with another particle, but we will leave this to the interested reader to pursue if they so wish.

4.2 Liquid in a Storage Tank

Our second example again involves a dynamical system. Assume we have a verticle-sided tank, whose base area is A, and suppose there is some water in the tank. The state of the tank can be specified by giving the depth, h, of water in the tank together with the rate, r, at which the height of the surface of the water is changing. The schemas that specify the state of the tank, and its initial state, are given below. The predicate in the state schema expresses

The Wider Uses of Z in Mathematical Modelling

the rather obvious dynamical relationship that connects the variables h and r. Both are modelled as appropriate partial functions of time.

―― $StateOfTank$ ――――――――――――――――
$h : T \nrightarrow L$
$r : T \nrightarrow LT^{-1}$
$A : L^2$
――――――――――――――――――――――
$\forall t : T \mid t \geq t_0 .$
$r(t) = \frac{d}{dt} h(t) \wedge A > 0 \wedge h(t) \geq 0$
――――――――――――――――――――――

―― $InitStateOfTank$ ――――――――――――――
$\Delta StateofTank$
――――――――――――――――――――――
$\forall t : T \mid t \geq t_0 .$
$h'(t) = 0 \wedge r'(t) = 0 \wedge A' = A_0$
――――――――――――――――――――――

If water begins flowing into the tank at time $t?$, at a rate $in?$, then the filling operation can be described by the following schema.

―― $Fill$ ―――――――――――――――――――
$\Delta\ StateofTank$
$in? : T \nrightarrow L^3 T^{-1}$
$t? : T$
――――――――――――――――――――――
$t? > t_0$
$dom(in?) = \{\tau \mid \tau \in T \wedge \tau \geq t?\}$
$\forall t : T \mid t \geq t_0 .$
$(t < t?) \Rightarrow r'(t) = r(t) \wedge A' = A$
$(t \geq t?) \Rightarrow r'(t) = in?(t)/A' \wedge A' = A$
――――――――――――――――――――――

Emptying of the tank, through a circular orifice of area $a?$, in the bottom of the tank, starting at time $t?$, can now be specified via the use of Bernouilli's equation and a supplied coefficient of discharge $Cd?$.

┌─ *Empty* ─────────────────────────────────
│ Δ *StateofTank*
│ $a? : L^2$
│ $out! : T \rightarrowtail L^3 T^{-1}$
│ $Cd? : \mathbb{R}$
│ $t? : T$
├───
│ $t? > t_0$
│ $a? > 0$
│ $Cd? > 0$
│ $a? < A$
│ $\forall t : T \mid t \geq t_0.$
│ $(t < t?) \Rightarrow r'(t) = r(t) \wedge A' = A$
│ $(t \geq t?) \Rightarrow out!(t) = Cd? \times a? \times \sqrt{(2 \times g \times h'(t))} \wedge$
│ $\qquad\qquad r'(t) = -Cd \times a? \times \sqrt{(2 \times g \times h'(t))}/A' \wedge$
│ $\qquad\qquad A' = A$
└───

Once again, various other operations, such as the simultaneous filling and emptying of the tank, could be specified.

4.3 Heat Conduction in a Metal Bar

To change the emphasis slightly, we now consider a steady state problem involving a one-dimensional, distributed-parameter system. The system is a metal bar of small square cross section, whose width is w and whose length is ℓ. At any distance, x, from one end of the bar a temperature θ and heat flux q will be defined in the bar. If the coefficient of heat conduction is k, then the state of the bar can be specified as follows, along with its initial state.

┌─ *StateOfBar* ────────────────────────────
│ $\theta : L \rightarrowtail K$
│ $q : L \rightarrowtail J L^{-2} T^{-1}$
│ $k : J L^{-1} T^{-1} K^{-1}$
│ $w, \ell : L$
├───
│ $w > 0 \wedge \ell > 0 \wedge k > 0$
│ $\forall x : L \mid 0 \leq x \leq \ell.$
│ $-k \frac{d}{dx} \theta(x) = q(x)$
└───

The Wider Uses of Z in Mathematical Modelling

```
┌─ InitStateOfBar ─────────────────────────────────
│ Δ StateOfBar
│ ┌──────────────────────────────────────────────
│ │ $w = w_0 \wedge \ell' = \ell_0 \wedge k' = k_0$
│ │ $\forall x : L \mid 0 \leq x \leq \ell'$.
│ │ $\theta'(x) = \theta_0$
│ │ $q'(x) = 0$
```

In these schemas K is the set of absolute temperatures and J is the set of heat values. Initially we have defined the bar to be at a steady temperature of θ_0. The quantities w_0, ℓ_0 and k_0 are the defining constants for the values of w, ℓ and k respectively. If we now subject the end of the bar, at $x = 0$, to a hot object at temperature $\theta end?$ then heat will flow along the bar away from this end. If the bar is uninsulated then heat will escape, by convective cooling, from its surface at a rate determined by the convective heat coefficient, $h?$, and the surrounding ambient air temperature, $\theta air?$. This heat conduction operation can be specified as follows.

```
┌─ HeatConduction ─────────────────────────────────
│ Δ StateOfBar
│ $\theta end?, \theta air? : K$
│ $h? : JL^{-2}T^{-1}K^{-1}$
│ ┌──────────────────────────────────────────────
│ │ $\theta end? > \theta air?$
│ │ $h? > 0$
│ │ $\forall x : L \mid 0 \leq x \leq \ell'$.
│ │ $\lim_{\delta x \to 0}(q'(x)w'^2 - q'(x+\delta x)w'^2 - 4w'\delta x h?(\theta'(x) - \theta air?)) = 0$
│ │ $\theta'(0) = \theta end? \wedge \theta'(\ell) = \theta air?$
│ │ $w' = w \wedge \ell' = \ell \wedge k' = k$
```

5 Comment

The above specifications are not the kind one normally sees. Nowhere in the Hayes specification case studies book [1], for example, are continuous systems specified. A common myth exists about Z, namely, that as it is based on discrete mathematics, it must therefore only be used to specify discrete systems such as software and algorithms. This is demonstrably not the case, since Z can be used to specify continuous systems as illustrated. Z promotes several positive features which are vital in modelling work:

- In the signatures of schemas all variables known to be of interest are clearly defined, and their exact mathematical nature specified. How often in modelling work do we fail to do this properly?

- In the predicate the precise connection between the variables is set out along with the assumptions being made. In modelling we often take for granted many assumptions and then overlook them. In a Z schema we cannot do this because the schema would not work. Schemas can be systematically checked for consistency, using precondition analysis, which automatically brings to light any assumptions that have been overlooked.

- Having to write a specification forces us to separate out the formulation stage of a problem from the solution stage, which has its advantages. All too often we write down equations and procede to solve them before realising that other key constraints have been violated.

- Writing schemas helps to break the problem into manageable parts and, in turn, this structures our thoughts about the problem and how it should be solved.

- Finally, schemas provide a definite focus of interest, which helps concentrate the mind when communicating ideas to peer mathematicians or to the lay person.

6   Experiences with using Z at Sheffield City Polytechnic

Z is taught on a range of courses at the Polytechnic. Bachelors and post graduate degree students in Computer Studies and Information Technology receive formal instruction in both Z and VDM because of the importance of these languages in modern software engineering. Students on the bachelors degree course in Computing Mathematics, and the Computing and Application Mathematics option on the Applied Science Degree, also learn Z and VDM for the same reason. But, additionally, they study Z in modelling classes for the wider reasons just outlined. Clearly the presentation is varied to suit individual needs, but the key aim of developing formal skills in model formulation and specification is the same.

In modelling workshops students tackle a range of problems from specifying software sytems for banking, library use, security systems, electronic mail, electronic diaries etc, to describing how vending machines and other physical systems work, to formulating logic puzzles for solution in PROLOG. Students typically work in groups and are expected to give group presentations, where they defend the specifications they have written or the models they have formulated. Word-processed reports, aimed at the lay person in some instances, are also produced and assessed. By and large students enjoy this activity. They see it as different, creative and exciting, and a welcome change from the rather mundane problem solving exercises that all too frequently they are given.

Observations that can be made, based on our experience of operating these kinds of workshops in the past few years, are the following.

Students are happiest when they are being creative and writing specifications. Reasoning about what they have written, in a formal way, is difficult for them. Proof ideas tend to come only slowly.

Students prefer to specify systems where they are in charge of defining the rules. They are happiest, therefore, modelling libraries, banks etc, as opposed to physical systems where additional knowledge of physics or mechanics is needed.

Their ability to communicate about specifications, in writing, seems to be enhanced. The structure of the specification seems to help structure the reports that are produced.

It is surprising just how little discrete mathematics is needed before students can begin to write Z specifications. There is no reason why first year students, in their second and third terms, should not make a start on formally specifying simple systems such as vending machines or banks, based only on a knowledge of sets, functions, and logic.

Finally students enjoy writing specifications, and working in groups, where they can use each other as sounding boards. They all agree that Z helps focus the mind on the essential features of the problem and helps develop a disciplined approach to problem formulation.

7 Conclusions

To summarise, then, we feel that Z does have a valuable role to play on mathematics degrees in modelling and problem solving work. The notation is easily learnt and should present few problems to mathematicians who have a good grasp of set theory and predicate logic. It offers the mathematician all the advantages of precision and rigour that it offers the software engineer, and we feel strongly that it should not remain an approach to problem formulation to which only software engineers are exposed.

References

[1] Hayes, I., (ed), *Specification Case Studies*, Prentice Hall, (London),1987.

[2] Jones, C., B., *Systematic Software Development Using VDM*, Prentice Hall, (London), 1986.

[3] McLone, R., "The Training of Mathematicians", SSRC Report, 1973.

[4] Spivey, J., *The Z Language: A Reference Manual*, Prentice Hall, (London), 1988.

Scene Analysis via Galois Lattices

M. Andrew
Royal Sussex County Hospital, Brighton
D. K. Bose & S. Cosby
I.T. Research Institute, Brighton Polytechnic, BN2 4GJ

Abstract

The paper will discuss the implementation of knowledge based analysis to gated blood pool studies of the human heart. In particular, it will consider the use of Galois lattices in the identification and classification of 'abnormal' areas of the left ventricle of the heart.

1 Introduction

Our aim is to describe a method which allows one to study a digital image and give conceptual descriptions of significant regions. The method is based on the Galois lattice structure of binary relations, and it will be illustrated by considering the problem of interpreting gamma camera images of the human heart. These images arise in gated blood pool studies of the heart.

2 Gated Blood Pool Studies of the Heart

The human heart has four main chambers, namely the left and right ventricles and the left and right atria (Figure 1). These chambers act in such a way as to form a double pump. The pumping capability and structure of the left ventricle is crucial in determining a cardiac diagnosis. A common non-invasive testing of the left ventricle is achieved by using a radioactive tracer. When administered to a patient it results in the blood pool of the heart emitting gamma rays. These gamma rays are then counted by a gamma camera and their position in the heart noted. Since the heart is in motion while the rays are being counted one would get a blurred image. To overcome this, one first of all calculates an average time for the cardiac cycle from an E.C.G (electrocardiogram) trace of the patient over a 30 second period. This time interval is then divided into 16 equal sub-intervals. The *R-wave* (usually the dominant signal) of the E.C.G triggers the gamma camera and all the counts in the first time sub-interval are put into a matrix. This matrix keeps track of the number of counts arriving from a point of the heart and for all the points covered by the camera. The counts in the second sub-interval are put into a second matrix, and so on. When the next R-wave is received the counts in the first interval are added to

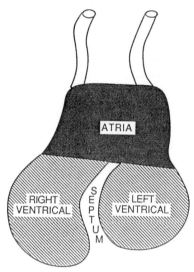

Figure 1. A simplified diagram of the heart

the first matrix and so on and this process is repeated until each of the 16 matrices have sufficient counts so that a reliable set of images are obtained.

Thus each matrix is an image of the heart at a certain interval of its cycle, and an element g_{ij} represents the number of counts (or grey level) at the ij^{th} pixel on the face of the camera corresponding to a certain point in the heart. The 16 matrices therefore correspond to 16 'snapshots' of the heart during its cycle. Each matrix is 64*64 and the size of a pixel is a 4 mm square.

If we consider this sequence of 16 digital images and pick a pixel in the first image and note its grey value, then note its value in the second image and so on until the 16th image, we then have a sequence of counts associated with that pixel. This sequence can be thought of as the sampled values of a certain curve, called the *Time Activity Curve* or TAC, associated with that pixel.

3 Objective

The collection of 64*64 TAC's one for each pixel and the pixel coordinates then represent all the data available to us from the Gated Blood Pool Study. Our aim is to use this information together with appropriate clinical knowledge to help interpret the behaviour and structure of the left ventricle. More specifically, our objective is to take the image corresponding the fully expanded ventricles and highlight in it those regions in the left ventricle which are clinically interesting. Our highlighting will consist of a logical predicate description of the region and an interpretation in medical terms of this predicate. Standard available methods allow us to identify a *region of interest* (R.O.I) enclosing the left ventricle in each image.

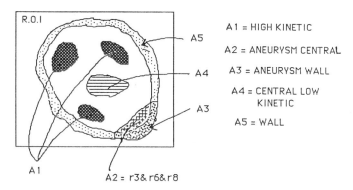

Figure 2. Left ventrical description

Figure 2 gives an example of what we would like to achieve, namely in the left ventricle we have interesting areas A_1 to A_5. Each of these areas has a medical description associated with it. For example, A_2 is a central aneurysm of the left ventricle. Also we 'explain' how this classification was obtained by describing the unique logic predicate that is satisfied by every pixel in this area. For example, let r_3 denote the predicate EF > 20, where EF denotes the *Ejection Fraction* and is defined by :

$$EF = (MAX\ TAC - MIN\ TAC)/\ MAX\ TAC.$$

Let r_6 denote each neighbour has a grey scale value within $\pm$ 2 of the pixel in question, and let r_8 denote the first Fourier coefficient of the TAC of the pixel in question lies between 10 and 20. Then A_2 could have the predicate description r_3 & r_6 & r_8, where & means 'logic AND'. In other words every pixel of A_2 and no other satisfies r_3 and r_6 and r_8. Having got the logic description of A_2 one can now interpret this medically and come up with the description ANEURYSM CENTRAL. We shall discuss later how this interpretation is done.

4 A Method based on Galois lattices.

To achieve our objective we want a method which will segment the left ventricle into regions, each region will have a predicate describing it, and the predicate should be as detailed as possible so as to assist in the medical interpretation of that region. Thus we want a method which will generate from the TAC data, regions and predicates such that each region $A = \{p_1, p_2, \ldots\}$ (say) has a predicate r (say) such that each pixel p_i satisfies r and no pixel not a member of A satisfies r. Further, r should be as detailed as possible i.e if q is any other predicate then at least one pixel in A will not satisfy r & q.

Obviously, the base set of predicates from which the predicate descriptions of the regions will be generated must be chosen so that these descriptions can be interpreted medically. To help us choose these predicates we turn to the set of rules given to us by the medical experts for classifying regions in the left ventricle. These rules are determined empirically from past experience of analysing gated heart studies as well as deductions about the shape of the TAC's due to the physical conditions. A highly simplified example of such a rule is :

IF	mean intensity of TAC is high,
AND	amplitude of TAC is significant,
AND	2_{nd} Fourier coefficient of TAC is large,
AND	phase of TAC is approximately 180 degrees out of phase with the right ventricle,
AND	pixel has a similar count to some but not all of its neighbours,
THEN	pixel is an aneurysmic wall pixel.

These rules allow one to pick out the 'base' predicates which are important for classification e.g 2nd Fourier coefficient of TAC is large. For the sake of illustration let us now assume we have a 'base' set of predicates $R = \{r_1, r_2, \ldots, r_6\}$ as shown in Table 1.

μ	r_1	r_2	r_3	r_4	r_5	r_6
p_1	0	0	1	1	0	0
p_2	0	1	1	1	0	0
p_3	0	1	1	1	0	1
p_4	0	1	0	1	0	1
p_5	0	1	1	1	0	0
p_6	1	1	1	1	0	0
p_7	0	0	0	0	0	1
p_8	1	1	0	1	1	1
p_9	1	1	1	1	0	1
p_{10}	0	0	0	0	1	1
p_{11}	1	1	0	1	0	1
p_{12}	0	1	0	1	1	1
p_{13}	0	1	0	0	1	1
p_{14}	0	1	0	1	0	0

Table 1. $r_1, \ldots, r_6$ are predicates satisfied by 'pixels' $p_1, \ldots, p_{14}$, $r_1 \equiv$ MAX OF TAC > 160, $r_2 \equiv$ MAX OF TAC > 100, $r_3 \equiv$ MIN OF TAC > 120, $r_4 \equiv$ MIN OF TAC > 80, $r_5 \equiv$ EF > 30, $r_6 \equiv$ EF > 20.

We now raster scan the left ventricle of a patient pixel by pixel and test for each pixel whether it satisfies a predicate. This way we can build up a logic, or 0,1, matrix between the pixels of the left ventricle and the predicates r_1, r_2, ... , r_6. Note however many pixels will have identical rows in the matrix. These pixels will be grouped together (an equivalence class) as one pixel or entry in the matrix and a record will be kept of the pixels in the group. Table 1 is the result of such a scan, note for example p_8 in Table 1 consists of 11 pixels distributed in the left ventricle as shown in Figure 4, and each of these pixels satisfies the predicates r_1, r_2, r_4, r_5, and r_6. Let P denote the set of 'pixels' p_1 to p_{14}.

Recall what we want is for the binary relation μ represented by the logic matrix to generate "maximal" entities of the form

$$[\{p_i, p_j, \ldots\}; R_k]$$

where each pixel $p_i, p_j, \ldots$ satisfies predicate R_k and if we expand the set $\{p_i, p_j, \ldots\}$ then this satisfaction of R_k by all the pixels is no longer true. Further we want R_k to be as 'detailed' as possible. This can be done using Birkhoff's Galois lattice description of a binary relation and the extension given in Barbut and Monjardet [1] (see also [5] in this volume), and is achieved as follows.

The relations μ and μ^{-1} induce maps

$$F : P \to 2^R \quad \text{and} \quad G : R \to 2^P$$

where $F(p) = \{r \in R : p \text{ is } \mu\text{-related to } r\}$ and $G(r) = \{p \in P : p \text{ is } \mu\text{-related to } r\}$. For example $F(p_{13}) = \{r_2, r_5, r_6\}$ and $G(r_3) = \{p_1, p_2, p_3, p_5, p_6, p_9\}$ in Table 1.

We are interested in properties shared by pixels so it is natural to extend F from $P \to 2^R$ to a mapping $f : 2^P \to 2^R$ by defining

$$f(A) = \cap_{p \in A} F(p) \quad \text{for} \quad A \subseteq P.$$

For example with reference to Table 1

$$f(\{p_3, p_{13}\}) = F(p_3) \cap F(p_{13}) = \{r_2, r_3, r_4, r_6\} \cap \{r_2, r_5, r_6\} = \{r_2, r_6\}.$$

One can similarly extend G to $g : 2^R \to 2^P$ by

$$g(B) = \cap_{r \in B} G(r) \quad \text{for} \quad B \subseteq R.$$

We note f and g form a Galois connection between 2^P and 2^R, i.e.

if $A_1 \supseteq A_2$ then $f(A_1) \subseteq f(A_2)$
if $B_1 \supseteq B_2$ then $g(B_1) \subseteq g(B_2)$
with $A \subseteq gf(A)$ and $B \subseteq fg(B)$.

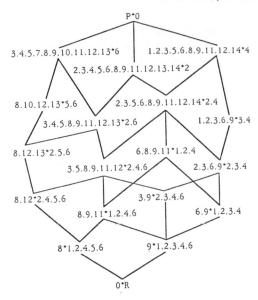

Figure 3. Galois Lattice for μ

Thus $fg : 2^R \to 2^R$ and $gf : 2^P \to 2^P$ are closure operators, i.e.

$gf(A) \supset A$ (Extensive)
$gf(gf(A)) = gf(A)$ (Idempotent)
$A_1 \subset A_2$ implies $gf(A_1) \subset gf(A_2)$ (Isotone).

(Similarly for fg). $A \subseteq P$ is defined to be *closed* iff $A = gf(A)$ $B \subseteq R$ is defined to be *closed* iff $B = fg(B)$. Hence we can define the closure spaces

$$C(P) \stackrel{\text{def}}{=} \{A \subset P : A = gf(A)\}$$
$$C(R) \stackrel{\text{def}}{=} \{B \subset R : B = fg(B)\}.$$

These are complete lattices with the usual set inclusion ordering. Hence we can get our 'maximal entity' data structure by considering the pair $(A, f(A))$ for each A a member of $C(P)$. This too has a lattice structure where

$$(A_1, f(A_1)) \leq (A_2, f(A_2)) \quad \text{iff} \quad A_1 \subseteq A_2 \text{ and } f(A_1) \supseteq f(A_2).$$

This lattice is the Galois lattice of a binary relation and algorithms for finding it from the logic matrix in an efficient and incremental way can be found in Ho [4] and Norris [7]. We have implemented Ho's version and used it to find the Galois lattice for μ given by Table 1. Our results are displayed in Figure 3 where, for example, the node 8,12,13 * 2,5,6 represents the 'maximal entity' [$\{p_8, p_{12}, p_{13}\}$; r_2 & r_5 & r_6].

5 Results

The Galois lattice obtained can be used to lead to medical descriptions of interesting areas of the left ventricle in various ways. The simplest is when a logic expression in a rule appears in a node of the Galois lattice. In this case one can use the rule to give a medical label to the region appearing in that node. For example, if there was a rule that said r_2 & r_5 & r_6 (from Table 1) implied a lower edge pixel of the left ventricle then since $[\{p_8, p_{12}, p_{13}\}; r_2$ & r_5 & $r_6]$ is a node the region $\{p_8, p_{12}, p_{13}\}$ is the lower edge of the left ventricle.

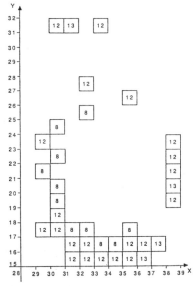

Figure 4. Node 8, 12, 13 * 2, 5, 6 approximately represents the lower edge of the left ventricle.

Figure 4, which is based on an actual scan, illustrates what this region really looks like and one can see it is roughly the lower edge of the left ventricle. This is an artificial example, in real cases one would have a larger base set of predicates and more complex and accurate rules. However, the rules used are not infallible and what can happen is that the logic expressions in a rule do not appear in any node of the Galois lattice. An obvious deduction is that the conclusion of that rule does not apply to the patient whose Galois lattice is being studied. Since the rules are not infallible we want to try and make sure abnormalities suggested by them have not been missed. To help do this we search for a node in the Galois lattice whose logic expression is in some sense 'close' to the expression in the rule. We then test the region described by it for geometrical and other information consistent with the conclusion of the rule. If these tests are positive we then tentatively classify it according to the rule. This procedure is also carried out on the neighbours in the lattice of the original node provided they are 'close enough' to the logic expression of the rule. For example, refering to Table 1 if a rule is r_2 & r_5 & r_6 & r_3 implies a lower edge pixel, then a close node is $[\{p_8, p_{12}, p_{13}\}; r_2$ & r_5 & $r_6]$. If we tested the 'shape' of $\{p_8, p_{12}, p_{13}\}$ as shown in figure 4 we would find it is consistent with the conclusion of the rule and so we would put this area forward (tentatively) as the lower edge of the left ventricle. A similar analysis on a 'neighbour' node $[\{p_8, p_{10}, p_{12}, p_{13}\}; r_5$ & $r_6]$ would not support the conclusion of the rule and so would be rejected. It is worth pointing out that the 'shape' analysis is partly carried out using morphological operations (see Haralick [3] for an introduction). Notice that this procedure opens up the possibility of the system learning and suggesting new rules. Another way the system can learn is if a certain logic expression appears in a node of the galois lattice for a large number of patients classified as having a certain left ventricular disorder. It would then be worthwhile for the experts to study this node to see whether it could be used for reinforcing or suggesting a diagnosis. We are at the moment exploring this aspect of the system.

There is a system developed by Niemann [6], based on fairly standard expert system ideas, that is in clinical use. Our system differs from his in that it uses in a crucial way information about the shape of the TAC's and not just the sampled values of the TAC,s. Also our system has the capability of suggesting classifications when the rules do not quite fit, as well as learning from examples. However, our system is still being modified and extended and we cannot yet give a proper clinical comparison with Niemann's.

References

[1] Barbut, M. & Monjardet, B., *Ordre et classification algebre et combinatoire*, Hachette Université, 1970.

[2] Birkhoff, G., *Lattice theory*, AMS Colloquim Publications **Vol XXV**, 1948.

[3] Haralick R. M., Sternberg, S. R., Zhuang, X., 'Image Analysis using Mathematical Morphology' *IEEE PAMI*, 4, 532-550, July 1987.

[4] Ho, Y-S, 'The planning process : stucture of verbal descriptions' *Environment and Planning B*, 9, 297-470, 1982.

[5] Johnson, J. H., 'The mathematics of complex systems', in *The mathematical revolution inspired by computing*, J.H. Johnson & M. J. Loomes (eds), Oxford University Press, (Oxford), 1991.

[6] Niemann, H., Bunke, H., Hofmann, I., Sagerer, G., Wolf, F., Feistel, H., 'A Knowledge Based System for Gated Blood Pool Studies' *IEEE PAMI*, bf 7, 246-259, 1985.

[7] Norris, E., 'Maximal rectangles' *Revue Romaine de Math Pure et Appliques*, **22**, 243-250, 1978.

The Mathematics of Complex Systems

Jeffrey Johnson
Centre for Configurational Studies
The Open University, Milton Keynes, MK7 6AA
JH_JOHNSON@UK.AC.OU.ACSVAX

Abstract

An account will be given of mathematical structures which may underlie all complex systems. These include a universal part-whole hierarchical construction, a canonical relational prism construction, the notion of system traffic on a structured backcloth, and the Fundamental Diagram of Complex Systems. The ideas will be illustrated with applications to social and physical systems, and new areas of mathematical research will be discussed.

1 Introduction

Complex systems admit certain universal mathematical constructions based on sets, relations, and numerical mappings. These underlie a mathematical methodology for the analysis and control of complex systems which has been developed over the last twenty years [3][4][5][6][18][21][22][25]. The basic ideas are that systems are composed of structured sets, that system activity is measured by numerical mappings defined on the structured sets, and that the relational structure of the sets constrains the dynamics of the system as expressed by its mappings. Complex systems tend to have many parts and are usually hierarchical. This means that complex systems need to analysed within a mathematical framework which can integrate low and high-level structure without losing information. Thus we will speak of a relatively static hierarchical backcloth which supports a relatively dynamic traffic of system activity.

2 A universal part-whole hierarchical construction

Consider a table with a top, e, and four legs a, b, c, and d. Then $\{a, b, c, d, e\}$ could be defined to be the *set of parts* of the table. It could be said that the table is *assembled* from the parts to satisfy the table relation, R. Let $< a, b, c, d, e; R >$ denote the elements assembled to form the table. Let the assembly process be denoted by a mapping a: $\{a, b, c, d, e\} \to\, < a, b, c, d, e; R >$. In general the assembly relation is an n-ary relation. For example the table is assembled under a 5-ary relation which puts a leg at each corner and makes sure no corner has more than one leg. Once sets of elements have been assembled to form structured objects, it is very common to give that object a name. Let h be the *hierarchical naming mapping* which takes

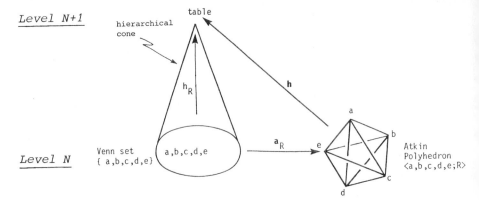

Figure 1. The fundamental diagram of hierarchical structure

a structured object to its *name*. For example, h:$< a,b,c,d,e; R >\rightarrow$ table. Let $h_R =$ h∘a be the *hierarchical aggregation mapping* which takes the set $\{a,b,c,d,e\}$ to the structured object **table** considered to be a single element at the next hierarchical level, $h_R(\{a,b,c,d,e\}) =$ **table**. Then $h_R^{-1}(\textbf{table}) \stackrel{def}{=} \{a,b,c,d,e\}$. These definitions are summarised graphically in Figure 1 where the ellipse represents the set in a *Venn diagram* and the structured set is represented by what I will call its *Atkin polyhedron*. Atkin polyhedra are useful for illustrating particular multidimensional connectivity properties in the same way that Venn diagrams are useful for illustrating particular set-theoretic properties. Connectivity is fundamental in analysing complex systems and it will be developed in more detail in Section 4.

Parts and wholes seem to cause endless confusion when analysed using vernacular languages. One amusing example comes from the journal *Cognitive Psychology*: "Simpson's Finger is part of Simpson. Simpson is part of the Philosophy Department. Simpson's finger is part of the Philosophy Department" [33]. This conundrum is easily resolved using this simple mathematics: Simpson's finger is an R_1-part of Simpson, Simpson is an R_2-part of the Philosophy Department, and Simpson's finger is an $R_1 \circ R_2$-part of the Philosophy Department. The word 'part' taken from vernacular English has been used to represent three distinct relations (each perfectly meaningful) and has introduced unnecessary ambiguity into the analysis. This careless use of language has lead to the consideration of the set $h_{R_2}^{-1} \circ h_{R_1}^{-1}(\text{PhilosophyDepartment})$, which is the set of the parts of the bodies of the members of the Philosophy Department. In most cases this is unlikely to be an important or useful set, and so the analysis has been sidetracked. The mathematical formulation allows that the set exists implicitly in the Universe until it is activated within the analysis by some reference which makes it explicit.

The diagram in Figure 1 is useful shorthand for the assembly of elements to structures in complex systems. In reality the process may be more subtle. An *n-ary relation* will be defined to be an *n*-valued proposition taking values in a *truth set* such as $\{True, False, \ldots\}$. In general the truth set may have a temporal quality, but that will not be developed here. We will say $x_1, x_2, \ldots, x_n$ are *R-related* if and only if $R(x_1, x_2, \ldots, x_n)$ is a meaningful proposition which can practically be assigned a value from the truth set by a truth decision mapping T. When $T : R(x_1, x_2, \ldots, x_n) = True$ the polyhedron $< x_1, x_2, \ldots, x_n; R >$ will be said to *exist* in any system whose class of relations contains R.

Finally it must be observed that R acts on ordered n-tuples rather than sets of elements. Let U be the mapping which 'forgets' the order on of the elements in the n-tuple $(x_1, x_2, \ldots, x_n)$ and makes all permutations of $x_1, x_2, \ldots, x_n$ equivalent. U will also 'collapse' any copies of an element in the n-tuple. In general the number of terms in $U(x_1, x_2, \ldots, x_n)$ can be less than n. For example $U(x_1, x_1) \stackrel{def}{=} \{x_1, x_1\} = \{x_1\}$. To complete the commutative diagram in Figure 2, the formal arrows O_s are defined with one for each $(x_1, x_2, \ldots, x_n)$. Each O_s produces a unique n-tuple $(x_1, x_2, \ldots, x_n)$. Thus each assembly a_R factors through an O_s which orders its constituent parts appropriately, and an *alpha-assembly* α_R which respects more precisely the ordering required by the relation R.

This formulation preserves the simplicity of Figure 1 with the more subtle requirements of n-ary relations represented in Figure 2. Category theorists will realise that the various sets, n-tuples, polyhedra, and truth sets can be viewed as the objects in a category (possibly a topos) whose morphisms are the various arrows in Figure 1 and Figure 2, together with all their compositions and identities. The implications of this will not be considered here, but the use of category theory for the study of complex systems is well established in the literature (e.g. [11][29]). The mappings a_R and $O_s \circ R$ are related to the notion of polymorphism as discussed by Burtsall [9] and Hyland [16] in this volume.

Figure 1 illustrates the idea of a *hierarchical cone*. This is defined to be a pair $(x^{N+k}, h_R^{-1}(x^{N+k}))$, where x^{N+k} is the *apex* of the cone and $h_R^{-1}(x^{N+k})$ is its *base*. In Figure 1, for example, $base(\text{table}) \stackrel{def}{=} h_R^{-1}(\text{table}) = \{a, b, c, d, e\}$ is the base of a cone with apex **table**. Hierarchical cones are useful for discussing non-partitional hierarchies, and establishing properties such as the *Nested Base Rule* for repeated applications of hierarchical aggregations [13]

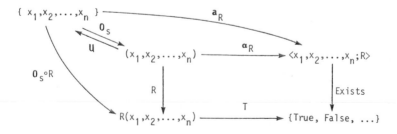

Figure 2. The commutative n-ary relation assembly diagram

3 Hierarchical set definition and the intermediate word problem.

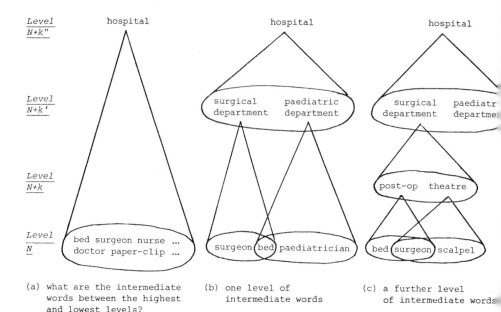

Figure 3. The intermediate word problem

In his book 'The blind watchmaker' Dawkins [10] says: "If we wish to understand how a machine or living body works, we look to its component parts and ask how they interact with each other. If there is a complex thing that we do not yet understand we can come to understand it in terms of simpler parts that we do already understand." In practice this often means *looking for* component parts, making them explicit in the analysis by defining and naming them, and seeking their relationships with other explicitly named parts. When encountering an unknown complex system one usually has to construct intermediate levels of named parts between the highly aggregate whole, and the disaggregate atoms.

Consider a management consultant trying to understand the structure of a large hospital. In the first instance there is the word *hospital* which names the whole assembly, and a preliminary look around will show that the system contains *doctors, nurses, beds, thermometers, blankets,* and so on. The analyst will invariably be faced with the problem that the word *hospital* describes the system at too high a hierarchical level, while *doctor, nurse,* etc. describe parts of the system at too disaggregate a level. Here the analyst faces the *Intermediate Word Problem* [13] which requires that words be found which name parts of the system between the more obvious highest and lowest levels. Very soon the analyst will find words which reflect the structure of the particular institution: *surgical department, paediatric department, administrative department*, and so on.

Figure 3 shows how the intermediate word problem amounts to constructing hierarchical cones, and nested sequences of hierarchical cones. This process of disassembly is a relatively easy part of an analysis because it does not require the n-ary assembly relations to be specified. The cone construction in Figure 3 corresponds to the trees and lattices that are frequently drawn to represent organisations and other complex systems. However defining the hierarchical vocabulary is just the beginning of trying to understand complex systems. This must be followed by an explicit assembly process, and an understanding of how this structure constrains system activity. For further details of hierarchical set definition as a knowledge elicitation procedure see [21].

4 Relations, graphs, hypergraphs, simplicial families, and complexes.

Relations are ubiquitous in complex systems and give rise to a number of well known mathematical structures. These will be defined briefly in this section and the next section will show how they relate to each other through the March prism construction. Let λ be a binary relation between two sets A and B. In the notation of Section 2, the expression $<a, b; \lambda>$ means that a is λ-related to b in the system. More commonly the reference to the defining relation is omitted from related pairs, and they are written $<a, b>$. The set of ordered pairs $\{<a, b> \in A \times B \mid a$ is λ-related to $b\}$ can be defined to be the set of *edges* in a *graph* with vertex set $A \cup B$. When $A \cap B = \emptyset$ the graph is *bipartite*. The elements of A and B are called the *vertices* of the graph. Let $\lambda(a)$ be the set of all members of B which are λ-related to a, for $a \in A$. Let $\lambda^{-1}(b)$ be the set of all members of A which are λ-related to b, for $b \in B$. Let $H_A = \{\lambda(a) \mid \text{all } a \in A\}$ and let $H_B = \{\lambda^{-1}(b) \mid \text{all } b \in B\}$. Relaxing Berge's restriction [7] that edges must be non-empty, H_A is a *hypergraph* with *edges* the sets $\lambda(a)$ and *vertices* the members of B. H_B is the *dual hypergraph* with *edges* the sets $\lambda^{-1}(b)$ and vertices the members of A. See also [32].

Let $\mathbf{H}_A$ be a hypergraph with all the edges of H_A, together with all intersections pairwise, triplewise, and so on. Similarly let $\mathbf{H}_B$ be the hypergraph with all the edges of H_B and all their intersections. Consider the mappings $\gamma_A : \mathbf{H}_A \to \mathbf{H}_B$ and $\gamma_B \mathbf{H}_B \to \mathbf{H}_A$ with $\gamma_A(e_A) = \bigcap_{b \in e_A} \lambda^{-1}(b)$ and $\gamma_B(e_B) = \bigcap_{a \in e_B} \lambda(a)$ for all $e_A \in \mathbf{H}_A$ and $e_B \in \mathbf{H}_B$.

Both γ_A and γ_B are order-reversing mappings, $e_A \subseteq e'_A$ iff $\gamma_A(e_A) \supseteq \gamma_A(e'_A)$ and $e_B \subseteq e'_B$ iff $\gamma_B(e_B) \supseteq \gamma_B(e'_B)$. Thus they form a *Galois connection* ([8] page 124). From this it follows that the set lattices of $\mathbf{H}_A$ and $\mathbf{H}_B$ are isomorphic. The pairs $(e_A, \gamma_A(e_A))$ form a lattice with the partial order $(\alpha, \beta) \leq (\alpha', \beta')$ iff $\alpha \subseteq \alpha'$ and $\beta \supseteq \beta'$. I will call this the *Ho-Galois lattice* [15]. Andrew *et al* make use of the Ho-Galois in their paper in this volume [1].

Two hypergraph edges are defined to be q-*near* if their intersection contains more than q common elements. q-nearness is a *tolerance relation* (reflexive and symmetric) on the set of edges with more than q elements. Its closure is the q-*connectivity* relation which partitions $\mathbf{H}_A$ and $\mathbf{H}_B$ into equivalence classes of q-*connected components*. A listing of the q-connected components of the hypergraphs $\mathbf{H}_A$ and $\mathbf{H}_B$ is called a Q-*analysis* of the relation λ (in the literature this term applies almost exclusively to simplicial complexes). In this set-theoretic context Q-analysis can be viewed as a clustering procedure. This may be useful as a means eliciting information about the system under consideration but may not be a useful end in itself [21][22][25].

In algebraic topology a structured set of abstract vertices, $< v_0, \ldots, v_p >$ is called a p-*simplex*, and often denoted σ or σ_p. Its *geometric realisation* is a p-dimensional polyhedron. σ is a *face* of σ' if all its vertices are also vertices of σ'.

Depending on the system under investigation, the members of $\lambda(a)$ are related to each other under an induced n-ary relation (it is possible that the binary relation λ is a projection of an n-ary relation), and as such $\lambda(a)$ can be considered to be the vertex set of an Atkin polyhedron, denoted $\sigma(a)$. Where there is no ambiguity the Atkin polyhedron with vertex set $\lambda^{-1}(b)$ is denoted $\sigma(b)$. $F_A \stackrel{def}{=} \{\sigma(a) \mid$ all $a \in A\}$ and $F_B \stackrel{def}{=} \{\sigma(b) \mid$ all $b \in B\}$ are called the *conjugate simplicial families* of the relation λ in which the $\sigma(a)$ and $\sigma(b)$ are viewed as abstract simplices. The *conjugate simplicial complexes* K_A and K_B are obtained by taking F_A and F_B with all their faces respectively. The *intersection closures* of F_A and F_B, denoted $\mathbf{F}_A$ and $\mathbf{F}_B$ respectively, are formed from the simplicial families and all their pairwise, triplewise, etc. intersections. $\mathbf{F}_A$ and $\mathbf{F}_B$ will be called the *canonical simplicial families* of the relation λ.

Two simplices are defined to be q-*near* if they share a q-dimensional face, i.e. they share at least $q + 1$ vertices. Two simplices are q-*connected* if there is a chain of pairwise q-near simplices between them. Thus q-nearness is a tolerance relation and q-connectivity is the equivalence relation formed by taking its transitive closure. The q-connectivity relation partitions a family of simplices with dimension $\geq q$ into q-*connected components*. A listing of the q-connected components for the simplicial families F_A and F_B is called a Q-*analysis* of the relation λ [4][5][6].

5 Stars, hubs, and the canonical March prism construction

The canonical simplicial families $\mathbf{F}_A$ and $\mathbf{F}_B$ are related by a Galois connection similar to that between the hypergraphs of λ. For every simplex σ in $\mathbf{F}_A$ there is a unique simplex σ' in $\mathbf{F}_B$ such every vertex of σ is λ-related to a vertex of σ', and σ' is not a face of any other simplex with this property. We will write $\gamma_A : \sigma \to \sigma'$ and $\gamma_B : \sigma' \to \sigma$. The simplicial families are partially ordered by the face relation, and the mappings γ_A and γ_B form a Galois pair of order reversing mappings: $\sigma_1 \leq \sigma_2$ in $\mathbf{F}_A$ iff $\gamma_A(\sigma_1) \geq \gamma_A(\sigma_2)$ in $\mathbf{F}_B$. Since γ_A and γ_B are both 1-1 mappings, they establish a 1-1 correspondence between the simplices in $\mathbf{F}_A$ and those in $\mathbf{F}_B$.

Given two simplices σ and σ', the polyhedron $\sigma.\sigma'$ is defined to have all and only the vertices of σ and σ'. In other words, $\sigma.\sigma'$ is the *prism* formed on the union

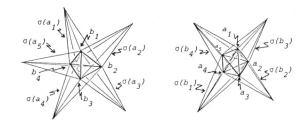

(a) the star-hub pair $\{\sigma(a_1),\sigma(a_2),\sigma(a_3),\sigma(a_4),\sigma(a_5)\} \leftrightarrow <b_1,b_2,b_3,b_4>$ paired with the star-hub pair $\{\sigma(b_1),\sigma(b_2),\sigma(b_3),\sigma(b_4)\} \leftrightarrow <a_1,a_2,a_3,a_4,a_5>$.

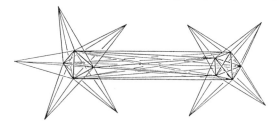

(b) the March prism $<a_1,a_2,a_3,a_4,a_5;b_1,b_2,b_3,b_4>$ between two stars.

Figure 4. Star-hub pairs and the March prism between them

of the vertices of σ and σ'. If $\sigma \in \mathbf{F}_A$, $\sigma.\gamma_A(\sigma)$ is defined to be the *March prism* spanning $\mathbf{F}_A$ and $\mathbf{F}_B$; these prisms were first suggested by March in 1983 [28]. The set of all March prisms of a relation will be called the *March family* of the relation. The members of the March family and their faces be called *spanning simplices*.

Proposition 5.1 The set of edges of the bipartite graph of a relation between the disjoint sets A and B is exactly the set of 1-dimensional spanning simplices in the March family.

Let σ be any simplex in $\mathbf{F}_A$. Then the set of simplices $\{\sigma(a) \mid \sigma \leq \sigma(a)\}$ is defined to be the *star* of σ, denoted $star(\sigma)$. (This is stricter than the usual definition in topology which requires only that two simplices share at least one vertex.) The intersection of all the simplices in star is called the *hub* of the star. By this construction stars and hubs are *maximal*, i.e. the star is defined to contain *all* the polyhedra having σ as a face, and the hub (which may be bigger than σ) is the largest shared face of the polyhedra in a (maximal) star. Stars are defined in a similar way for $\mathbf{F}_B$. Stars and hubs in the canonical families are uniquely paired:

Proposition 5.2 $\{\sigma(a_1), \sigma(a_2), \ldots\}$ is a star with hub $< b_1, b_2, \ldots >$ in $\mathbf{F}_B$ iff $\{\sigma(b_1), \sigma(b_2), \ldots\}$ is a star with hub $< a_1, a_2, \ldots >$ in $\mathbf{F}_B$.

Figure 4 shows a March prism between two star-hub pairs, a structure is that absolutely fundamental in complex systems . Complex systems are usually made up of many sets and many relations between them, and the star-hub-prism construction is ubiquitous. The assembly mapping of Figure 1 establishes a relation between hypergraphs and complexes. Furthermore, the ideas of hypergraphs, complexes, Ho-Galois lattice, March prisms, and networks are all part of a single coherent structure [21].

6 System activity as traffic on a structured backcloth.

The combinatorial structures defined in the previous sections are demonstrably present in those systems that we consider to be complex. It remains to be shown that this structure is *significant* in understanding system dynamics and controlling complex systems through predictive theories. We follow Atkin's suggestion [3] that these algebraic structures act as a kind of relatively static *backcloth* which supports a relatively dynamic *traffic* of system activity represented by numerical (and other) mappings defined on the underlying sets and polyhedra. Atkin's idea was that the connectivity of the backcloth constrains the behaviour of the traffic mappings, both in terms of their codomains and dynamic properties.

Definition 6.1   A finite *system backcloth* is a class of sets together with a class of relations on those sets. All the polyhedra defined by these relations will be called the *geometric realisation* of the backcloth. The term *backcloth* will be used for the the system backcloth and its geometric realisation. When a system backcloth is hierarchically structured we will refer to is as the *hierarchical backcloth*.

Definition 6.2   Any time-dependent mapping on a system backcloth is called *system traffic* on the system backcloth. Most mappings will take values in number systems such as the integers or rationals, but this need not be the case. If π is a traffic mapping, we will use the notation π_t to refer to the action of π at time t. For example, $\pi_t \sigma$ will be the value of π on the polyhedron σ at time t.

Definition 6.3   Let π and μ be time-dependent mappings on a system backcloth. Let $\delta\pi_{01}$ be a change in the mapping π between times t_0 and t_1, and let $\delta\mu_{12}$ be a change in the mapping μ between times t_1 and t_2. $\delta\pi_{01}$ and $\delta\mu_{12}$ will be called *traffic activity* on the system backcloth. If the relational backcloth remains unchanged during traffic activity the backcloth is said to be *relatively static* with respect to the *relatively dynamic* traffic.

Definition 6.4   Let σ_i and σ_j be simplices in a system backcloth. Let π be a mapping on σ_i and let μ be a mapping on σ_j. Suppose traffic activity $\delta\pi_{01}$ on σ_i results in traffic activity $\delta\mu_{12}$ on σ_j only if there exists a simplex σ which is a face of both σ_i and σ_j. Then the traffic activity on the backcloth is said to be governed by a *transmission mechanism*. We say system activity is *transmitted* through the system backcloth.

Definition 6.5 A system backcloth with traffic activity governed by transmission mechanisms will be called a *backcloth-traffic system*. This term also applies to systems in which changes in the backcloth geometry induce changes in the system traffic.

In his book on cybernetics, Ross Ashby says "Science today stands on something of a divide. For two centuries it has been exploring systems that are either intrinsically simple or that are capable of being analysed into simple components. The fact that such a dogma as "vary the factors one at a time" could be accepted for a century, shows that scientists were largely concerned in investigating such systems as allowed this method; for this method is often fundamentally impossible in the complex systems." [30].

In our terms Ross Ashby is saying that the traffic in complex systems depends on *all* the vertices of the supporting polyhedra, and sometimes on *whole faces* of the supporting polyhedra. The polyhedra represent 'wholes' or 'Gestalt phenomena' which cannot be understood by independent study of their atoms. As such many systems which are considered to be complex will be backcloth-traffic systems:

Definition 6.6 (tentative)   A backcloth-traffic system is *complex* if:

(a) it has assembly relations at two or more hierarchical levels

(b) changes in traffic values at *Level N+k* depend on a transmission mechanism, and these result in changes in traffic values at *Level N+k+1*.

(c) There exist traffic mappings π and p-simplices $\sigma = <v_0, \ldots, v_p>$ such that $\pi\sigma$ cannot be expressed in terms of the $\pi <v_i>$, i.e. one cannot "vary the factors one at a time".

In general complex systems are characterised by sets with very large numbers of elements, hence the need for hierarchical structure[1]. Apart from their size and the combinatorial complexity of their backcloth structure, complex systems are characterised by almost everything affecting almost everything else through transmission mechanisms. In some cases the transmitted traffic falls below a threshold at which it might be considered to be *noise*. However, we are learning that even small changes in traffic values can lead to wide divergence in behaviour, as in the case of chaotic systems [2].

Figure 5 shows what I will call the Fundamental Diagram of Complex Systems. It shows two intersecting polyhedra at *Level N* and a transmission mechanism on the mapping π. This transmission of change is manifest at the next hierarchical

---
[1] The absolute *a priori* existence of hierarchical structure is matter of debate. It can be argued that it is a cultural artefact imposed on complex systems which *redefines* them in a way which makes them comprehensible and controllable.

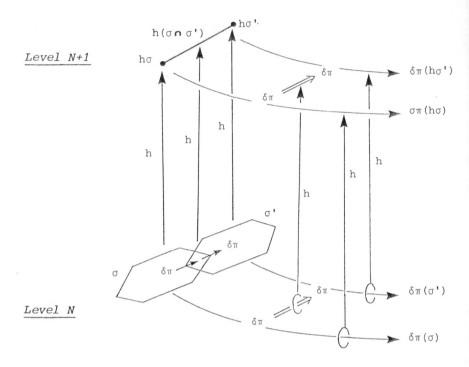

Figure 5. The Fundamental Diagram of Complex Systems

level, *Level N+1*, and appears to take place on the 1-simplex $\mathbf{h}(\sigma \cap \sigma')$. The point of this diagram is that in order to understand this change at *Level N+1* it may be necessary to understand the transmission process in terms of the specific supporting structure at *Level N*.

Statistical correlations provide an example of high level traffic relationships. Their value lies in alerting the scientist to the possible existence of system mechanisms at the more detailed lower levels. Sadly the challenge to follow these indicators to construct mathematical theories is often passed by in favour of vernacular accounts.

7 q-transmission in simplicial complexes

For simplicity, consider a traffic mapping π as a time-sequence of mappings π_t. Suppose there is a transmission mechanism between all q-near simplices, i.e. there is a transmission mechanism between the simplices if and only if they are q-near. Then, by definition, a change $\delta\pi_{t_0}$ on a simplex σ_0 at time t_0 will be transmitted to all its q-near neighbours at time t_1. Similarly the change will be transmitted to all their q-near neighbours at time t_2. In this way one can define *q-transmission fronts* of simplices through which the changes are propagated [17]: *the transmission of change is constrained by the connectivity of the system backcloth.*

Although transmission fronts are defined in terms of q-nearness, in almost all complex systems a single shared vertex is *sufficient* for system changes to be transmitted. In such cases many of the ideas of graph and network theory can be directly applied.

However in some complex systems the existence of a connection is *necessary* but not *sufficient* for transmission to occur. This is because the parts of complex systems are frequently not *homogeneous* and the behaviour of traffic is *vertex specific*, i.e. it is not always the case that one channel for transmission is equivalent to another. The significance of the dimension of a polyhedron in complex systems is often obscured by the structure being elevated to a single vertex at the next hierarchical level. Thus many systems which are currently represented by networks have a much more complex connectivity at lower hierarchical levels. Furthermore, the mechanisms governing a system may not exist unambiguously at the more aggregate levels modelled by networks, but must be stated in terms connectivities between polyhedra at lower levels.

8 Time as a combinatorial structure

In a remarkable departure from conventional thinking, Atkin [6] suggested that the formation of multidimensional polyhedron should be used to denote the passing of time. Atkin called the formation of a p-dimensional polyhedron a *p-event*, and suggested that events might be related to clocktime in a way proportional to the combinatorial number of their faces.

This idea is attractive because it accords with the common experience that adding one more thing does not result in a linear increase in complexity. Furthermore it has a logic which overpowers the wishful thinking of politicians and planners. For example, whatever the schedules say, a new road will support a traffic of vehicles only when it is ready (it's various structures have formed), and the hospital can only support a traffic of patients when it is built, equipped, and staffed.

Those engaged in trying to design and control complex systems ought to have some idea of the structures and mechanisms in those systems. However, they should also have some idea of the natural 'rhythms' of the complex systems they build. The assertion that certain things will take certain amounts of clock time may contradict the reality of the system's internal structure.

Atkin argued that physical events occur in physical time, and that the time of physics is measured by physical events such as the swing of the pendulum or the oscillation of the crystal. So why should social time be measured by physical events? Not only did Atkin suggest a new way of measuring social time though social events (polyhedra), but he suggested a combinatorial way of moving between social and physical time [6][19].

9 Examples

9.1 Parts and wholes in human perception

The human brain is one of the most complex systems known to man, and the process of human perception is not well understood. In his book 'The man who mistook his wife for a hat' Sachs [31] describes a man capable of seeing a pin on the floor, but making elementary blunders when trying to recognise faces and scenes. When the man is given an object he says it is "about six inches in length ... a convoluted red form with linear green attachment ... [which] could be an inflorescence or flower". On smelling it he exclaims "Beautiful! ... An early rose. What a heavenly smell".

Biological vision is far from well understood and we do not know if people with normal vision would see the rose as a Gestalt, or if they would see the parts and assemble them as { *convoluted red form, linear green attachment* } →< *convoluted red form, linear green attachment*; R > → rose. Certainly the assembly process does not happen automatically for this man, but he is able to invoke a process of reasoning which can assemble the parts. In fact it appears that, despite severe malfunctioning of the visual part of his brain, the man is using another part of the brain which performs intellectual tasks to compensate by reasoning about the visual atoms which he can see clearly.

9.2 Computer Vision as a Complex System Problem

Abstracting explicit information from digital pictures without human interpretation is a complex system problem. Typically one has hundreds of thousands of vertices (*pixels*) at the lowest hierarchical level, with each mapped to one or more integers (*greyscale traffic*). Also, there is an *adjacency relation* on the pixels.

At the highest hierarchical level there are usually a few named objects which one wants to recognise and locate, for example, the *eyes* in a face. This problem is much more difficult that one might think because eyes in digitised faces are highly textured mixtures of greys; the intuition that there should be coherent polygons corresponding to the pupil, iris, and whites is incorrect.

Defining appropriate intermediate words (named structures) is fundamental to computer vision. One approach is to assemble pixels into polygonal configurations by defining relations based on their adjacency and greyscale similarity as *pseudo-homogeneous polygons*[20]. The geometry of (parts of) the edges of these polygons can be used for low-level pattern recognition in a hierarchy of assembly which uses the relational structure of features to give robust pattern recognition [20].

Of course, these pseudohomogeneous polygons are not appropriate intermediate structures for *textured* features such as eyes. A new approach to this problem, also based on the mathematics of this paper, proceeds by defining *n-ary relations* on sets of pixels as follows.

Let $g(p_{ij})$ be the greyscale of the pixel in the ith row and j^{th} column of a digital image. Let pixel p_{ij} be R_0-related to the pixel beneath it, $p_{i,j-1}$ if $g(p_{ij}) \geq g(p_{i,j-1})$.

The Mathematics of Complex Systems

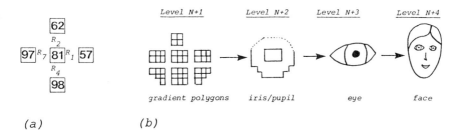

(a) (b)

Figure 6. (a) An R_{4127} pixel configuration (b) pixels → polygons → features → objects → ....

Similarly, $p_{ij} R_1 p_{i+1,j}$ if $g(p_{ij}) \geq g(p_{i+1,j})$, $p_{ij} R_2 p_{i,j+1}$ if $g(p_{ij}) \geq g(p_{i,j+1})$, and $p_{ij} R_3 p_{i-1,j}$ if $g(p_{ij}) \geq g(p_{i-1,j})$. Also define $p_{ij} R_4 p_{i+1,j}$ if $g(p_{ij}) \leq g(p_{i+1,j})$, $p_{ij} R_5 p_{i,j+1}$ if $g(p_{ij}) \leq g(p_{i,j+1})$, $p_{ij} R_6 p_{i-1,j}$ if $g(p_{ij}) \leq g(p_{i-1,j})$, and $p_{ij} R_7 p_{i+1,j}$ if $g(p_{ij}) \leq g(p_{i+1,j})$. Then we say p_{ij} and its four neighbours are $R_{\alpha\beta\gamma\delta}$-related iff $p_{ij} R_\alpha p_{i+1,j}$, $p_{ij} R_\beta p_{i,j+1}$, $p_{ij} R_\gamma p_{i-1,j}$, and $p_{ij} R_\delta p_{i+1,j}$. For example, the existence of $< p_{ij}, p_{i,j-1}, p_{i+1,j}, p_{i,j+1}, p_{i-1,j}; R_{0123} >$ means that P_{ij} is a locally 'bright' pixel, while $< p_{ij}, p_{i,j-1}, p_{i+1,j}, p_{i,j+1}, p_{i-1,j}; R_{4567} >$ means that p_{ij} is a locally 'dark' pixel. The existence of $< p_{ij}, p_{i,j-1}, p_{i+1,j}, p_{i,j+1}, p_{i-1,j}; R_{4127} >$ means that the image is getting darker in a North-East direction at pixel p_{ij} (Figure 6(a)). And so on.

There are sixteen of these 5-ary relations, R_{0123}, R_{0127}, R_{0163}, etc. Remarkably, when pixels are classified according their 5-ary relation with their neighbours, they form coherent polygons. In particular, in eyes one finds a polygon P_{0123} corresponding to the 'highlight', a polygon P_{4567} corresponding to the pupil beneath it, and polygons P_{4127} and P_{2345} corresponding to the lower left and right of the iris with a lower-centre iris polygon P_{2457} between them. When these *Level N+1 gradient polygons* are assembled by relations capturing their relative positions they provide sufficient information (intermediate structure) to recognise eyes (*Level N+3* in faces (*Level N+4*), as shown in Figure 6(b). These algebraic structures may be fundamental primitives in computer vision [24].

9.3 Finite Element Analysis

The mathematical approach to analysing complex systems set out in this paper is exemplified by the standard use of finite element methods in the design of complex physical systems such as engines, power stations, bridges, aircraft and motor cars.

Consider finding the temperature distribution over a two-dimensional section of a complex structure in an engine. It is well known from physics that the two-dimensional steady-state temperature of an object is given by the equation $\tau \partial^2 \phi / \partial x^2 + \tau \partial^2 \phi / \partial y^2 = 0$, where ϕ is the temperature at point (x, y) and τ is a constant. However, the complex shape of objects in engineering and their related boundary conditions often make it impossible to solve for ϕ directly.

The Finite Element method divides the object into finite number of small regular geometric parts, *elements* such as triangles and rectangles in two dimensions and tetrahedra and cuboids in three dimensions. Typically the temperature across an element is approximated by interpolation between the values on the vertices. The problem then becomes that of finding the temperatures at the vertices.

A standard result from the calculus of variations says that any function ϕ which satisfies the two-dimensional heat equation is also a function which minimises the integral expression $E = \int \int \frac{1}{2}[\tau(\partial\phi/\partial x)^2 + \tau(\partial\phi/\partial y)^2 dx dy - \int_C \phi'_C \phi ds$ where ϕ'_C is the function $\partial\phi/\partial N(x,y)$ which expresses the rate of change of ϕ normal to the boundary curve C. In this case ϕ'_C expresses a heat transmission mechanism in the system which acts across the shared one-dimensional faces of the two dimensional elements.

By substituting the polynomial interpolation approximations of ϕ for each element into the above equation, one obtains a set of equations linear in its node values. These can be assembled over all the elements to give a system of linear equations in the all node values. On substituting according to the boundary conditions this system of equations can be solved by Gaussian elimination or other computational techniques. As the mesh of elements is refined the approximate solution for ϕ converges to the correct solution. For an overview of the mathematical details of the Finite Element method see [14] while Zienkiewicz's book is a standard reference on the subject [34].

It could be said that Finite Element Analysis *extends* a very limited idealised physics to one which can be used to describe and predict the behaviour of physical systems of great geometric complexity.

The way the heat is transmitted is determined by the geometry of the backcloth and the connectivity of the triangular or tetrahedral elements. In general, the complexity of physical systems often comes from their topology and geometry: it matters that the backcloth has holes, since the physical geometry of those holes determines the boundary conditions. Atkin's thesis that the backcloth structure of complex systems *constrains* the behaviour of system activity traffic is well illustrated by the Finite Element, as is the power of hierarchical stucture.

9.4 Transmission mechanisms in a manufacturing system

Consider a manufacturing company c_1 which assembles machines m_1 and m_2, job skills s_1 and s_2, and raw materials r_1 and r_2 in a workshop w_1 in such a way that the company can manufacture the product g_1 In other words the company assembles the simplex $< m_1, m_2, s_1, s_2, r_1, r_2; R >$ at, say, *Level N*. The structure is named as w_1 at *Level N+1*. During a given period the workshop experiences a traffic of system activity expressed by mappings such as *production*: $w_1 \to$ quantity of g_1 produced, *costs*: $w_1 \to$ unit cost of producing g_1, *profit*: $w_1 \to$ profit made by producing g_1, and so on. These mappings at *Level N+1* are closely related to other mappings at *Level N* such as *costs*: $m_1 \to$ capital cost of machine m_1, *costs*: $s_1 \to$ labour costs of skill s_1, *quantity*: $r_1 \to$ quantity of raw material used in producing g_1, and so on.

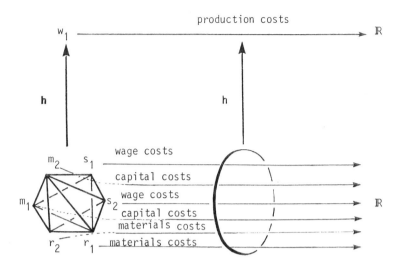

Figure 7. Hierarchical aggregation of mappings over a structure

The situation is illustrated in Figure 7 which shows how the mappings at *Level N* are hierarchically assembled to produce another mapping at *Level N+1*. In this case the assembly is a simple summation, but in general it will be more complex.

Consider another company which is manufacturing product g_2 in its workshop w_2, $w_2 = \mathbf{h} < m_3, m_4, s_2, s_3, r_1, r_3; R' >$. Suppose this company becomes very successful and $profit(w_2)$ takes a high value. Then the managers of this company may decide to reward their workers by sending some of this *Level N+1* profit down to *Level N* in the form of a pay rise.

Although the managers of company c_2 are only interested in their part of the manufacturing system, their actions may have repercussions elsewhere in the system. In particular the workers at company c_1 with skill s_2 may become dissatisfied that they are paid less than workers with the same skill at company c_2. There are many possible outcomes of this including striking for more money which might make the manufacturing polyhedron collapse and incapable of further production traffic. Alternatively the managers of company c_1 may agree to pay the workers with skill s_2 more to stay in line with the labour market, but this may result in increased prices. In time increased prices may result in loss of market, and possibly the sales traffic will grind to a halt. In turn this may result in a traffic of redundancies.

This discussion makes the rather obvious point that system activity in company c_2 has caused system activity in company c_1 because their manufacturing polyhedra are connected. In fact the two workshops share the face $< s_2, r_1 >$ so that any change in either company on this simplex may result in changes on the other.

In complex systems such as manufacturing industry many of the particular mechanisms for transmitting changes are relatively easy to understand and state as they apply locally at the common shared faces of the structure. However the global effects of many interacting mechanisms are frequently not understood, and this presents new mathematical problems.

9.5 Road Traffic Systems

Road traffic systems provide a subtle interplay between human and physical systems. They have a level of complexity which has defeated theorists, planners, and engineers to date. Surprisingly, many road traffic theorists have been content to assume that road systems can be modelled as equilibrium systems. The idea is that the volume of traffic builds up to a maximum at 'peak hour' and then decreases. The reality is, of course, different.

Thirty five years ago Lighthill and Whitham gave a model of traffic dynamics on long crowded roads [27]. The dynamics depend on the fact that as the *concentration*, k vehicles per unit length of road, increases the mean speed, v, of the vehicles decreases. For any given flow of vehicles past a point, say n vehicles per hour, with $n = kv$ there are two possibilities: k is small and v is large, or k is large and v is small (Figure 8(a)). Thus the flow-concentration and the flow-speed curves are multivalued (except at maximum flow). Lighthill and Whitham's theory explained the *shock waves* caused by jumping from a high-speed low-concentration state to a low-speed high-concentration state, the cause of these shock waves such as bottlenecks, and the ways they move. This is one of the few theories of road traffic behaviour which gives successful predictions. Macroscopic theories based on 'dynamic equilibrium' are essentially static, and therefore incompatible with the dynamic theory. Lighthill and Whitham's calculus-based *micro-level* theory has not been successfully extended to large road systems.

Road traffic planning desperately needs a dynamic macroscopic theory. This requires a mathematical representation of large complex road systems (backcloth) which supports the dynamics of vehicle (traffic) behaviour of over large areas.

Our complex systems approach to this problem goes back to first principles [26]. At the lowest level, *Level N*, roads are made up of two-dimensional *segments* with the three-dimension space above them. Segments have places at which vehicles can enter and leave them, called *entrance* and *exit* gates. A *link* across a segment is a possible vehicle path from an entrance gate to an exit gate (figure 8(b)). Two segments which share a gate are *contiguous*. A *physical route* is a sequence of contiguous segments, and an *abstract route* is a sequence of contiguous links.

Every segment s_i can be represented by a polyhedron $< g_{in_1}, g_{in_2}, \ldots, g_{out_1}, g_{out_2}, \ldots; R_{segment_i} >$. Links can be represented as $< g_{in}, g_{out}; segment_i >$, meaning vehicles can enter segment i over gate g_{in} and leave over gate g_{out}. Thus the one-dimensional spanning simplices between the input gates and the output gates of a segment are links.

Although this representation may seem more complicated that the usual network representation it is able to represent the *real* features of road system dynamics. A

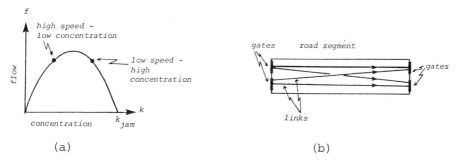

Figure 8. (a) the flow concentration relationship, (b) links and gates in a segment at *Level N*.

number of *observable* mappings can be defined on the structures. They include the *flow* over a gate, $f(g)$ vehicles/hour; the number of vehicles on each link and segments, $k(\ell_i)$, $k(s_i)$, where, for example, $\Delta k(\ell_i) = \Sigma f(g_{out}) - \Sigma f(g_{in})$ is the change in concentration over time interval $\Delta t = t' - t$. Thus this representation will support the microlevel theory of shock wave dynamics. It will also support other essential microlevel ingredients of road traffic control such as the setting of traffic lights, which attempt to control the whole system by controlling some of the gate flows, $f(g_{in})$ and $f(g_{out})$.

The representation can be extended to road systems of any size by a hierarchical scheme based on hierarchical sets of *zones*. Somewhat simplistically, *Level N* routes crossing *Level N* zones are hierarchically mapped to *level N+1* links. These support a traffic of vehicle flows, concentration, travel times, etc., and contiguous sequences of *N+1* links can form *N+1* routes. In turn *N+1* routes traversing *N+1* zones hierarchically aggregate into links at *level N+2* [18].

An important part of this theory concerns the state-change when links become jammed, i.e. $f(g) \approx 0$ at an entrance gate g. Then the traveltime on the link depends on dynamic behaviour elsewhere in the system. In fact one can define a *congestion dynamics* in which congestion moves through the road system. *The way that congestion moves through large complex road systems is highly constrained and largely determined by the multidimensional connectivity of the backcloth.*

An extensive theory of road traffic systems has been developed using the mathematics of this paper [26]. The representation of road systems is based upon algebraic hierarchies of links and routes which allow road systems of any size (e.g. Britain) to be modelled holistically. Also this representation supports a hierarchical shortest path algorithm [18].

9.6 Neural networks and multi-processor computational machines

Let $x_0, x_1, ..., x_n$ be the inputs to a one-output processing element. Then we can say the processing element assembles these inputs to produce an output $y = \mathbf{h} <$

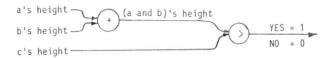

Figure 9. A 3-ary relation resolved into two binary relations and computed by two processing elements

$x_0, x_1, ... x_n; R_y >$. In conventional neutral networks h sums the input traffic values and applies a threshold operator to obtain an output traffic value, but in more general multi-processor computers h may be anything. In multi-layer neural networks the output of some processing elements are the inputs of others. In the terms of this paper, any multi-processor computer consists of a system backcloth of simplices representing processors with their shared vertices representing connections between processors. This backcloth supports a numerically encoded traffic of information. For example, the connection weights of neural networks are traffic on the vertices.

In applications of combinatorial mathematics to complex systems computers will be needed to see if any particular n-ary relations hold. Calculating n-ary relations presents new mathematical problems concerning their reducibility. Let an n-ary relation be *reducible* if is it logically equivalent to a combination of m-ary relations, $m < n$. For example, Fenton in this volume [12] gives the example that "a, b, and c are 3-ary related iff a is taller than b standing on c's shoulders". If the output of a processing element is a number, this 3-ary relation can be computed by two binary processors(Figure 9).

Complex geometric n-ary relations can often be resolved into binary relations. For example $< L_1, L_2, L_3, L_4;$ "*square*" $> \equiv\, < L1, L2;$ "*same length*" $> \wedge < L_1, L_3;$ "*same length*" $> \wedge\, ... < L1, L2;$ "*perpendicular*" $> \wedge < L_2, L_3;$ "*perpendicular*" $> \wedge ...$. This suggests a number of mathematical questions:

- When can n-ary relations be reduced to m-ary relations, $m < n$?
- When are reductions unique?
- Binary processors are 2-simplices; do 0-connected families made up only of 2-simplices have any interesting special properties?
- To what extent is finding reductions of n-ary relations algorithmic, and to what extent is it a creative process?
- Is finding reductions the equivalent to discovering theorems about the particular complex system under investigation?

In the absence of feedback, multilayer neural networks act as black box processor with information feeding forward from layer to layer. But there is also the possiblity of information being transmitted *sideways* due to the connectivity of the processing elements [23].

This sideways, or *horizontal* transmission may be very important in the analysis of neural architectures since combinatorial considerations make it impossible for every neural processing element at one level to be connected to every neural processing element at the next. In large networks some degree of horizontal *disconnection* is essential. Also disconnection may be highly advantageous since subsystems may become differentiated and undesirable information cross-talk avoided. Further details of these ideas can be found in [23].

10 Mathematical problems in complex systems

In this paper I have presented mathematical ideas which I claim underlie many or all complex systems. Many mathematical problems and questions remain.

10.1 The logic of relations

The existence of simplices in the representation of complex systems depends on the propositional functions of n-ary relations being decidably true. In Section 9.6 we asked *under which circumstances can n-ary relations be reduced to m-ary relations, $m < n$?* This question which seems to rest on logically equivalent classes of propositions. However this equivalence is more subtle than the usual *True/False* classification, since in some way it relates to an equivalence of information encoded in the relational propositions.

10.2 Hierarchical structure

The mathematical theory presented here is quintessentially hierarchical. The Fundamental Diagram of Complex Systems in Section 6 presents the explicit question "how does traffic at a given level relate to traffic at lower levels". Statistical theories of casuality can be interpreted as traffic at relatively high hierarchical levels. Seen this way, one can ask *how high-level statistical theory can be integrated with lower level deterministic theory in the study of complex systems?*

10.3 The combinatorial connectivity of simplicial complexes in relation to transmission of system traffic

Section 7 suggests that many complex systems can be considered to be multidimensional networks. In this context we can ask *what are the abstract mechanisms of transmission and how they are constrained by system connectivities?* I think new structures are required to reflect the connectivity of complexes from a global viewpoint, if we are to understand complex systems we may have to discover (or invent) new system-wide mathematical structures.

10.4 Finite Element Analysis

Sections 9.3 shows how differential equations can be intimately tied up with the backcloth geometry in complex systems. Certainly the differential calculus has been a very successful in the physical sciences, and engineering requirements have extended this to systems with complex backcloth topology through methods such as Finite Element analysis. By comparison very little is known of the traffic of complex social systems. Even relatively simple systems such as roads are unpredictable beyond the general observation that there was a jam here today so we 'predict' another tomorrow. *Is it possible that some analogue of Finite Element Analysis can be developed for these complex systems?*

10.5 Simulation

We do not know if it is possible to have a function ϕ which, for example, predicts the traffic flow at every point in a complex road system. One approach to this and many other complex systems is simulation based on various behavioural and dynamic hypotheses.

Simulation is analogous to computing a few cases to see what a function looks like. In general it is impossible to simulate for all possible initial states. Simulation offers two advantages: incorrect hypotheses may be detected because of their empirically incorrect consequences, and interpolation may give insights into a global behaviour. *To what extent is simulation a mathematical discipline, and to what extent is it likely to be only a step towards rigorous theory?*

10.6 Control of chaos by connectivity

In some simulations we use discrete steps to approximate continuous processes, while in others the system itself is inherently discrete. As is now well known, simulations of complex systems such as the weather can expose chaotic behaviour[2]. Any particular simulation may give poor long term predictions because of the sensitivity to initial conditions.

However, discrete events such as the changing of traffic lights, the closing down of a factory, or the eruption of a volcano may induce discontinuous jumps in their system trajectory (perhaps to another family of mappings). Research into chaos has undoubtedly informed our view of what is knowable or predictable, but it must be remembered that chaos is a theory of *traffic* on a static *backcloth*. Many complex systems also have a *discrete backcloth dynamics* which constrains this and other complex traffic behaviour. This leads to the question: *can chaos be controlled by disconnection-induced discontinuities?*.

11 Conclusion: Domain Dependent Dynamics

The fundamental idea underlying our mathematical approach to complex systems is that they have n-ary relational structure and that this structure constrains system behaviour. In particular it has been argued that the connectivity of backcloth polyhedra constrains the behaviour of system traffic: the traffic on $< x'_1, x'_2, \ldots ; R_1 >$ may be transmitted to $< x''_1, x''_2, \ldots ; R_2 >$ if they share the face $< x_0, x_1, \ldots, x_q >$. In other words, the dynamic behaviour of the traffic is determined by the *intersections of the domains* of the n-ary relations R_1 and R_2. In turn this supposes there is something special about the shared q-face, which could be represented by a relation $R_{1 \cap 2}$. In effect we are talking about a *Domain Dependent Dynamics* in which dynamic traffic behaviour is constrained by the connectivity of a changing backcloth domain. For this we need new mathematics.

References

[1] Andrew, M., Bose, D. K., Cosby, S., 'Scene analysis via Galois Lattice', in *The Mathematical Revolution Inspired by Computing*, J. Johnson & M. Loomes (eds), Oxford University Press (Oxford), 1991.

[2] Arrowsmith, D. K., 'The mathematics of chaos', in *The Mathematical Revolution Inspired by Computing*, J. Johnson & M. Loomes (eds), Oxford University Press (Oxford), 1991.

[3] Atkin, R. H., 'From cohomology in physics to Q-connectivity in social science', *I.J. Man-Machine Studies*, 4, 139-167, 1972.

[4] Atkin, R. H., *Mathematical Structure in Human Affairs*, Heinemann Educational Books (London), 1974.

[5] Atkin, R. H., *Combinatorial Connectivities in Social Systems*, Birkhäuser Verlag (Basel), 1977.

[6] Atkin, R. H., *Multidimensional Man*, Penguin (Hardmondsworth), 1981.

[7] Berge, C., *Graphs and hypergraphs*, North Holland (Amsterdam), 1973.

[8] Birkhoff, G., *Lattice Theory*, American Math. Soc., 1967.

[9] Burstall, R., 'Computer Assisted Proof for Mathematics: an introduction using the LEGO proof system', in *The Mathematical Revolution Inspired by Computing*, J. Johnson & M. Loomes (eds), Oxford University Press (Oxford), 1991.

[10] Dawkins, R., *The blind watchmaker*, Peguin (London), 1988

[11] Ehresmann, A. C., Vanbremeersch, J-P., 'A mathematical model for complex systems, II. Trial and error dynamics with hierarchical modulation', U.F.R. de Mathematiques, Université de Picardie, 80039 Amien, France, 1987.

[12] Fenton, N., 'The mathematics of complexity in software engineering', in Johnson J. H. & Loomes M. J. (eds) *The mathematical revolution inspired by computing*, Oxford University Press, 1991.

[13] Gould, P., Johnson, J.H., Chapman, G.P. *The structure of television*, PION (London), 1984.

[14] Goult, R. J., 'Finite Element Analysis', in *Computer Aided Design*, J. Rooney & J. P. Steadman (eds), Pitman (London), 1987.

[15] Ho., Y-S, 'The planning process: structure of verbal description', *Environment & Planning B*, 9, 397-420.

[16] Hyland, J. M. E., 'Computing and Foundations', in *The Mathematical Revolution Inspired by Computing*, J. Johnson & M. Loomes (eds), Oxford University Press (Oxford), 1991.

[17] Johnson, J. H., '*q*-transmission in simplicial complex', *I. J. Man-Machine Studies*, **16**, 351-377, 1982.

[18] Johnson, J. H., 'Hierarchical backcloth-traffic simulation', *Planning & design* **13**, 415-436, 1986.

[19] Johnson, J. H., 'A theory of stars in complex systems', in *Complexity, language, and life: mathematical approaches*, J. Casti & a. Karlqvist (eds), Springer (Berlin), 1986.

[20] Johnson, J. H.,'Pixel parts and picture wholes', in *From pixels to features*, J-C Simon (ed), North Holland (Amsterdam), 1989.

[21] Johnson, J. H., 'Expert Q-analysis', *Planning & Design*, **17**, 221-224, 1990.

[22] Johnson, J. H., 'Interpretation and hierarchical set definition in Q-analysis', *Planning & Design*, **17**, 277-302, 1990.

[23] Johnson, J. H., 'Neural Complexes: multidimensional structure in neural computing', British Telecom Neural Network Project, Research Report 3, Open University, Milton Keynes, MK7 6AA, 1990.

[24] Johnson, J. H., 'Gradient polygons: fundamental primitives in hierarchical computer vision', *Proc. Symposium in Honour of Prof. Jean-Claude Simon*, AFCET, University of Paris, 156 boulevard Péreire-75017, Paris, October 1990.

[25] Johnson, J. H., 'The rules of Q-analysis', *Planning & Design*, **17**, (in press), 1991.

[26] Johnson, J. H., 'The dynamics of large complex road systems', in *Transport Planning & Control*, J. D. Griffiths (ed), Oxford University Press, 1991

[27] Lighthill, J., Whitham, G. B., 'On kinematic waves: II. Theory of traffic flow on long crowded roads', *Proc. Royal Society, Series A* **229** 317-345.

[28] March, L., 'Design in a universe of chance', *Environment & Planning B* **10**, 471-484, 1983.

[29] Ratteray, C., 'Sketching an evolutionary hierarchical approach for knowledge based system design', *Proc. EUROCAST 89*, Springer Lecture Notes in Computer Science, 1989.

[30] Ross Ashby, W., *Cybernetics*, Methane (London), 1984 reprint of 1956 original.

[31] Sachs, O., *The man who mistook his wife for a hat*, Picador (London) 1986

[32] Seidman, S. B., 'Structure induced by collections of subsets: a hypergraph approach', *Mathematical Social Sciences*, **1**, 381-396, 1981.

[33] Winston, M, Chaffin, R., Hermann, D., 'The taxonomy of part-whole relations', *Cognitive Science*, **11**, 417-444.

[34] Zienkiewicz, O. C., *The finite element method*, MgGraw Hill, 1977.

PART III

Mathematics in Computing

The Mathematics of Complex Computational Systems

Stephen B. Seidman
Department of Computer Science and Engineering
Auburn University, Alabama 36830, USA

Abstract

For several centuries, mathematics has been associated with the modelling of complex systems. The continuous mathematical tools that were first developed for modelling astronomical systems have been applied to a wide variety of complex physical systems, and they have also had a major influence on the evolution of mathematical thought. More recently, discrete mathematical tools have been created in response to the need for models of complex social systems. The introduction of digital computers has given rise to an entirely new class of computational complex systems, found in the highly structured assemblies of electronic components that make up computer hardware, and equally in the software that governs computer operations. The process of model development first involves the use of existing mathematics, followed by the construction of new mathematical tools, which are finally abstracted and incorporated into mathematics. Although this process is beyond observation for physical systems, and is difficult to observe for social systems, it can be directly examined for computational systems. The initial and final stages of the modelling process for computational systems are typified by the use of methods taken from mathematical logic and by the incorporation of automata theory into mathematics. Examples to illustrate the second stage will be drawn from mathematical models for concurrency and for the design of software development environments.

1 Introduction

For thousands of years, the study of increasingly complex systems has occupied a central place in human societies. The need to understand such systems has led to the development of tools that can deal with the levels of complexity that have been encountered. These tools are almost exclusively mathematical in nature, and it is possible to regard the development of mathematics as having been driven in large part by the need to model complexity.

The most straightforward approach to the study of a complex system is to observe and record its behaviour. Such observation and record-keeping can be regarded as *descriptive modelling*, and it can be found in the astronomical and calendrical records of ancient societies, in the more contemporary compilation of economic and social statistics, and in the performance estimates currently used to compare and evaluate supercomputers and parallel computers. Although it can be argued that descriptive modelling requires a nontrivial understanding of the system being modelled, descriptive models are extremely limited in their ability to provide predictions of the future state of a system from information about present and past states.

In order for a system model to provide useful predictions of future behaviour, the model must itself generate a certain level of complexity. This complexity may be provided directly by the mathematical tools used to build the model, and if the complexity obtained in this way yields a useful description of system behaviour, the modelling process terminates. Otherwise, the existing mathematical tools must be assembled into a more elaborate model, which can generate more complexity than is present in the model's individual components. While such emergent complexity is still likely to reflect only a small portion of the complexity of the original system, it may be sufficient to provide a useful system model. In certain circumstances, the emergent complexity produced by a model may make it an object of study in its own right. If this is so, the properties of the model will be explored in great detail, and the model's essential characteristics will eventually be abstracted into new mathematical constructs. The modelling process therefore mediates the conversion of existing mathematics into new mathematics.

The effects of this modelling process can be observed throughout the history of mathematics, e.g. the development of geometry from descriptive models of the physical universe, the development of the calculus from geometrical models of astronomical systems, and the development of some significant subfields of discrete mathematics from relational models of social and economic systems. The actual operation of the process, however, is not so easy to observe. It is impossible to recreate the particular steps by which geometry evolved out of descriptive models, nor is it easy to trace the development of the calculus through the historical record. Although the development of discrete mathematics is more recent, its transition from mathematical model to mathematical discipline is already receding rapidly into the past.

The operation of the modelling process is best observed by identifying a class of complex systems that are presently being modelled for the first time. In recent years, the introduction and development of computers has given rise to a completely new class of complex systems. While the previously encountered systems have arisen from nature or from human society, complex computational systems were constructed by humans as complex systems. If the computers that embody and incorporate these systems are to be successful, mathematical models must be developed that can describe and predict their behaviour. The effort to model these systems has been under way for several decades, and models of complex computational systems can be used to illustrate all stages of the modelling process: the application of existing mathematics, the construction of system models from components drawn from existing mathematics, the exploration of the complexity derived from these models, and the abstraction of successful models into new mathematics.

2 The Development of Mathematical Models of Complex Systems

The first encounters of agricultural civilisations with complex astronomical and meteorological systems led to informal though sophisticated methods for recognising and recording the periodic regularities of these systems, as well as less common deviations from the observed regularities. The recorded information was put to predictive use for such applications as irrigation control and navigation. As experience with these complex systems increased, so did the complexity of the observed and recorded phenomena, and this complexity was eventually abstracted to form the foundations of Babylonian, Egyptian, and Greek mathematics (see [13] for an overview of the history of mathematics). The development of these tools in turn made possible the sophisticated modelling of astronomical systems done by such Hellenistic mathematicians as Archimedes and Ptolemy.

Similar interactions between the study of complex systems and the development of mathematics can be observed at many points in history. For example, the increasingly detailed studies of astronomical systems carried out in the sixteenth century led to the realisation that although the Ptolemaic model of the solar system had given an efficient description of the previously known behaviour of the solar system, the incorporation of newly observed phenomena required elaboration of the model. In particular, Kepler was able to propose three laws of planetary motion, corresponding to higher-level regularities that should be consequences of any satisfactory model. Although it was apparent that the Ptolemaic model and its elaborations were not capable of yielding a description of planetary motion with this predictive power, it was by no means clear how to proceed. Copernicus had replaced the geocentric circular planetary orbits of the Ptolemaic model by heliocentric circular orbits, but this change was not a result of a deeper understanding of the nature of the planetary system. In order to achieve such an understanding, it was necessary both to consider the system from a new perspective and to develop new mathematical tools. The new perspective was provided by the development of physics by Newton, who studied the motion of objects under the influence of forces, and the tools were provided by Newton's creation of the differential and integral calculus.

The effectiveness of these new mathematical tools in predicting the behaviour of planetary systems led to their application to the study of other physical systems. The success of these applications in turn encouraged the rapid development and elaboration of the differential and integral calculus that took place during the eighteenth and early nineteenth centuries. During this period, the mathematical tools were closely linked to their prospective physical applications. Toward the end of the nineteenth century, however, the mathematical superstructure that had been erected on top of the calculus began to be regarded as a purely mathematical entity, worthy of study (by mathematicians) for its own intrinsic interest. This perception gave rise to the new mathematical disciplines of real and complex analysis, which proved to be of as much use in seemingly unrelated areas of mathematics as they had been in modelling physical systems. Although many mathematicians were still sensitive to the need for models of newly discovered complex physical systems, such models could generally be constructed by applying existing mathematical techniques directly. Indeed, the process of constructing such models became known as *applied mathematics*. Even if existing mathematical techniques could not be applied to a particular system in a straightforward way, a model could often be built on top of these techniques, and such models were relatively easily abstracted and reabsorbed into mathematics. A typical example of this process is the use of trigonometric series by Fourier for the study of heat, and the transformation of this tool into what is now called Fourier analysis.

It is hard to identify the point in the past when humans first became aware that the social system in which they lived was a complex system whose behaviour could be studied and possibly predicted. Descriptive models, in the form of census data and mortality statistics, were available by the seventeenth century, and their use led to the development of models of growth and mortality in human populations. These models used tools taken from the calculus, and their essential features were absorbed into mathematics in the eighteenth century among the foundations of probability and statistics.

The use of tools taken from the calculus is found in a wide range of social system models. While some of these models have been quite successful (models for human population systems and economic systems), others have been less so (models of the arms race). All of these models share the use of continuous mathematical techniques, which assume that the members of the (usually very large) set being modelled are essentially indistinguishable, and that there is no interaction between the members of the set. Such an assumption is valid in modelling the behaviour of the molecules in a gas, the behaviour of the money supply of an economy, or even individuals' susceptibility to death from various causes, but it is no longer valid if the model is to describe, say, the evolution of conflict in a small group. As a consequence, the mathematical tools used to model most social systems are discrete and relational in nature [22].

Discrete and relational mathematical tools are typified by the theory of graphs, whose first ideas were described by Euler in 1735. Although graph theory was developed further in the nineteenth century, and applied by Cayley to the study of molecular configurations, it is only in recent years that it has reached maturity, partly under the stimulus of the development of graph-theoretic models for social systems. For one example, Harary, Norman, and Cartwright [7] developed a natural graph- theoretic model for small human groups that incorporated the psychological theory of cognitive dissonance. This in turn led to significant results on signed graphs that have been absorbed back into graph theory. Sociologists and anthropologists studying small human groups have also built graph-theoretic models that mathematically capture such concepts as group centrality and group cohesion, and some of the mathematically formulated structural insights gained from these models have also been absorbed into graph theory [22].

The development of discrete and relational models was also encouraged by the demand for models that could describe the behaviour of economically significant systems that were composed of a finite number of distinguishable, interacting units. A typical example of such a system is a network of pipelines, where each pipeline has a given maximum capacity. The need to develop a model that will allow the determination of the maximum flow from the source to the sink of the network has given rise to a large volume of research, and many of the underlying graph-theoretic tools have been absorbed into mathematics [24]. More generally, much of the material that is now considered to be part of the mathematical discipline of operations research (and also part of the discrete applied mathematician's toolkit) was first considered as a discrete model of a complex economic system.

The discrete models that have been discussed so far have all been concerned with systems of dyadic relations. Such systems are made up of relations between pairs of model elements, and a relation can be regarded as a set of (possibly ordered) pairs. These models do not generate sufficient complexity to model many interesting social systems effectively. For example, a social system based on overlapping committee memberships can not be modelled by a dyadic relational model. In recent years, non-dyadic relational models have been constructed to model such social systems [22]. The mathematical tools used to build these models are usually taken from the theory of hypergraphs, where a hypergraph is a collection of subsets of a set of vertices, just as a graph can be seen as a collection of pairs of vertices. The mathematical tools used to construct these models are currently undergoing further development, and their incorporation into pure or applied mathematics is yet to come.

3 Complex Computational Systems

The development of digital computers and their accompanying software has taken place essentially simultaneously with the development of mathematical models of computational systems. While the first digital computers can be associated with the work of Turing on the formalisation of computability and that of von Neumann on self-reproducing automata, it was only after computers became relatively common that serious attention was paid to the need for mathematical models of the phenomena produced by these new machines.

As we have seen, the first stage of the modelling process involves the use of techniques drawn from existing mathematics. This can be illustrated here by the theory of abstract computability, which had been developed in the 1930's within mathematical logic by such researchers as Turing, Gödel, and Post [20]. Since this theory was able to describe the behaviour of a abstract machine (by means of Turing machines) and to define precisely which functions could be computed by algorithms (using recursive function theory), the introduction of computers made it an active and significant area of research within both mathematics and theoretical computer science.

As experience with computers grew, it was soon realised that computability theory could not be used to develop higher-level, modular descriptions of abstract machines. This observation gave rise in the 1960's to the study of abstract automata as a new discipline within computer science, and these have played an important role in computer science during the past two decades [12]. In one important application, they have been used to develop recognisers for formal languages, and thus they have formed part of the theory that supports the development of compilers, parser generators, and other programming language tools. Abstract automata also form the basis for the theory of Petri nets [18], which are widely used as models of distributed and concurrent computer systems.

The theory of abstract automata was developed along algebraic lines, and its basic constructs (quotients, products, homomorphisms) are analogous to those used in the study of more traditional algebraic objects. Researchers in automata theory were able to use algebraic techniques to develop powerful structure theorems, such as the Krohn-Rhodes theorem ([12], p. 141). As a consequence, automata theory is often regarded as a branch of abstract algebra, and thus it is an example of the incorporation of a new mathematical model into mathematics.

The rapid development of computers led to the discovery of complex systems that could not be modelled effectively by existing mathematical techniques. An important example is the development of multitasking operating systems in the late 1960's, which made the modelling of *concurrency* an important problem, and this problem has become even more significant with the recent development of parallel computers and distributed computer systems. One approach to modelling concurrency was taken by C. A. Petri, whose development of Petri nets from automata theory actually antedates the recognition of the central importance of *concurrency* in computer science [18]. In Petri nets, concurrency is seen as a relation on the components of an existing system. By contrast, in the Communicating Sequential Processes (CSP) model of Hoare [11] and the Calculus of Communicating Processes (CCS) model of Milner [15], concurrency is seen as a fundamental operator that can be used to build complex processes from simpler processes. While Petri nets were proposed in the context of an already existing mathematical discipline, both CSP and CCS were developed as new abstractions of observed behaviour. Although CCS has mathematical roots in the formal semantics of programming languages, CSP is relatively independent of previously developed formalisms. CSP is therefore an excellent example of a newly developed mathematical model.

The following incomplete and oversimplified discussion of CSP is intended to give the flavour of the model. The fundamental constituents of CSP are *processes* and *events*. A process may engage in an atomic event drawn from its *alphabet*; the process $STOP_A$ has alphabet A but does nothing. CSP provides constructors that allow complex processes to be built out of simple processes. Among them are sequential composition $(P;Q)$ and concurrent composition $(P||Q)$ of processes. A process can be preceded by an event $(a \rightarrow P)$, and this *prefix* constructor can be applied recursively. CSP also provides process interleaving, two forms of nondeterministic process composition, as well as the concealment of some of the events in the alphabet of a process.

Abstractly, a CSP process is defined by its alphabet, its *divergences*, and its *failures.* The failures of a process describe the circumstances under which a process may fail to continue execution after some initial behaviour, while the divergences of a process describe the initial behaviours of a process after which its behaviour is unpredictable. The need to include unpredictable behaviour in the formal description of a process is due to nondeterminism, which can arise explicitly from the use of nondeterministic composition operators and implicitly from the concealment of events.

Hoare uses this abstraction to prove many useful formal properties of processes and process constructors, including a clear distinction between deterministic and nondeterministic processes. The resulting properties can be used to investigate the presence or absence of deadlock in particular systems ([11], p. 145, [21]). Although the CSP model of concurrency arose from the need to model operating systems, it was very influential in the design of the Ada tasking model, and it has provided the basis for the development of the transputer and the occam programming language [3], so that it is currently at the center of much important work in the design of parallel computer systems and parallel programming.

The CSP formalism developed by Hoare is an excellent example of *applied mathematics*: the use of mathematical ideas to model newly perceived complex systems. In order to prove that recursive process construction fits naturally into CSP, Hoare does use concepts and results from the theory of partially ordered sets that are also used in the semantics of programming languages ([11], pp. 92-99, 132)). With this exception, CSP does not use any sophisticated tools that were developed elsewhere in theoretical computer science or in mathematics. Although CSP is well recognised as a topic in theoretical computer science, the novelty of its formalism is likely to delay its further abstraction into an explicitly mathematical discipline.

Despite the recent nature of these models of concurrency, the fifteen or twenty years that separates us from their introduction makes it difficult to observe the operation of the modelling process that was involved in their creation. In order to examine the first stages of that process more closely, it is necessary to look at a mathematical model that is only now being developed.

4 A Mathematical Model for Software Development Environments

Although our discussion of complex computational systems has so far been limited to systems arising from computer hardware, it is clear that software systems are also extremely complex. Programs to control extremely complex systems (such as NASA's mission control system) can require over a million lines of code, and the development of reliable software systems of this order of magnitude is a topic of major current interest, usually called *software engineering*. It is generally accepted that software systems should be built hierarchically of modular components. The significance of hierarchy and modularity in software design can be derived from the software engineering principles of data abstraction and information hiding [14].

Software systems of this scope must be produced by teams of programmers, and such teams must be supported by additional software that allows the archiving and restoration of reusable software components. A software system that is intended to support large-scale programming efforts is generally called a *software development environment*, and many such environments have been built ([8], [9], [10]). Although these environments differ substantially from one another in their intended application domains, they all support the development of hierarchical and modular software systems.

Since software development environments can expedite the construction of reliable software, attention should be focused on the production of these environments, which are themselves large and complex software systems. This process would be greatly simplified if the essential nature of a software development environment were abstracted into a mathematical model that formalised the essential features of hierarchy and modularity. The general form of such a model can best be seen by examining one particular software development environment.

The environment to be described was developed to support coarse-grained dataflow programming. In general, a dataflow program is a graph whose nodes represent computations and whose edges represent the flow of data between the nodes. The computations in the nodes can be simple arithmetic or comparison operations (fine-grained dataflow), or more complex computations (coarse-grained dataflow). Fine-grained dataflow programs are generally produced by compilers from programs written in a more conventional style. They may be executed on a specifically designed dataflow parallel computer [6], or they may be used to aid in code optimisation for sequential or parallel computers [1]. Coarse-grained dataflow programs are generally written specifically as dataflow programs, and they are intended for execution on parallel computers ([2], [17]). The software development environment to be discussed here ([4], [5]) allows a user to build dataflow graphs whose nodes correspond to signal processing algorithms for execution on AT&T's EMSP dataflow computer [2].

A user's interaction with the environment begins with an empty graph, to which nodes can be added by importing signal processing algorithms (*primitives*) from a *primitive library*, which contains a description of the external interface needed for each algorithm. The flow of data between the nodes can be established by adding directed edges between the nodes, which must be consistent with the requirements of the external interfaces. At any point, a graph that is under construction can be packaged and stored in a file as a *subgraph*. The subgraph's external interface will be inferred from the external interfaces of the nodes of the underlying graph and from the pattern of edges linking the nodes. Such a subgraph can also be added (as a *subgraph* node) to a graph that is under construction. Figure 1 illustrates how (appropriately simplified) environment commands can be used to build a graph. The ability to encapsulate, store, and reuse subgraphs shows that this environment supports both hierarchy and modularity.

Figure 1(a) shows primitive algorithms named A and B. They are imported into a graph construction setting by the commands BUILD(A) and BUILD(B), as shown

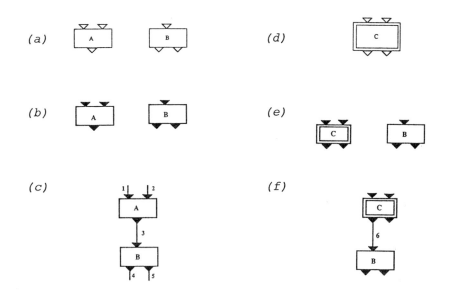

Figure 1. Coarse-grained dataflow environment operations

in Figure 1(b). Rectangles drawn with thin lines represent primitives, while rectangles drawn with thick lines represent graph nodes. Figure 1(c) shows the dataflow graph that has been produced by the commands ASSIGN CHANNEL(A,1), ASSIGN CHANNEL(A,2), ASSIGN CHANNEL(A,3), ASSIGN CHANNEL(B,3), ASSIGN CHANNEL(B,4), and ASSIGN CHANNEL(B,5). Figure 1(d) shows the subgraph C that is obtained by encapsulating the graph of Figure 1(c). Double-walled rectangles are used to represent subgraphs and subgraph nodes. Note that only the external channels of the graph of Figure 1(c) appear as part of the external interface of subgraph C. In Figure 1(e), this subgraph is imported as a subgraph node into a construction setting, along with a node based on algorithm B. Finally, in Figure 1(f), the commands ASSIGN CHANNEL(C,6) and ASSIGN CHANNEL(B,6) are used to link these nodes.

The key features of this environment are:

1. the importation of primitive or subgraph nodes

2. the characterisation of nodes by interfaces

3. the linking of interfaces on distinct nodes

4. the encapsulation and storage of a group of nodes as a subgraph

5. the determination of a subgraph interface from the node interfaces and the graph structure

These features embody the environment's approach to hierarchy and modularity, and they can be used in the construction of a model that abstracts and formalises the essential features of many software development environments.

Informally, the software construction process in a software development environment is modelled as a *construction setting*, which is initially empty [16]. Modularity is supported by providing the user with a library of *templates*, any of which may be instantiated and added to the construction setting as *nodes*. The external interface of a library template is specified as a set of *ports*, which become node *slots* when a template is instantiated as a node. Nodes in a construction setting can be linked together by assigning *labels* to corresponding slots. Hierarchy is implemented by allowing an entire construction setting to be encapsulated as a template and stored in the library. The external interface of the new template is inferred from the labelling of the slots that make up the interfaces of its component nodes.

The general model can be applied to a particular software context by specifying the rules that govern the assignment of labels to slots. For example, the model's templates and nodes correspond naturally to the primitives and nodes used in the coarse-grained dataflow graphs described above. The label assignment rules can be used to guarantee, for example, that an edge must link an output slot on one node to an input slot on another, and that edges can only be used to link slots that have corresponding data types.

The model is described formally using the set-theoretic specification language **Z** [23]. Its components are drawn from three *basic sets*: the set **Temp** of templates, the set **Port** of ports, and the set **Node** of nodes. Z schemas are used to establish the associations between these sets that constitute the definition of a construction setting. For example, the association between a template and its ports is specified by the schemas **Template_Association** and **Port_Association**, shown in Figure 2. Notice that the functions **interfaces** and **owner** are used to associate a template with its ports. The symbol $\rightarrowtail$ is used in **Z** to denote a *partial* function, and $F_1 S$ denotes the set of *finite* subsets of a set S. The template library can be identified with the domain of **interfaces**.

Template_Association
interfaces : Temp --|--> $\mathbb{F}_1$Port
disjoint {interfaces(t) : t ∈ dom interfaces}

Port_Association
Template_Association
owner : Port --|--> Temp
dom owner = ∪{interfaces(t) : t ∈ dom interfaces}
∀t : dom interfaces • interfaces(t) = owner^{-1}(t)

Figure 2. Z schemas for **Template_Association** and **Port_Association**

Every template port has a set of attributes, which can be used to control the ways in which nodes based on the templates can be linked. These attributes are described by a partial function attr from the Cartesian product **Port** × **AttrIndex** to **Attr**, where **Attr** is the set of all attributes, and **AttrIndex** is used to organise the attributes into classes, which will vary according to the application area. For coarse-grained dataflow graphs, the attribute classes include TYPE, with values EDGE and VARIABLE, and DIRECTION, with values IN and OUT. Alternatively, this association can be described by the Z schema **Port_Attribute** (see [16] for details).

The association of nodes with templates is described by the **Node/Template Association** schema, which requires the existence of a partial function **node_parent** that maps each node to its corresponding template. The domain of **node_parent** is the collection of all currently instantiated nodes. The schema **Slot_Association** describes how the external interface of a node is derived from that of its underlying template. Mathematically, the set **Slot** is defined to be the subset $\{(n, p)\}$ of **Node** × **Port** for which owner(p) = node_parent(n). The corresponding pullback diagram is shown in Figure 3.

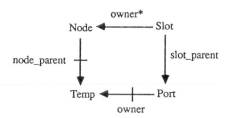

Figure 3. Relationships of Slots, Ports, Nodes, and Templates

Constr_Setting
Node/Template Association
Slot_Attribute
label : Slot --|--> Label
type : Label --|--> (NonStruct ---> Attr)
ran label   dom type
$\forall$ a:NonStruct, s:dom label • type(label(s))(a) = attr*(s)(a)

Figure 4. Z specification for **Constr_Setting**

The slots created in this way inherit the attributes of their corresponding ports, and this inheritance is described by the schema **Slot_Attribute**.

After two or more nodes have been instantiated, they can be linked by associating a label from the basic set **Label** with a collection of slots belonging to these nodes. Each label has a set of *non-structural* attributes that correspond to a subset **NonStruct** of **AttrIndex**. These associations are established formally by the functions **label** and **type**. To ensure that the slots linked by a label are compatible, the attributes shared by a slot and its label must match. It is now possible to define a construction setting formally with the Z schema shown in Figure 4.

A construction setting can be modified in several ways. First, a slot can be assigned a label, and the label previously assigned to a slot can be modified or deleted. Second, a node based on a library template can be added to a construction setting. Finally, a template can be created from a construction setting and stored in the template library. The model specifies operations that perform each of these actions. While full details can be found in [16], Figure 5 shows the Z schema for the operation **EXT**, which creates a new library template from the current construction setting.

The labels which are currently in use in a construction setting are called **Channels**; **External Channels** are those **Channels** that have been used only once, and are therefore intended for communication with the external environment. The

EXT
ΔConstr_Setting
t? : Template
P? : F_1Port
t? $\notin$ dom interfaces
$\forall$t : dom interfaces • P? $\cap$ interfaces(t) = $\varnothing$
$\exists$h : p? >--->> ExternalChannels
interfaces' = interfaces $\oplus$ {t? |---> P?}
attr' = attr $\oplus$ {p |---> attr*(x) | (p$\in$ P? $\wedge$ label(x) = h(p)}

Figure 5. Z specification for **EXT**

EXT operation uses the **External Channels** of a construction setting to define the ports of the new template. The symbol Δ**Constr_Setting** in the schema indicates that the existence of a **Constr_Setting** is both a precondition and a postcondition of the operation. The components of the modified **Constr_Setting** are indicated by primes, and the double-headed arrow is used to represent a bijection.

The model can be tailored to particular development environments by choosing relevant attributes and attribute classes, and by modifying the definition of a construction setting to include regulation of the way in which these attributes govern the valid assignment of labels to slots. It may also be necessary to correspondingly alter the specifications of the operations that modify construction settings.

The Z specification language is well suited to describe the dynamically changing basic sets and functions needed for the model. The model serves three important purposes:

1. it provides a useful abstraction of the concepts of hierarchy and modularity in software development

2. it identifies the principles underlying the construction of a wide variety of existing software development environments

3. it allows the formulation of interesting mathematical questions about the structure of the model

In particular, it is now possible to investigate the completeness of the model by asking whether a given construction setting can be obtained from another by a sequence of model operations. The categorical flavour of this question seems well suited to the abstraction of the mathematical model.

The mathematical tools that were used to construct the model were taken from set theory and category theory, but they they were used in a relatively unusual way that was dictated by the system that was to be modelled. Although the model generates complexity that corresponds well to the system complexity, the formal consequences of that complexity have not yet been fully explored.

5 Summary and Conclusions

In this paper, the process of constructing mathematical models of complex systems has been described as taking place in four stages:

1. the use of existing mathematical tools to create system models

2. the extension and combination of existing mathematical tools to create more complex system models

3. the exploration of the emergent mathematical complexity of these models

4. the abstraction and absorption of this complexity into new mathematics

Although this process has played a major role in the development of mathematics for thousands of years, its operation in the past is beyond our observation. By examining the construction of models for complex computational systems, however, the operation of the modelling process can be observed directly. Examples have been presented to illustrate each of the four stages outlined above. The first stage is typified by the application of computability theory to actual computers, the second by the set-theoretic model of software development environments, the third by the CSP model of concurrency, and the fourth by automata theory.

The conclusion that can be drawn from these examples is that there is indeed a mathematical revolution inspired by computing, which is similar in significance to the previous mathematical revolutions inspired by the modelling of physical and social systems.

References

[1] A. V. Aho and J. D. Ullman, *The Theory of Parsing, Translation, and Compiling. Volume 2: Compiling*, Prentice-Hall, Englewood Cliffs, 1973.

[2] N. H. Brown, 'The EMSP Data Flow Computer', in *Proceedings of the 17th Annual Hawaii International Conference on System Sciences*, Honolulu, 1984, pp. 39-48.

[3] A. Burns, *Programming in occam 2*, Addison-Wesley, Wokingham, 1988.

[4] T. E. Gerasch, M. D. Rice, and S. B. Seidman, 'A Programming Support Environment Based on a Formal Model for Coarse-Grained Dataflow Computation', Technical Report TR-6-86, Department of Computer Science, George Mason University.

[5] T. E. Gerasch, M. D. Rice, and S. B. Seidman, 'Graphical Programming Languages for MIMD Computation', Technical Report TR-10-87, Department of Computer Science, George Mason University.

[6] J. R. Gurd, C. C. Kirkham, and I. Watson, 'The Manchester Prototype Dataflow Computer', *Communications of the Association for Computing Machinery*, **28**, 111-129, 1985.

[7] F. Harary, R. Norman, and D. Cartwright, *Structural Models: An Introduction to the Theory of Directed Graphs*, Wiley, New York, 1965.

[8] P. Henderson, ed., *Proceedings of the ACM SIGSOFT/SIGPLAN Software Engineering Symposium on Practical Software Development Environments*, Pittsburgh, 1984.

[9] P. Henderson, ed., *Proceedings of the ACM SIGSOFT/SIGPLAN Software Engineering Symposium on Practical Software Development Environments*, Palo Alto, 1986.

[10] P. Henderson, ed., *Proceedings of the ACM SIGSOFT/SIGPLAN Software Engineering Symposium on Practical Software Development Environments*, Boston, 1988.

[11] C. A. R. Hoare, *Communicating Sequential Processes*, Prentice-Hall, Englewood Cliffs, 1985.

[12] W. M. L. Holcombe, *Algebraic Automata Theory*, Cambridge University Press, Cambridge, 1982.

[13] M. Kline, *Mathematical Thought from Ancient to Modern Times*, Oxford University Press, New York, 1972.

[14] R. L. Kruse, *Data Structures and Program Design* (second edition), Prentice-Hall, Englewood Cliffs, 1987.

[15] R. Milner, *A Calculus of Communicating Processes*, Springer-Verlag, Berlin, 1980.

[16] M. M. Pett, M. D. Rice, and S. B. Seidman, 'A Formal Framework for the Design of Development Environments', in *Proceedings of the Fifth International Workshop on Software Specification and Design*, 1989, 284-286.

[17] R. D. Rasmussen, G. S. Bolotin, N. J. Dimopoulos, B. F. Lewis, and R. M. Manning, 'Advanced General Purpose Multiprocessor for Space Applications', in *Proceedings of the 1987 International Conference on Parallel Processing*, pp. 54-57.

[18] W. Reisig, *Petri Nets: An Introduction*, Springer-Verlag, Berlin, 1985.

[19] M. D. Rice and S. B. Seidman, 'Construction and Formal Specification of Graphical Programs', Technical Report TR-11-87, Department of Computer Science, George Mason University.

[20] H. Rogers, *Theory of Recursive Functions and Effective Computability*, McGraw-Hill, New York, 1967.

[21] A. W. Roscoe and N. Dathi, ' The Pursuit of Deadlock Freedom', Technical Monograph PRG-57, Oxford University Computing Laboratory.

[22] S. B. Seidman, 'Relational Models for Social Systems', *Environment and Planning B : Planning and Design*, 14, 135-148, 1987.

[23] J. M. Spivey, 'Understanding Z : A Specification Language and its Formal Semantics', Cambridge University Press, Cambridge, 1988.

[24] R. E. Tarjan, *Data Structures and Network Algorithms*, Society for Industrial and Applied Mathematics, Philadelphia, 1983.

A Euclidean Basis for Computation

Dan Simpson
Department of Computing, Brighton Polytechnic
Moulsecoomb, Brighton, BN2 4GJ

Abstract
Given a few simple ideas from Euclid the theory of computation is constructed. The construction provides a simple intuitive motivation for the Church-Kleene approach to computation.

1 Motivation

A number of theories of computation are known and it is well established that they all lead to systems which are equivalent in power with respect to the functions which may be computed (see for example [1]). Here we present a new construction which attempts to explain the recursive functions. However, the construction is intuitively appealing as it is built on very simple foundations which all mathematicians should be able to agree upon. It is appealing to computer scientists as it is, in some sense, constructive.

The study was motivated by a comment of Bertrand Russell who wondered what a mathematician would do if stranded on a desert island. Given only the simple tools available in such a setting and some very simple ideas on using these tools, the theory presented here may be obtained. To make the explanation simpler some notation is used before it is derived, however this is not required.

2 The Pre-requisites

We shall assume that we have available a straight edge, a compass, something on which to make marks and something with which to make these marks. We shall use paper and ink for these last two tools although sand and a stick would be better and this would not imply continuity as does paper and ink. We need a very simple idea of ratio, simple in that we only need consider ratios of objects with the same dimensionality, lines with lines and, although not strictly required, area with area. We need the idea of a map, but again we only relate items of the same type.

Given these we use a few well known constructions without developing them here. We shall assume that a perpendicular can be erected from any point on a given line, that the mid-point between any two points may be found and that given the length of a side a square may be drawn.

We shall present the construction as known not as discovered but during the discovery three types of action are needed. These are:

1. the ability to make a choice between things (otherwise a wrong track in the development could never be discarded)

2. the ability to do one thing and then another (or the development would be paralysed after the first step)

3. the ability to do one thing a number of times and to know when to stop doing it (otherwise the development would have to stop short of being anything interesting or would take forever to complete).

3 The Construction

Given a straight edge, we make two arbitrary marks on it. Although we shall use the word distance, these marks should not be thought to imply that we have invented the idea of distance. But let us say that these marks are d apart.

We now proceed to draw marks and invent the graph of a function, although we are only marking points in a map we construct. From an arbitrary point on an arbitrary straight line (call it the axis) mark a distance d (conventionally to the right, but it doesn't matter). At this point create a perpendicular (conventionally up). We now have to choose how far up this perpendicular to mark a point. From this chosen point draw another line parallel to the axis, continue in this way until boredom sets in. For simplicity we assume the horizontal distance is always 'd' and the choice to be made is how far (defined as a ratio to d) we wish to proceed vertically. Note that we always go 'up'; to go down at any point would violate our wish to keep things as simple as possible.

The choice for our vertical distance falls to always using a constant amount or using a variable amount. Going up a constant amount leads to a family of linear functions. Going up an arbitrary amount at each step will lead to polynomial functions. If at each step we go up the same distance as the height we are currently at we obtain exponential functions. We can now (using bisection) interpolate *ad nauseum* but not *ad infinitum*. Logarithms can be developed by analogy with exponentials. In this way we can simply, but not quite arbitrarily, obtain the classes of functions important in complexity analysis. Interpreting our horizontal move as adding an item of data and our vertical move as amount of work to be done, this is not surprising. But we did not wish to start with this idea and these functions but rather to build them from our simple tools. In fact the functions we have built are not the standard functions we may think them to be, because we do not have an idea of continuity.

4 Using the Graphs

We started our constructions from an arbitrary point and it is tempting to choose this point to represent zero and choose an arbitrary point (say at distance d from zero) to represent one. In fact this would suffice for the presentation here but we

A Euclidean Basis For Computation

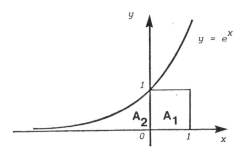

Figure 1. Given $A1 = A2$, zero and one are defined

can do a little better than this. Later papers will try to show why the word better is used.

We choose zero and one as follows. Take the curve e^x and draw a square on the axis so that it just reaches the curve. The square is situated so that its area is equal to the area under the curve to its left. As we are working *ad nauseum* and not *ad infinitum* a course grain equality is sufficient and may be defined as putting the grains of sand in the two areas in a one to one relation (i.e. remove one from each area at a time until the equality is good enough). Note that we are not counting, simply using ratio on areas. We choose the point on the lower left corner to represent zero and the lower right to represent one (see Figure 1).

We now define our three base functions and two composition strategies.

Our first function is the successor which has its domain and range as marks along an arbitrary line. We define SUC (point representing zero) as the point representing one (see Figure 1). To construct the SUC (point representing one) we mark, using ratio, a further point on the line and label it 2. In this way we can build the set N. Continue in this way and then define a function SUC on N such that $SUC(n) = n + 1$.

Our second function $zero(n) = 0$, $n \in \mathbb{N}$. Which may again be interpreted using Figure 1 for the representation of the natural numbers.

The third function needs a further basic idea: that of making a choice. The select function chooses an item from a tuple of integers:

$select(0, 0) = 0$
$select(n, k)$ projects the nth item from a list of k numbers, where $0 < n \leq k$.

We now obtain our two compositional strategies. Doing one thing followed by doing another is simple function composition. We note that applying function composition to SUC has the properties of the natural numbers we want and allows us to interpret +1 as addition and not simply taking a step.

Doing one thing after the other and knowing when to stop is primitive recursion. Assume g is a primitive recursive function already obtained, we define f as

$$f(0) = k, \quad f(n+1) = g(n, f(n))$$

Having developed the natural numbers, our base functions and composition strategies, we can now define the class of primitive recursive functions:

Definition The class PRF of *primitive recursive functions* is the smallest class on the natural numbers such that

1. Each base function is in PRF

2. PRF is closed under composition and primitive recursion.

This class is exactly the one which is given in the standard texts. However it is hoped that the construction helps to explain why the class is structured as it is. Our definition of PRF is restricted to functions of only one variable. However, we have already used the idea of a tuple and the definitions are easily generalised to functions on $\mathbb{N}^n$.

5 Evaluation

We have now defined the class of primitive recursive functions but for these to form a basis for computation we need to define an evaluation strategy. To do this we must introduce a further idea which is that of being able to replace a function point by its value. This is the well known computing idea of referential transparency. Evaluation then follows the standard method.

Our construction of a function was always done in a pointwise fashion, we may thus ensure that our function will be defined for all natural numbers. However once constructed, it is useful to be able to present the definition in a different, more economical way. An evaluation mechanism is then needed to ascertain the complexity of the calculation. If we then start to define functions using a non-pointwise constructive method we loose the certainty that the function may be evaluated at all in a finite time. However constructing functions as done above does ensure that those in PRF are indeed computable.

6 Mu Functions

All the functions defined so far are computable and indeed total. For a long time it was thought that all total functions were members of PRF but it is now known that this is not the case. We now turn our attention to asking why.

The standard function to consider at this point is due to Ackermann and is defined by the system.

$$\begin{aligned} A(0, x) &= x + 1 \\ A(n+1, 0) &= A(n, 1) \\ A(n+1, x+1) &= A(n, A(n+1, x)) \end{aligned}$$

A Euclidean Basis For Computation

If we calculate (by induction) the first few values, we find

$A(1, x) = x + 2$
$A(2, x) = 2x + 3$
$A(3, x) = 2^{x+3} - 3$
$A(4, x) = 2^{2^{\cdot^{\cdot^{16}}}} - 3$, where the height of the exponential is $x + 1$.

This function is total but is not primitive recursive. The reason being that the function grows faster than any primitive recursive function may grow. We may see why this is if we consider our ratio method of defining functions. When a function is constructed, the next value for the ordinate will depend on a certain ratio to the previous ordinates. The person building the function will press his stick into the sand a certain number of times. When we come to calculate the function we may not know how many times this was done, but we are sure it was done. It will be noted in passing that this is why members of PRF are equivalent to count programs.

If we look at the definition of Ackermann we can see that this restriction has been violated as the height of the exponential depends on x. We may evaluate A using our previous methods but we may not define it using them. To include such a function in our theory of computation, we must move away from our idea of definition by ratio.

It appears from looking at the definition of A that we have two choices. The value of $A(4, x)$ leads us to consider doing some calculation to help determine the value of the ordinate. We are able to do this now as we have defined N and operations on it using our simple ratio idea. Alternatively, looking at the system of equations which define A we notice that the third line of the definition violates the definition for primitive recursion given above which required f and g to be distinct functions. Of course this may not give us any extra power as it may be that we could reformulate the equations; but it seems worthy of investigation.

If we are to allow arbitrary calculation in the value of a function f, we need to consider definitions of the form $f(x) = y$ such that $P(g(x, y))$ is true, and consider alternatives for the predicate P and function g. For g we may require that it is primitive recursive, total or arbitrary. We require g to be defined everywhere and so reject the idea of using a partial function but we also wish to include all computable functions and so choose g to be total. The choice of P is slightly arbitrary because it will affect the particular g we choose to match it in the definition of f. A simple yet sufficiently powerful choice is $g(x, y) = 0$. But $g(x, y)$ may be zero for many and various choices of y so we choose the nearest such y to zero. This means we may set y to zero evaluate g and continue to increase y until $g(x, y) = 0$; but of course this may never occur for any y.

We may now define the class of mu-recursive functions MRF as

1. Every primitive recursive function is in MRF.

2. If g is a total function in MRF, then the function f defined as above by minimisation from g. $f(x) = \mu y[g(x, y) = 0]$ is in MRF.

If we consider relaxing the definition of primitive recursion so that a function may be defined in terms of itself we are allowing exactly the same type of calculation for the ordinate of a given $f(x)$ as we allowed in our first relaxation of a rule and the class of functions defined is exactly MRF.

7 Other Functional Theories

In the above we talked about computable and computation when we should have talked about the computation of number theoretic functions only. However, having developed the theory to this point we can appeal to Gödel numbering [2] to allow us to generalise to any symbolic system.

We may investigate other functional theories of computation by considering different alternatives for the predicate P and function g which we discussed when considering mu-minimisation.

8 Observations

The major aim of this paper was to use only a very simple set of basic tools to explain why the theory of computation is as it is. At the very least I hope that some future students may be less mystified about the topic than I was when I first met it.

As a by-product of the development it was possible to give a fairly rigorous basis for the natural numbers without the need for the idea of a set.

When considering how to calculate the value of a function at a point we considered all possibilities from using a constant value up to using an arbitrary subcalculation. Of course someone may come up with a different concept of evaluation, consider different domains or may find a flaw in the current work. But I am convinced by the Church/Turing Thesis.

Ackowledgement: I am deeply indebted to Dr Arthur Charles Allen for numerous discussions on function theory. The importance of the exponential as an area for investigation is due to him.

References

[1] Denning, P. F., Dennis, J. B., Qualitz, J. E., *Machines, Languages and Computation*, Prentice Hall, (Englewood Cliffs), 1978

[2] Hyland, J. M. E., 'Computing and Foundations', in Johnson and Loomes (eds) *The Mathematical Revolution Inspired by Computing*, Oxford University Press (Oxford), 1991.

An Extension of Turing Machines

Claudio Sossai
Via Vlacovich 20 - 35124 Padova, Italy

Abstract

A extension of Turing's definition of a machine is defined which has the usual Turing Machine as a special case. Turing's machine chooses between two possible instructions depending on the whether the scanned symbol has value 0 or 1. The machine defined here can work with an extension of Boolean valued arithmetic, in which numbers are objects that are associated with more than one standard number with different degrees of truth. This definition describes a time-dependent non-monotonic computation agent that may learn from its computing process. This shows how new mathematics needs to be developed for computing and its applications.

1 Introduction

The last few years have seen a growing interest in all those techniques by which one can simulate deduction by means of computational process. In practical applications, one can see that applied reasoning has some peculiarities which are not completely described by mathematical reasoning, or better, by its description in terms of mathematical logic. In particular, it has been observed that quite often concrete or applied reasoning differs in at least two aspects from reasoning as described by classical logic:

1. deductions in situation of partial information

2. non-monotonicity.

Drawing from my personal experience, I met a typical example of such situations when I constructed an expert system for renal function monitoring. Frequently I found the following situation: let p be a proposition that asserts the value of some medical test and let q be a proposition that asserts the value of some other medical tests. Very often the physician derives the diagnosis B from p and derives the diagnosis $\neg B$ from $p \wedge q$. This is a typical example of the above situation in the sense that in the first case the physician must reason with partial information, and the first and second case together give an example of what will be called non-monotonic reasoning.

It is well known that a programmed computer is equivalent to a Turing machine, and quite close to the deductive structure of classical logic; therefore, either one considers the kind of reasoning used by physicians as absolutely arbitrary and with no scientific interest, or one meets severe difficulties in describing the physician's way of reasoning on a computer.

This paper investigates if and how the notion of computability can be properly extended to deal in a natural way with two peculiarities of applied reasoning: deduction in conditions of partial information and non-monotonic reasoning. In this context we give a definition of machine that is an extension Turing's, where here extension means that if we particularise a parameter the Turing definition appears again.

A Turing machine chooses between two possible instructions depending on the value of the scanned symbol, which may be 0 or 1. In the definition given below the machine can work with an extension of Boolean valued arithmetic, where numbers are objects which are equated to more than one standard number with different degrees of truth.

For this reason in general the machine needs a certain amount of information to make a choice, and hence three different kinds of computation may arise:

1. The machine never finds situations of uncertainty. In this condition it works like a Turing machine.

2. The machine finds situations of uncertainty, but it has the information to make a choice. In this case it makes some operations that are combinations of Turing operations.

3. The machine has not enough information to decide in situations of uncertainty. In this case it makes a random choice between the choices that are possible. It is interesting to note that to every random choice there corresponds a proper extension of the information previously owned by the machine.

This definition of machine describes a time dependent non-monotonic computation agent that may learn from its computing process.

It may happen that, if the machine starts the calculation with a piece of information p, it gives an output α. If it starts with a piece of information $p \wedge q$, that is *an extension* of p, it gives an output β different from α. We will refer to such a process as being *non-monotonic*.

To be able to describe a computing process with such properties, an environment is needed within which to speak about uncertainty, and partial information, and hence define what computability means under such conditions.

2 Uncertainty

The usual environment in which one can describe uncertainty is the Boolean (sigma) algebra of possible events. It is well known that the set of truth values $2 = \{0, 1\}$, where 0 stands for false and 1 for true, can be a Boolean algebra under the operations $\neg$, $\wedge$ and $\vee$. It is possible to extend the semantics of classical logics to a logic with uncertainty by interpreting propositions in a Boolean algebra different from 2. Since the language of set theory is a sufficient tool for describing mathematics, it is natural to extend standard set theory to probabilistic set theory and use this extension to express our ideas and describe a computing process with uncertainty. There is an elegant theory with all the tools we need, namely Boolean-valued models of set theory. An exposition of this theory may be found in [1], but here we give an informal summary of the ideas that we shall need.

A set A in the universe of sets V may be identified with its characteristic function c_A defined on all $x \in V$ by:

$$c_A(x) = \begin{cases} 1 & \text{if } x \in A \\ 0 & \text{if } x \notin A \end{cases}$$

So we can identify the universe of sets V with the class of functions defined by transfinite recursion on the ordinal number α in the following way:

$$V_\alpha^{(2)} = \{f : function(f) \wedge range(f) \subseteq 2 \wedge (\exists \beta)[\beta < \alpha \wedge domain(f) \subseteq V_\beta^{(2)}] \wedge range(f) \subseteq 2\}$$

and

$$V^{(2)} = \{x : (\exists \alpha)(x \in V_\alpha^{(2)})\}$$

Now if we substitute a complete Boolean algebra B for 2 in the previous definition, we get a new definition that describes a new universe, named $V^{(B)}$, within which we can interpret the logical sentences of set theory.

It is well known that a suitable language for set theory is a first order language built up from the atomic formulas $x \in y$ and $x = y$. So we can associate with every sentence ϕ of the language of set theory an element of the Boolean algebra B that describes its truth value and we shall denote this element by $b \in B$ with the formula $b = \|\phi\|$. Let us briefly repeat how this is done :

1. for the atomic formulas $u, v \in V^{(B)}$

$$\|u \in v\| = \bigvee_{x \in dom(v)} [v(x) \wedge \|x = u\|]$$

$$\|u = v\| = \bigwedge_{x \in dom(u)} [u(x) \to \|x \in v\|] \wedge \bigwedge_{y \in dom(v)} [v(y) \to \|y \in u\|]$$

2. for all other formulas

$$\|\phi \wedge \psi\| = \|\phi\| \wedge_B \|\psi\|$$
$$\|\neg \phi\| = \neg_B \|\phi\|$$
$$\|(\exists x)\phi(x)\| = \bigvee_{u \in V^{(B)}} \|\phi(u)\|$$

Where $\wedge_B, \neg_B$, and $\bigvee$ are the operations *and, not* and *or* on the complete Boolean algebra B.

In this model, logical axioms as well as all axioms of set theory have truth value 1 and this property is preserved by the rules of inference.

3 Information

Let B be a complete Boolean algebra. A subset C of B is said to be *dense* in B iff 0 does not belong to C and for every element $b \in B$, with $0 < b$, there is an element $c \in C$ such that $c \leq b$. A dense subset of B is also called a *set of information* for B.

A central idea in information theory is that if a is an event with probability less than event b, then the occurrence of a gives us more information than the occurrence of b. In our model of set theory we will not use measures and so we cannot speak of probability; however, it is well known that in a Boolean (sigma) algebra if $a \leq b$ then, as soon as we define a measure μ, necessarily $\mu(a) \leq \mu(b)$.

Let p be an element of a set of information and let ϕ be a sentence. There is a well known relation between information and sentences that captures among others the idea of information theory: the forcing relation (written $p \Vdash \phi$). In this context the forcing relation may be defined in the following way:

$$p \Vdash \phi \text{ iff } p \leq \|\phi\|.$$

This relation has many nice properties including the following:

1. $(q \leq p \wedge p \Vdash \phi) \rightarrow q \Vdash \phi$; this captures the above idea of information theory
2. $(\forall p)(\exists q \leq p)(q \Vdash \phi \vee q \Vdash \neg \phi)$; this says that there is enough information
3. $(p \Vdash \phi) \rightarrow \neg(p \Vdash \neg \phi)$; this shows the soundness of the definition

Since we want to define a computable relation between information and sentences, first of all we need a complete Boolean algebra with a set of informations in which Boolean operations are computable. Let G be 2^N, the set of all functions from the integers to $2 = \{0, 1\}$. Let K be the set of functions from finite subsets of the integers to 2 partially ordered by inverse inclusion (i.e. for all elements h, k of K, $k \leq h$ iff $k \supseteq h$). For every $p \in K$ we put

$$N(p) = \{f \in G : p \subset f\}$$

Subsets of the form $N(p)$ form a base for the product topology on G when 2 is assigned the discrete topology. Each $N(p)$ is a clopen (i.e. closed and open) set in this topology, in particular it is a regular open set, and it is easy to verify that the map $p \mapsto N(p)$ is an order isomorphism of K onto a dense subset of $RO(G)$, the set of the *regular open elements* of G. Therefore $< RO(G), N >$ is a Boolean completion of K, and the latter is (up to isomorphism) a set of information for $RO(G)$. So if we take $B = RO(G)$ as a Boolean algebra, all the information we need is given us from the characteristic functions of the finite subsets of the integers, which are computable objects. We call C the set of such informations and from now on when we will speak of a Boolean algebra B we will mean the *complete Boolean algebra* $B = RO(G)$. We will also use the following convention to simplify the notation: for instance, if $f \in K$ is defined as

$$domain(f) = \{1, 4, 5\}, \text{with } f(1) = 0, f(4) = 1, f(5) = 1,$$

An Extension of Turing Machines 215

then we will indicate f as $f = [\bar{1}, 4, 5]$. Identifying f with its isomorphic image $N(f)$ in B, with this notation it is easy to see that:

$$[1, 2, \bar{6}] \wedge [\bar{3}, 8] = [1, 2, \bar{3}, \bar{6}, 8]$$

$$\neg [1] = [\bar{1}]$$

In this framework, we thus can express the two crucial ideas of uncertainty and information in a homogeneous mathematical environment sufficiently strong to describe our extension of the Turing machines.

4 An Arithmetic with Uncertainty

Recursive functions, Turing machines and computability have their natural environment in the natural numbers, and it is known that the set of finite ordinals is a model in set theory of the integers; so we can describe an arithmetic with uncertainty using the set of finite ordinals in a Boolean valued extension of set theory.

What are such *integers with uncertainty*? It is known that it is possible to define integers in set theory with a restricted formula $INT(x)$ and so integers in our model, which we will call $N^{(B)}$, are all the elements x of $V^{(B)}$ such that $\|INT(x)\| = 1$. It is also possible to construct them explicitly.

First of all we can describe a copy of the integers in our model in this way: given an integer number n we have a copy of n, named $\hat{n}$, so defined

$$\hat{n} = \{<\hat{j}, 1> : j \in n\}$$

The set of such numbers behaves just like the natural numbers and so from now on we shall identify $\hat{n}$ with n.

A partition of unit is a family $\{a_i\}_{i \in I}$ of elements of B such that :

1. $a_i \wedge a_j = 0$ for every $i, j \in I$,
2. $\bigvee_{i \in I} a_i = 1$

If we have a partition of unit $\{a_i\}_{i \in I}$ and a family of integers $\{u_i\}_{i \in I}$ we can define a new element u, called a *mixture* and written $u = \Sigma_{i \in I} a_i \cdot u_i$, in the following way:

$$dom(u) = \bigcup_{i \in I} dom(u_i)$$

$$u(z) = \bigvee_{i \in I} [a_i \wedge \|z \in u\|], \text{ for } z \in dom(u)$$

We will usually write $u = b \cdot 1 \oplus \neg b \cdot 0$ for $u = \Sigma_{i \in I} a_i \cdot n_i$ when $I = \{1, 2\}$, $a_1 = b$ and $a_2 = \neg b$.

It is possible to show that all the extended integers are mixtures of standard integers and they have the properties of the standard integers. For example, we can see how the sum works in this extended arithmetic. We can follow the traditional way to define the sum, i.e. for $a \in N^{(B)}$ let $a' = a \cup \{a\}$ and then

$$a + 0 = a$$
$$a + b' = (a + b)'$$

With this definition we can show that

Theorem. If $\Sigma_{i \in I} a_i \cdot n_i$ and $\Sigma_{j \in J} b_j \cdot m_j$ are two elements of $N^{(B)}$ then

$$[\![\Sigma_{i \in I} a_i \cdot n_i + \Sigma_{j \in J} b_j \cdot m_j = \Sigma_{i,j}(a_i \wedge b_j) \cdot (n_i + m_j)]\!] = 1$$

Corollary. Let $\Sigma_{i \in I} a_i \cdot n_i$ be any element of $N^{(B)}$ where $I = \{0, 1, ..k\}$. Then there exist j_i and c_i, $i \leq k$, such that:

$$\Sigma_{i \in I} a_i \cdot n_i = \sum_{i=0}^{k} \left(\sum_{l=1}^{j_i} (\neg c_i \cdot 0 \oplus c_i \cdot 1) \right)$$

where big Σ is the summation symbol. This corollary says that every extended integer is the sum of numbers that are partial like the integer number 1.

5 Extended Machines

We are now able to define extended machines. Let D be the Boolean algebra generated by the set of informations C, so that $C \subset D \subset B$. To see the relation with the Turing machines, it is convenient to repeat the standard definition, let:

$Q = \{q_i : i \in N\}$ be the set of internal states
$2 = \{0, 1\}$ be the set of scanned symbols
$O_p = \{R, L\} \cup 2$ be the set of operations

(where R means go to the right, L go to the left, 1 write 1 and 0 write 0).

The set of Turing machines, TM, is defined as:

$TM = \{\tau : \tau$ is a mapping from a finite subset of $Q \times 2$ into $O_p \times Q\}$

According to our philosophy, the first thing that we must do is to extend the Boolean algebra 2 to the algebra D and so let:

$$O'_p = \{R, L\} \cup D.$$

This, with a restriction that we shall explain below, gives us the definition of *extended Turing machines*, ETM:

$ETM = \{\pi : \pi$ is a mapping from a finite subset of $Q \times D$ into $O'_p \times Q\}$

An ordered pair $<< q_h, b >, < o, q_k >> \in \pi$ is called an instruction of the machine π and we shall write it in the form $q_h b o q_k$; furthermore, we will call q_h the initial state and q_k the terminal state.

Then the restriction is that any machine π must contain at most two different instructions with the same initial state; that is, if $q_n b_1, ..., q_n b_m \in$

$dom(\pi)$ are all pairs with initial state q_n then $m = 2$ and $b_1 = \neg b_2$. We shall call an element of C an *informational state*.

An extended machine will simply be an element of ETM that satisfies the restriction above. It is important to note that at this point the computing process is not yet defined. In particular, every machine will be provided with a set of possible information, as we will see below.

6 Running

Extended machines work with a tape that has an infinite number of cells and each cell may contain an element $b \in D$. An element b of a cell is interpreted by the machine as the number $b \cdot 1 \oplus \neg b \cdot 0$. Note that when the value of b is $1 \in D$, then the number represented on the cell is the number 1, while for $b = 0$ it is the number 0. This means that if we restrict to the Boolean algebra 2, the tape will be exactly as in the standard Turing machines.

At every moment of the computation the machine will be positioned on a cell that we will call a *scanned symbol* or *scs*. The pair of instructions beginning with the same initial state will be called the *macro instruction*.

We can describe the computing process as composed by two kinds of operations performed by the machine, the choice operation and the execution operation. In the *choice operation*, the machine chooses the instruction to execute inside the macro instruction to which it is pointing. The *execution operation* uses the chosen instruction to execute the operation indicated by the instruction.

7 Operations

First let us describe the choice operation. Let

$$q_i b...$$

$$q_i \neg b..$$

be the macro instruction to which the machine is pointing and let α_1, α_2 be defined in the following way:

$$\alpha_1 = \|b \cdot 1 \oplus \neg b \cdot 0 = scs \cdot 1 \oplus \neg scs \cdot 0\|$$

$$\alpha_2 = \|\neg b \cdot 1 \oplus b \cdot 0 = scs \cdot 1 \oplus \neg scs \cdot 0\|$$

We can say that α_1 represents the degree of likeliness between the number (represented by) b and the number (represented by) the scanned symbol; similarly, α_2 is the degree of likeliness between $\neg b$ and the scanned symbol. We shall show that $\alpha_2 = \neg \alpha_1$.

Lemma. If $a, b \in B$, then $\|a \cdot 1 \oplus \neg a \cdot 0 = b \cdot 1 \oplus \neg b \cdot 0\| = a \leftrightarrow b$.

Proof: To simplify notation, we call $n(b)$ the number represented by b on the tape, i.e. $n(b) = b \cdot 1 \oplus \neg b \cdot 0$. It is easy to see that $dom(n(a)) = dom(n(b)) = \{\emptyset\}$. Applying the definition of equality we have:

$$\|n(a) = n(b)\| =$$

$$= (n(a)(\emptyset) \to \|\emptyset \in n(b)\|) \wedge (n(b)(\emptyset) \to \|\emptyset \in n(a)\|)$$

$$= (((a \wedge \|\emptyset \in 1\|) \vee (\neg a \wedge \|\emptyset \in 0\|)) \to \|\emptyset \in n(b)\|)$$
$$\wedge (((b \wedge \|\emptyset \in 1\|) \vee (\neg b \wedge \|\emptyset \in 0\|)) \to \|\emptyset \in n(a)\|)$$

$$= (((a \wedge 1) \vee (\neg a \wedge O)) \to b) \wedge (((b \wedge 1) \vee (\neg b \wedge O)) \to a)$$

$$= a \leftrightarrow b.$$

Corollary. If $a, b \in B$, $\alpha_1 = \|b \cdot 1 \oplus \neg b \cdot 0 = scs \cdot 1 \oplus \neg scs \cdot 0\|$ and $\alpha_2 = \|\neg b \cdot 1 \oplus b \cdot 0 = scs \cdot 1 \oplus \neg scs \cdot 0\|$ then $\alpha_2 = \neg \alpha_1$.

Proof: Using the previous lemma we have $\alpha_1 = b \leftrightarrow scs$ and $\alpha_2 = \neg b \leftrightarrow scs$ and using the tautologies $[(a \leftrightarrow b) \vee (\neg a \leftrightarrow b)]$ and $\neg[(a \leftrightarrow b) \wedge (\neg a \leftrightarrow b)]$ we have that $\alpha_1 \vee \alpha_2 = 1$, $\alpha_1 \wedge \alpha_2 = 0$ and therefore $\alpha_2 = \neg \alpha_1$.

At every moment the machine possesses a certain amount of information, i.e. it knows $p \in C$. We assume that when a machine starts for the first time it possesses no information, i.e. the information possessed by the machine is $1 \in B$. If p is the information possessed by the machine, only three mutually exclusive cases are possible :

1. $p \leq \alpha_1$
2. $p \leq \alpha_2 = \neg \alpha_1$
3. $p \not\leq \alpha_1 \wedge p \not\leq \alpha_2$.

If the first case occurs, we have from the definition of forcing that

(1) $p \Vdash b \cdot 1 \oplus \neg b \cdot 0 = scs \cdot 1 \oplus \neg scs \cdot 0$.

For the same reason if the second case occurs we have

(2) $p \Vdash \neg b \cdot 1 \oplus b \cdot 0 = scs \cdot 1 \oplus \neg scs \cdot 0$.

If the first case occurs the machine chooses the first instruction to execute and if the second case occurs the machine chooses the second instruction.

If neither the first nor the second case occurs, this means that $p \wedge \alpha_1 = p_1 > 0$ and $p \wedge \alpha_2 = p_2 > 0$. It is easy to see that $p_1 < p$, $p_2 < p$ and that

$p_1 \Vdash b \cdot 1 \oplus \neg b \cdot 0 = scs \cdot 1 \oplus \neg scs \cdot 0$
$p_2 \Vdash \neg b \cdot 1 \oplus b \cdot 0 = scs \cdot 1 \oplus \neg scs \cdot 0$.

In this case the machine chooses randomly between p_1 and p_2; if p_1 is chosen, the state of information is extended from p to p_1, and similarly if p_2 is chosen.

So if it chooses to extend the information p with p_1, it will execute the first instruction, and the second otherwise. It is worth noting that in this way a proper extension of the information previously owned by the machine corresponds to every random choice.

It is easy to see that no information is needed to make a choice in the case of the Turing machine, where $b = 1$ or $b = 0$ and $scs = 1$ or $scs = 0$.

When the machine has chosen the instruction to execute, the control is passed to the execution operation. Let $q_h a_h O_h q_k$ be the instruction chosen by the choice operation; then the execution operation works in the following way:

if $O_h = L$, it moves the tape one cell left
if $O_h = R$, it moves the tape one cell right
if $O_h = b$, it writes b in the scanned cell.

The stop condition is given by a state to which no macro instruction corresponds.

8   An Example

We can see with a simple example how the extended machines can simulate some characteristics of concrete reasoning. Take the following simple machine

$$q_0([1,2] \vee [\bar{1},2])1q_1$$
$$q_0([1,\bar{2}] \vee [\bar{1},\bar{2}])1q_0$$

and give it input 1, i.e. let the tape be ...|0|1|0|... at the start, and let the scanned symbol be |1|.

Let p be the set of information of the machine. Initially we can give the machine the information $p = [1]$. Then the value 1 on the tape is viewed by the machine as representing the number $1 \cdot 1 \oplus 0 \cdot 0$ and so the operations performed by the machine are:

compute the degree of likelihood between the number on the tape and the number on the first instruction:

$$\alpha_1 = \|([1,2] \vee [\bar{1},2]) \cdot 1 \oplus ([1,\bar{2}] \vee [\bar{1},\bar{2}]) \cdot 0 = 1 \cdot 1 \oplus 0 \cdot 0\| = \|\phi\|$$

do the same thing for the second instruction:

$$\alpha_2 = \|([1,\bar{2}] \vee [\bar{1},\bar{2}]) \cdot 1 \oplus ([1,2] \vee [\bar{1},2]) \cdot 0 = 1 \cdot 1 \oplus 0 \cdot 0\| = \|\psi\|$$

Using the definition of equality and that of mixture it is easy to see that $\alpha_1 = [1,2] \vee [\bar{1},2]$ and $\alpha_2 = [1,\bar{2}] \vee [\bar{1},\bar{2}]$.

Since $([1] \not\leq \alpha_1)$ and $([1] \not\leq \alpha_2)$ we have $not(p \Vdash \phi)$ and $not(p \Vdash \psi)$ and so the machine must make a random choice. We can suppose that the choice made is $p \wedge \alpha_1 = [1,2]$ (i.e. an extension of the information owned by the machine with [2]) and then the machine chooses the first instruction, writes 0 on the tape and then stops.

Now let $q = [\bar{2}]$ and we give to the machine the information $I = p \wedge q = [1,\bar{2}]$. In this case, since $[1,\bar{2}] \leq \alpha_2$, the machine chooses the second instruction and the machine does not stop.

So at the end of this example we can say that the machine with information p may stop, but if we extend p to $p \wedge q$ the machine does not stop.

9 Conclusions

The situation described in the above example seems very similar to that of the physician mentioned at the beginning of the paper: if he has information p then the diagnosis is B but if he has information $p \wedge q$ his diagnosis is $\neg B$. The only difference between the physician and the machine seems to be that the machine reached the conclusion "stop" through a random choice. But if we look a little more closely into the reasoning of the physician we find that when he reached the diagnosis B with information p he knew neither q nor $\neg q$, but if he has opted for B he implicitly made a random choice, i.e. $\neg q$, (otherwise he would have concluded $\neg B$ because we know that with information $p \wedge q$ he would diagnose $\neg B$) which is what the machine did. If we start from the fact that man does not now have all information about the universe, random choices and uncertainty are not something outside knowledge, on the contrary random choices are mechanisms essential to knowledge; while uncertainty appears to be the energy of the knowledge. In particular it is very difficult to describe a learning machine without the possibility of random choices. Computing may be viewed as a link between experience and theoretical mathematics and in this link the experience of non-monotonic reasoning in conditions of partial information may offer new insight to theoretical mathematics: for example this extension of the idea of computing. This is a possible perspective on *"The Mathematical Revolution Inspired by Computing"*.

References

[1] Bell, J.L. *Boolean valued models and independence proofs in set theory* Second edition, Oxford University Press, (Oxford), 1985.

[2] Rogers, H. *The theory of recursive functions and effective computability*, McGraw-Hill, (New York), 1967.

[3] Halmos, P.R. *Lectures on Boolean algebras*, Van Nostrand, (New York), 1963.

[4] Halmos, P.R. *Measure theory*, Van Nostrand, (New York), 1950.

[5] Takeuti, G. Zaring, W.N. *Introduction to axiomatic set theory*, Springer, (Berlin), 1971.

Algorithmic Languages and the Computability of Functions

Newcomb Greenleaf
Department of Computer Science, Columbia University
New York, N. Y. 10027, U.S.A., newcomb@cs.columbia.edu

Abstract

This paper concerns the *algorithmic paradigm* for mathematics and the revolution in which it confronts the *logical paradigm*. In particular, we consider the way in which language helps to establish the paradigm, and how the paradigm confronts the phenomena of *non-computable functions*. Algorithmic interpretations are given of the *busy beaver function* and the *Cantor diagonal method*. Functions normally regarded as non-computable can be seen as computable when the range of the function is suitably extended.

1 Algorithmic Languages

Among the many revolutions which the computer is bringing to mathematics is a *linguistic revolution* in which the algorithm concept and languages for algorithms are becoming central in mathematical thinking and practice. The word *algorithm* acquired its modern meaning only half a century ago. The first Oxford English Dictionary referred to it only as an "erroneous refashioning" of *algorism*[1]. According to the new OED of 1989, the first modern use of the word was in the 1938 number theory text of Hardy and Wright. The term has now become so central in our thought that Knuth has proposed *algorithmics* as the best name for the discipline of computer science [19]. This suggests that, after the revolution, mathematics may come to be seen as *pure algorithmics*. As the computer comes to dominate our life, there may be no escape from the computational metaphor. If, as some would have it, our minds and bodies are algorithmic, can our mathematics be far behind? It is our thesis that this revolution should be welcomed, because computing has given mathematics powerful new languages of algorithms, which can refine and sharpen our thinking about process and change.

The study of phenomena associated with algorithms has opened up many new areas of mathematics. Here we consider how the algorithmic revolution can change the way that we view mathematics as a whole.

---

[1] *Algorists* used the ten digits for computing (instead of an abacus). This art was brought to Europe by the Latin translation of a text by al-Khwârizmî, the eponymous ninth century mathematician of Baghdad.

1.1 Algorithms and Proofs

In the terms of Kuhn [20], the computer has brought forth a new *algorithmic paradigm* for mathematics, which stands in opposition to the standard view, based on logic and set theory.[2] The logical and algorithmic paradigms confront each other on this basic issue: what is the relationship between *algorithm* and *proof*? In computer science proofs are used to verify algorithms. Any algorithm must be supported by some form of proof to be believed. The proof often consists of a very informal argument buttressed by testing, but many workers in program verification argue that a program should be a *proof that can be compiled* [10]. Mathematicians use algorithms in their proofs, and many proofs are totally algorithmic, in that the triple [*assumption, proof, conclusion*] can be understood in terms of [*input data, algorithm, output data*]. Such proofs are often known as *constructive*, a term which provokes endless unfortunate arguments about ontology.

We are seeing the emergence of a new concept, of which proof and algorithm are but two aspects. Michael Beeson recently put it nicely (in the context of a discussion of Prolog), "The flow of information seems now to be logic, now to be computation. Like waves and particles, logic and computation are metaphors for different aspects of some underlying unity. [3] " I have no good candidate for a name for this underlying unity (neither *verified algorithm* nor *constructive proof* does it justice).

For our purposes, this superficial analysis of the relationship between algorithms and proofs will do. But computation and logic interact in many more specific ways: the survey [3] covers areas such as automated deduction (theorem provers and proof checkers), logic programming (Prolog and equational logic), program verification, and the implementation of mathematics [9].

1.2 The Algorithmic Counter-revolution

In dialectic terms, we could say that the algorithmic revolution in mathematics is actually a *counter-revolution* to the *logical revolution* which occurred during the period 1850 to 1925, and that it is bringing forth a new *synthesis* of logic and computation. The logical revolution was driven by the new language of logic and sets provided by Boole, Cantor, Frege and their followers. Of course, mathematicians had always used logic and talked about sets, but had lacked an adequate language. The new language was used by Hilbert and his followers to dramatically move mathematics away from its algorithmic roots, both in theory and practice.[3]

Now mathematics has again been given a powerful new language, the language of *algorithms and data structures*, and with it a new vision of mathematical reality. Since Euclid, mathematicians have used algorithms, but only recently have

---

[2] While Kuhn himself saw mathematics as a realm of eternal truths and thus exempt from revolutions [20], Grabiner has argued persuasively that 'Mathematics is *not* the unique science without revolutions. Rather mathematics is that area of human activity which has at once the least destructive and still the most fundamental revolutions.' [15]

[3] Recall Gordan's anguished cry, "This is not mathematics, it is theology." [23]

systematic languages for algorithms been developed. And only very recently has an evident quorum of mathematicians, through their programming experience, become fluent in higher-level algorithmic languages. A Pascal-like pseudo-code has become a new *lingua franca*.[4]

Of course, we have had the languages of Turing machines and recursive functions for half a century. While these precise formalisms have made their mark on our thought, they are at the level of assembly or machine language, and, as Martin-Löf and others have shown us, mathematics is best regarded as a *very high level programming language* [21]. Knuth illustrated this point when he translated a proof (of the Weierstrass Approximation Theorem) from Errett Bishop's constructive mathematics into a Pascal-like form, and noted that the whole point of Bishop's work is that *every proof is an algorithm* [19,4].

Knuth's exercise can be repeated for any mathematical theorem. To understand the theorem in algorithmic terms, represent the assumptions as *input data* and the conclusions as *output data*. Then try to convert the proof into an algorithm which will take in the input and produce the desired output. If you are unable to do this, it is probably because the proof relies essentially on the *law of excluded middle*.

It was L. E. J. Brouwer[5] who first noted that the algorithmic vision forces us to examine our use of logic critically. In the logical paradigm for mathematics, the traditional test for correctness is *consistency*, and this is not sufficient for the needs of algorithms. In particular, the *law of excluded middle* is algorithmically valid only for decidable statements. Today we are witnessing an explosive development of new logics for the needs of computing [2,24,9].

The second half of this paper turns to the critical question of the computability of functions to show how algorithmic mathematics can interpret phenomena which are standardly referred to by the phrase "non-computable function," taking as main examples the *Busy Beaver Function* and the *Cantor Diagonal Algorithm*. Let's look first at a simple example of how the availability of the language of algorithms can change the shape of mathematics.

1.3 The Derivative Algorithm

To illustrate the significance of the new algorithmic literacy, consider the example of elementary differential calculus, which has traditionally been presented as a large collection of formulas for derivatives. Of course, these formulas were intended to be used as the base cases and recursive operations of a grand recursive *derivative algorithm*, which, for want of a proper language, was not made explicit (and therefore never formally verified). The base case formulas, such as the sine formula: D(sin

---

[4] There has been much discussion of the proper mathematics prerequisites for computing courses. Soon we may expect to see mathematics courses with a computer science prerequisite, since students master the language of algorithms through learning to program [16].

[5] The *intuitionism* of L. E. J. Brouwer (1881-1966) was a romantic counter-revolution which attempted to reject the new tools of logic and set theory and to return to an earlier uncorrupted innocence. In contrast, the new algorithmic mathematics is comfortable with a variety of levels of formalization. Dijkstra argues that computing scientists tend to use logic more systematically than do mathematicians [13].

$x) = \cos x$ give the derivative of a specific function. Others, such as the addition rule: $D(f + g) = Df + Dg$ describe the derivatives of more complicated functions in terms of derivatives of simpler functions. The student of calculus is expected not just to learn the various formulas, but to understand the operation of the recursive algorithm, in which formulas of the first type cover the base cases and formulas of the second type correspond to recursive calls.

Tomorrow's calculus texts will express the derivative algorithm recursively in a (formal or informal) algorithmic language, as is done in programming texts such as [1]. When the algorithm assumes its rightful place as the *primary explicit structure* of differential calculus, there is a welcome gain in clarity, but also a profound shift of meaning. The proofs of the various derivative formulas are now part of the *verification* of the algorithm. They prove the *correctness* of an algorithm rather than the *truth* of a theorem.

2 Are All Functions Computable?

If we take algorithms and data structures to be fundamental, then it is natural to define and understand *functions* in these terms. The phrase "non-computable function" then becomes problematic, and the understanding which sees *almost all* functions as non-computable becomes mysterious. If a function does not correspond to an algorithm, what can it be? There is no higher court corresponding to the set theory of logical mathematics.

Since there are evident advantages of simplicity and unity in defining functions in terms of algorithms, we shall take the stand that functions are, by definition, computable, and then test those phenomena which are standardly taken as evidence for the existence of non-computable functions, to see if we need to yield any ground.

Given a putative function f, we do not ask "Is it computable?" but rather "What are the *data types* of the domain and of the range?" This question will often have more than one natural answer, and we will need to consider both restricted and expanded domain/range pairs. Distinguishing between these pairs will require that we reject excluded middle for undecidable propositions. If you attempt to pair an expanded domain for f with a restricted range, you will come to the conclusion that f is non-computable.

2.1 Background

Let N denote the set of positive integers (which we shall simply call *integers*). For concreteness we shall work mainly with functions from N to N, and denote the set of all such functions by N^N. Unless modified by 'partial' the term *function* denotes a total function (defined for all $n \in N$), and unless modified by 'non-computable' it refers to something given by an algorithm. Under the Church-Turing thesis, this algorithm may be represented by a Turing machine. The text [6] gives a very careful description of the way in which numerical functions are computed by Turing machines. We use the simple and intuitive terms *decidable* and *enumerable* instead of the more traditional 'recursive' and 'recursively enumerable.' A set of integers is *decidable* if there is an algorithm or Turing machine for deciding membership, and *enumerable* if there is an algorithm or Turing machine for listing its members.

The pioneering researches of Turing, Church, and others showed that the functions defined by Turing machines (or equivalent formalisms) are typically partial, and their domains are typically undecidable (because of the undecidability of the *halting problem*). They also concluded that functions are typically non-computable, on the grounds that Turing machines can be enumerated, while functions cannot. Later, specific examples of non-computable functions were found, most notably the *busy beaver function*. We will present another interpretation of the busy beaver phenomenem, based on careful attention to the data types of domain and range, in which the function is indeed computable. Then we will consider the Cantor diagonal algorithm and questions of cardinality.

2.2 The Busy Beaver Function

The busy beaver phenomenem concerns Turing machines (TMs) whose tape alphabet consists of a single non-blank symbol '*'. A *beaver* is a TM which, when started on a blank tape, halts and computes an integer, known as its *productivity*. Two conventions are commonly used for what counts as the computation of an integer. The more restrictive requires that the machine halt on the leftmost * of a contiguous block on an otherwise blank tape. The less restrictive requires only that the machine halt and takes as productivity either the number of *'s on the tape or the number of steps of the computation. A k-state beaver is *busy* if, among all TMs with k states, it has greatest productivity. It does not matter which convention is taken, beavers turn out to be extremely busy. Already Rado had proved the following [22]:

Theorem 2.1 Busy Beaver Theorem.
Let f be any (total) Turing-computable function. Then there is an integer n such that for all integers $k \geq n$ there is a k-state beaver with productivity greater than $f(k)$.

An extremely careful proof is given in Chapter 4 of the text [6]. It will be sketched in the next section in a slightly different context. If we define the *busy beaver function* bb by taking $bb(k)$ to be the maximum productivity of any k-state beaver, then the theorem shows that bb grows faster than any Turing-computable function. Hence, under the Church-Turing Thesis, it appears that bb is a non-computable function. (I prefer the alternative interpretation given in the next section.)

But Rado's theorem gives no hint of the extraordinary complexity of computations performed by extremely small machines. While k-state busy beavers have been found for $k \leq 4$, computer searches are continually finding busier and busier 5-state beavers. In recent months a 5-state machine which halts with 4,098 symbols on the tape after running for 23,554,760 steps has been announced![6] For descriptions of this work, we particularly recommend Brady's fascinating article [7] and the entertaining account in the *Scientific American* column of Dewdney [12].

2.3 Reinterpreting the Busy Beaver Function

The busy beaver function bb becomes computable when its domain and range are properly defined. When the domain is taken to be N, the range will be the set of 'weak integers,' a superset of N which we shall define shortly. The standard proof then demonstrates that bb grows faster than any *integer-valued* function.

To determine the proper data type for $bb(k)$, consider what can be done by a suitable universal Turing machine. Given an input k, the UTM can first enumerate all k-state TMs. Then it can proceed to execute each k-state TM for longer and longer periods, starting each from a blank tape. Whenever a beaver is found, its productivity is placed on an output tape. We obtain $bb(k)$ as an enumerable set of integers, of cardinality bounded by the (very large) number of TMs with states $\{1\ldots k\}$.

Hence we define a *weak integer* to be an enumerable set X of positive integers which contains at least one and at most B elements, for some integer B. Intuitively, a weak integer X is an approximation from below, and every element $x \in X$ establishes a lower bound. It is crucial to understand that while we are given a bound B on the number of elements in a weak integer X, we do not necessarily have any bound on the values of these elements. Hence we do not generally have access to the entire list, but only to the algorithm or TM which enumerates it. So, in concrete terms, a weak integer is an algorithm or Turing machine which enumerates a set of integers of bounded cardinality. Given k, the function bb produces such a TM $bb(k)$ by a finite process. But there is no reason to expect that the enumeration by $bb(k)$ of the productivities of k-state machines will ever halt.

The situation here is analogous to that of real numbers. When we speak of a real number x, we generally have in mind a Cauchy sequence of rationals. But when an algorithm produces a real number, what it actually delivers as x is an algorithm for computing as much of the Cauchy sequence as we may wish to see [19]. Similarly, when we speak of a weak integer X, we have in mind an enumerable set of integers. But when an algorithm produces a weak integer, what it actually delivers is an algorithm or Turing machine for enumerating as much of that set as we may wish to see.

Keeping in mind that weak integers approximate from below, it is natural to define equality and order on the collection of all weak integers as follows. For weak integers X and Y:

- $X \leq Y$ means $(\forall x \in X)(\exists y \in Y)(x \leq y)$

---

[6] Note that these are 5-tuple machines, which simultaneously print and move. Other authors, like [6], work with 4-tuple machines which can either move or print (but not both at once). A 5-tuple machine with 5 states will generally convert to a 4-tuple machine with 8 or 9 states.

- $X = Y$ means $(X \leq Y) \wedge (Y \leq X)$
- $X < Y$ means $(\exists y \in Y)(\forall x \in X)(x < y)$

By associating each integer n with the singleton set n the integers become a subset of the weak integers. Clearly a weak integer X equals an integer x if and only if x is the maximum element of X).

Hence there are really two busy beaver functions. Rather than extend the range to **W**, the set of weak integers, we can shrink the domain to **D**, the set of integers at which bb takes integer values (**D** contains at least $\{1, \ldots, 4\}$).

$$\begin{array}{ccc} \mathbf{N} & \xrightarrow{bb} & \mathbf{W} \\ \cup & & \cup \\ \mathbf{D} & \xrightarrow{bb} & \mathbf{N} \end{array}$$

In this context, the usual arguments [6] now prove:

Theorem 2.2 The Busy Beaver Theorem.
*Let f be any total Turing-computable function from **N** to **N**. Then there is an integer n such that for all $k \geq n$*

$$bb(k) > f(k)$$

That is, the busy beaver function grows faster than any total *integer-valued* function.

> **Proof** Let f be a total, integer-valued function which is computed by a Turing machine T with n_T states. It is no loss of generality to assume that the function f is strictly increasing. While we don't generally know the busiest beaver, we do know that beavers can be very busy indeed, and this easily allows us to find a beaver M with n_M states that computes a number m larger than $n_M + n_T$. The composite machine MT, which has $n = n_T + n_M$ states, computes $f(m)$ from an empty tape, so $bb(n) \geq f(m) > f(n)$. Further, it is easily seen that this can be done for all $k \geq n$. □

This theorem confirms our intuition that the complexity of a computation is incomparably more sensitively linked to the size of the machine than to the size of the input. There can be no universal total machine which computes all (total) functions from **N** to **N**; no single machine can keep up with a sequence of ever larger machines. Note that, since we do have a universal machine for partial functions, this implies the unsolvability of the halting problem, for if we could decide the halting problem then we could carry out a brute force computation of $bb(k)$ as an integer by running all machines with k states which halt when started on a blank tape.

The theorem also shows that we can obtain faster growing functions by relaxing the data type of the range. Functions to the weak integers can grow faster than functions to the integers. Hence the weak integers cannot be identified with the integers, and this interpretation requires that we use intuitionistic rather than classical logic. Of course there are other good algorithmic reasons for preferring such a logic [9,8]. For a general discussion of intuitionistic extensions, see [25].

2.4 The Diagonal Algorithm

We shall consider the Cantor diagonal method (which we will call the diagonal algorithm) in the context of the set $\mathbf{N}^\mathbf{N}$ of all integer valued functions. The algorithm takes as input a sequence of such functions $\{F_k(n)\}$ and produces as output a function G different from each F_k. It was Cantor's genius to notice that this is achieved if we simply 'go down the diagonal' and construct G by a simple rule such as

$$G(n) = F_n(n) + 1.$$

As long as the functions F_k are total, this procedure is wholly algorithmic, and one implication is similar to that drawn from busy beavers: there does not exist a universal total machine. A single fixed Turing machine which takes two integer inputs k and n cannot, by fixing one input, imitate the behavior of an arbitrary machine which takes one integer input, when all functions are required to be total (which is hardly surprising, since Cantor also showed that two integers are really no better than one).

The diagonal algorithm is commonly used to point to the existence of non-computable functions in two different ways. It is used to directly construct specific non-computable functions. Used indirectly, it is the source of the theory of infinite cardinal numbers, which seems to imply that almost all functions are non-computable. We shall examine the direct argument here and briefly consider cardinality in the next section. The direct construction of a non-computable function by the diagonal algorithm is carried out with great care in Chapter 5 of [6]. The basic idea is very simple. The collection of all Turing machines which compute partial functions is arranged in a list $\{F_k\}$ so that $F_k(n)$ represents the value computed by the k-th TM when given input n. However, because the functions are partial, we must modify the diagonal algorithm:

$$G(n) = \begin{cases} 1 & \text{if } F_n(n) \text{ is undefined} \\ F_n(n) + 1 & \text{otherwise} \end{cases}$$

Certainly G is a function distinct from every Turing computable function, and it is very tempting to say that $G(n)$ is an integer, since it is an integer if $F_n(n)$ is defined or is undefined. Using excluded middle, we could conclude that logically it *must* be an integer. On the other hand, it is certainly not an integer in any computational sense, since we have no general way of finding its value. (Indeed, here lurks another proof of the undecidability of the halting problem). So the question arises, what is the data type of $G(n)$? Again, we must describe the proper superset of N. {*Warning*: this definition will seem artificial and even paradoxical to those unused to intuitionistic logic. But the artificiality really resides in the application of the diagonal algorithm to a sequence of partial functions.}

A *pseudo-integer* is a set X of integers satisfying:

- X contains at most one integer (i.e. if $x \in X$ and $y \in X$, then $x = y$),
- X is non-empty (in the sense that it is contradictory that $X = \emptyset$).

Let **P** denote the set of all pseudo-integers. If we identify each integer m with the singleton set $\{m\}$, then $\mathbf{N} \subset \mathbf{P}$, and G, as defined by the diagonal algorithm, is a (computable) function from **N** to **P**. If we want G to take integer values, then the domain must be restricted to the set of integers n for which $F_n(n)$ is either defined or undefined. This set, while it has empty complement, cannot be identified with **N** (again a paradoxical situation for those unused to the algorithmic niceties of intuitionistic logic). We should emphasize that, unlike weak integers, pseudo-integers are not in general enumerable. Indeed, an enumerable pseudo-integer is very close to being an ordinary integer. We can easily prove that an enumerable pseudo-integer equals an integer if we assume *Markov's Principle* ([5], page 63).

2.5 Cardinality as Shape.

Cantor argued that the diagonal algorithm showed that the set $\mathbf{N}^\mathbf{N}$ was larger than the set **N** of natural numbers. It is this last step which introduces into mathematics a supposed universe of non-algorithmic functions. We might well wonder how so simple an algorithm could transcend the computable.[7] Cantor did indeed show that there is a fundamental difference between the sets $\mathbf{N}^\mathbf{N}$ and **N**, but this difference can be understood not as a quantitative difference, but as a difference of quality or structure. Rather than call the set $\mathbf{N}^\mathbf{N}$ uncountable, it might better be called *productive*, because there are very powerful methods for producing elements of $\mathbf{N}^\mathbf{N}$, in particular for producing an element outside of any given sequence in $\mathbf{N}^\mathbf{N}$ [17]. This use of the term productive, taken from recursive function theory, grounds our understanding in algorithmic reality rather than idealistic fantasy.[8] Given any sequence of elements of $\mathbf{N}^\mathbf{N}$, we can find an element of $\mathbf{N}^\mathbf{N}$ outside the sequence, not because there are more functions than integers, but because of the structure of $\mathbf{N}^\mathbf{N}$.

It follows, of course, that there is no surjection from **N** to $\mathbf{N}^\mathbf{N}$. Let $\mathbf{N}^\mathbf{N}_{par}$ denote the set of all partial binary functions on **N**, with intensional equality. Then, under the assumptions of the Church-Turing Thesis, the set of all partial functions is just the set of Turing machines, which can be listed, so there is indeed a bijection from **N** to $\mathbf{N}^\mathbf{N}_{par}$. Since $\mathbf{N}^\mathbf{N}$ is a subset of $\mathbf{N}^\mathbf{N}_{par}$, this might be considered as evidence that **N** is *larger* than $\mathbf{N}^\mathbf{N}$, were one inclined to make a quantitative comparison

---

[7] This point was perhaps first made by Wittgenstein, who wrote of the diagonal algorithm: "Our suspicion ought always to be aroused when a proof proves more than its means allow it. Something of this sort might be called a 'puffed-up proof'." [27]

[8] The fantasy here is not the theory of sets, as elegantly elaborated in the concrete confines of Zermelo Frankel set theory, but rather the notion that, somewhere, there are really 'more' elements of $\mathbf{N}^\mathbf{N}$ than of **N**. It is worth remembering that Cantor lobbied the Vatican to recognize that the higher cardinals pointed the way to God [11,18].

between them. Cantor, and most mathematicians after him, considered sets as 'mere collections of elements,' which could differ only in quantity. We do not find this position algorithmically intelligible, since the extra structure of N^N plays an essential role in the diagonal algorithm.

Perhaps shape is a better metaphor than size for the difference between N and N^N which is revealed by the diagonal algorithm. As shown in Figure 1, we naturally have

$$N \subset N^N \subset N^N_{par}$$

but N and N^N_{par} have the same shape, which is distinct from that of N^N.

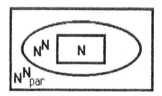

Figure 1.

In our discussions of the busy beaver function and the diagonal algorithm we produced demonstrations of the undecidability of the halting problem. The means that there is no halting function, if by that we mean a binary function f on N such that the n-th Turing machine halts when given a blank tape if and only if $f(n) = 1$. However, there are two obvious halting functions, obtained either by restricting the domain to consist of all Turing machines for which the halting problem is decidable, or expanding the range to a set Ω of intuitionistic truth values [14]. The set Ω was, of course, devised to extend the truth function to all propositions. So the idea of extending the range is not wholly new, though we have not seen it applied to the busy beaver function and the functions produced by the diagonal algorithm.

3 Conclusions

The computer has supplied mathematics with a new central concept, the *algorithm*, and with new languages for algorithms, indeed with a new algorithmic view of the nature of mathematics.

It is often felt that the existence of non-computable functions shows that mathematics necessarily transcends the algorithmic. I have tried to show that this is not so, that the phenomena standardly connected with non-computability can better be understood in purely algorithmic terms.

I am not at all able to propose a foundations or codification of algorithmic mathematics. As Tseytin has argued, we are just beginning to appreciate the implications of thinking algorithmically [26].

Thanks to Matthew Kamerman for his insistent curiosity about busy beavers and for suggesting *shape* as a good metaphor for cardinality, and to Jonathan Nash for bringing my attention to Wittgenstein's remarks about the diagonal algorithm.

References

[1] Abelson, H. and Sussman, G. J., *Structure and Interpretation of Computer Programs*, M. I. T. Press, 1984

[2] Apt, K. R., Ten years of Hoare's logic: a survey', *ACM Transactions on Programming Languages and Systems*, 431-483, 1985

[3] Beeson, M., 'Computerizing mathematics: logic and computation' in Herken, R. (ed) *The Universal Turing Machine: a Half-Century Survey* Oxford University Press, pages 191-225, 1988

[4] Bishop, E., *Foundations of Constructive Analysis*, McGraw-Hill, 1967 ([5] is a new edition.)

[5] Bishop, E. and Bridges, D., *Constructive Analysis*, Springer-Verlag, 1985, (revised edition of [4])

[6] Boolos, G. S., and Jeffrey, R. C., *Computability and Logic*, Cambridge University Press, 1980

[7] Brady, A. H., 'The busy beaver game and the meaning of life', in *The Universal Turing Machine: a Half-Century Survey*, Herken, R. (ed), Oxford University Press, pages 259-278, 1988

[8] Clarke, M. R. B. and D. M Gabbay, 'An intuitionistic basis for non-monotonic reasoning', *Non-Standard Logics for Automated Reasoning*, Smets, P. et. al (eds), Academic Press, 1988

[9] Constable R. L. et al., *Implementing Mathematics with the Nuperl Proof Development System*, Prentice-Hall', 1986

[10] Constable, R. L., 'Programs as proofs: a synopsis', *Information Processing Letters*, **16**, 105-112, 1983.

[11] Dauben, J. W., , *Georg Cantor: His Mathematics and Philosophy of the Infinite*, Harvard University Press', 1979

[12] Dewdney A. K., 'Computer Recreations', *Scientific American*, **252**, 20-30, April 1984 Reprinted in *The Armchair Universe*, Freeman, 160-171, 1988

[13] Dijkstra, E. W., 'Mathematicians and computing scientists: the cultural gap', *The Mathematical Intelligencer*, **8**, 48-52, 1986, Reprinted in *Abacus*, Summer 1987, 26-31

[14] Goldblatt, R., *Topoi: the Categorial Analysis of Logic*, North-Holland', 1979

[15] Grabiner, J. V., 'Is mathematical truth time-dependent?', *American Mathematical Monthly*, **81**, 354-365, 1974. Reprinted in *New Directions in the Philosophy of Mathematics*, T. Tymoczko (ed), Birkhauser, 201-213, 1986

[16] Greenleaf, N., 'Algorithms and proofs: mathematics in the computing curriculum', *ACM SIGCSE Bulletin*, **21**, 268-272, 1989

[17] Greenleaf, N., 'Liberal constructive set theory', in *Constructive Mathematics*, Richman, F. (ed), Springer Lecture Notes in Mathematics, **873**, 213-240, 1981

[18] Hallett, M., *Cantorian Set Theory and Limitation of Size*, Oxford University Press', 1984.

[19] Knuth, D. E., 'Algorithms in modern mathematics and computer science', *Algorithms in Modern Mathematics and Computer Science*, Ershov, A. P. and Knuth, D. E. (eds), 82-99, 1981 Springer Lecture Notes in Computer Science, **122**, Revised (from ALGOL to Pascal) and reprinted in *American Mathematical Monthly*, **92**,170-181, 1985

[20] T. S. Kuhn, *The Structure of Scientific Revolutions*, 'University of Chicago Press', 1970.

[21] Martin-Löf, P., 'Constructive mathematics and computer programming', in *Sixth International Congress for Logic, Methodology, and Philosophy of Science*, Cohen, L. J. et al. (eds), North-Holland, 1982. Reprinted in *Mathematical Logic and Programming Languages*, Hoare, C. A. R. and Shepherdson, J. C. (eds.), Prentice-Hall, 1986.

[22] Rado, T., 'On non-computable functions', *Bell Sys. Tech. Journal*, 887-884, 1962

[23] Reid, C., *Hilbert*, Springer-Verlag, 1970

[24] Smets, P., Mamdani, A., Dubois, D., and Prade, H. (eds), *Non-Standard Logics for Automated Reasoning* Academic Press', 1988

[25] Troelstra, A. S., 'Intuitionistic extensions of the reals', *Nieuw Arch. Wisk*, **28**, 63-113, 1980.

[26] Tseytin, G. S., 'From logicism to proceduralism (an autobiographical account)' in *Algorithms in Modern Mathematics and Computer Science*, Ershov, A. P. and Knuth, D. E. (eds), Springer Lecture Notes in Computer Science, **122**, 390-396, 1981.

[27] Wittgenstein, L., *Remarks on the Foundations of Mathematics*, Basil Blackwell, (Translated by G. E. M. Anscombe), 1956

The Parallel Computation Hypothesis and its Applications to Computer Science

V.J. Rayward-Smith
School of Information Systems.
University of East Anglia, Norwich, NR4 7TJ

Abstract

The class of NP-complete problems has been one of the most fruitful contributions of mathematical reasoning to the understanding of the complexity of problems. We briefly survey these results but our major area of interest in this paper is the development of a theory which isolates a subclass of P containing problems for which we do not expect a dramatic speed-up on parallel machines. We describe a common model for parallel computations known as the P-RAM and relate this model to current practical machines. We present some of the algorithms designed for a P-RAM. The parallel computation hypothesis asserts that the class of problems solvable in $f(n)^{O(1)}$ time by a P-RAM is equal to the class of problems solvable in $f(n)^{O(1)}$ workspace by a RAM. From this assertion we construct the desired class of problems viz the class log space complete for P. We present some of the problems in this class. The paper provides an illustration of how the mathematics of complexity classes has evolved to respond to the use of parallelism by computer scientists.

1 Introduction

Of fundamental interest to any mathematician must be the efficiency (or otherwise) of the algorithms (s)he uses. Techniques now exist which enable algorithms to be compared within a rigorous framework. Further, the theory of complexity enables us to isolate problems with similar characteristics and hence to isolate problems for which no really efficient algorithm can exist. This paper reviews this exciting area, paying particular attention to parallel algorithms. First, however, we review some standard notation used to describe algorithm performance.

The worst case time complexity function, $t_A(n)$, for an algorithm, A, determines the maximum amount of time that algorithm A can spend on processing an input of length n. This is machine dependent up to some constant multiplicative or additive factor in $t_A(n)$. To overcome this problem we introduce (big) O-notation.

Definition 1 If $f(n), g(n)$ are positive valued we say $f(n)$ is $O(g(n))$ iff there exist $c > 0$ and $N \geq 0$ such that $f(n) \leq cg(n)$ for all $n \geq N$.

Computer scientists have used this notation extensively in algorithm analysis, [1][16]. For example, to sort n integers stored in computer memory, sorts such as selection-sort and bubble-sort are $O(n^2)$ but heapsort is $O(n \log n)$. The commonly used algorithm, Quicksort, is $O(n^2)$ but it can be shown to have average time complexity of $O(n \log n)$; moreover, it will generally outperform heapsort. In this paper, however, we will restrict our attention to worst case performance. As well as O-notation, computer scientists also use Ω- and θ-notation.

Definition 2 If $f(n), g(n)$ are positive valued, we say that $f(n)$ is $\Omega(g(n))$ iff there exist $c > 0$, and $N \geq 0$ such that $f(n) \geq cg(n)$ for all $n \geq N$. $f(n)$ is $\theta(g(n))$ iff $f(n)$ is both $O(g(n))$ and $\Omega(g(n))$.

Definition 3

(a) An algorithm runs in *polynomial time* iff $t_A(n)$ is $O(p(n))$ for some polynomial, p.

(b) An algorithm runs in *exponential time* iff $t_A(n)$ is $\theta(k^n)$ for some $k > 1$.

(c) An algorithm runs in *polylogarithmic time* iff $t_A(n)$ is $O(p(\log(n)))$ for some polynomial, p, and some logarithmic base.

The vast majority of algorithms in use are polynomial-time algorithms since exponential-time algorithms rapidly become impractical as n grows. For example, if $t_A(n) = n^5$ and $t_B(n) = 2^n$ then if A takes a quarter of a second to execute with $n = 60$, B can be expected to require over 300 centuries of computer time. Computer scientists have long been aware of the need for efficient, polynomial-time algorithms but they are sometimes unable to find them, e.g. for the travelling salesman problem (TSP). The theory of NP-completeness which we briefly review in section two provides an equivalence class of problems such that if any one of these problems can be solved in polynomial time then they can all be solved in polynomial time. Currently this equivalence class contains several thousand known problems, many of which have practical application but none of which have polynomial-time algorithms. Researchers in complexity theory have also been able to define other classes of problems which exhibit similar characteristics, e.g. space requirements. For a gentle introduction to complexity theory, the reader is referred to [29]; a more thorough introduction is given in [10].

A recent hope is that a parallel architecture might yield significant speed-up for a large class of important problems. If the best sequential algorithm for solving some problem, Π, is A, with worst case time complexity function, $t_A(n)$, then with m parallel processors, we might optimistically hope to solve P with an algorithm of worst case time complexity, $t_A(n)/m$. We might even aim for a dramatic speed-up from polynomial to polylogarithmic time given enough processors. Some problems with polynomial time sequential algorithms yield to a dramatic speed-up but others seem to require an essentially sequential approach and parallelism offers no advantage. We discuss parallel algorithms in section three and the contribution of complexity theory in section four.

2 The Theory of NP-completeness

The theory of NP-completeness is traditionally presented in terms of Turing machines and is limited to decision problems, i.e. problems with YES/NO answers. The familiar Turing machine [34] provides a suitable model for a computing device. Alternatively, we could use a random access machine (RAM), derived from an unlimited register ideal machine [32], which consists of a read-only input tape, a write-only output tape, a program and a memory in the form of a sequence of registers, each of which is capable of holding an integer of arbitrary size. Worst case time complexity functions for an algorithm implemented on a RAM are polynomially related to that for a Turing machine implementation, see, for example [1]. The Turing machine is simpler to use in formal proofs although the RAM is easier to program and is closer to current sequential computers.

The restriction of our theory to decision problems is not as severe as may first appear. Optimization problems such as TSP have corresponding decision problems constructed using a bound value. Thus, instead of asking for the cheapest Hamiltonian cycle, we ask whether there is a cycle of cost less than some given value. If the decision problem cannot be solved in polynomial time, then this must also be true of the original problem.

If Π is a decision problem then its *domain*, D_Π, consists of those instances of that problem. A subset $Y_\Pi \subseteq D_\Pi$ consists of those instances of Π that have YES solutions. We denote the set of decision problems solvable in polynomial time on a Turing machine by P.

A nondeterministic Turing machine allows an arbitrary choice of move from certain configurations. A nondeterministic Turing machine solves a decision problem provided: (a) the input is in Y_Π implies there must be some choice of moves to result in a YES output, and (b) the input is not in Y_Π implies no choice of moves can result in a YES output. The set of decision problems solvable in polynomial time on a non-deterministic machine is denoted by NP. It is perhaps more convenient to view this class of problems as those for which a "guessed" solution can be checked out in polynomial time. Thus TSP $\in$ NP since we could guess an ordering of all the vertices and then, in polynomial time, check if this ordering corresponded to a cycle with total weight less than the given bound.

Clearly, P $\subseteq$ NP. The whole theory of NP-completeness is based upon the very reasonable hypothesis that P $\neq$ NP.

Definition 4 If Π_1 and Π_2 are decison problems, we say there is a *polynomial transformation* from Π_1 to Π_2, denoted by $\Pi_1 \propto \Pi_2$, iff there exists $f : D_{\Pi_1} \to D_{\Pi_2}$ such that

(i) f is computable on a deterministic Turing machine in polynomial time, and

(ii) $I \in Y_{\Pi_1}$ iff $f(I) \in Y_{\Pi_2}$.

Definition 5 A decision problem Π is *NP-complete* iff

(i) $\Pi \in$ NP, and

(ii) for all $\Pi' \in$ NP, $\Pi' \propto \Pi$.

An immediate consequence is that an NP-complete problem can only be in P if P = NP. The set of all NP-complete decision problems is denoted by NPC.

Cook [5] determined the following decision problem was NP-complete.

SATISFIABILITY (SAT)
INSTANCE: A set, U, of variables and a collection, C, of clauses over U
  (i.e. a collection of subsets of variables and their negations)
QUESTION: Is there a satisfying truth assignment for C?

The proof is based upon a clever construction of a set of clauses, C, from an arbitrary polynomial time, non-deterministic Turing machine, M, and input, x, such that C is satisfiable iff M accepts the input string x.

Proposition 1 If $\Pi \in$ NP and $\Pi' \in$ NPC, then $\Pi' \propto \Pi$ implies $\Pi \in$ NPC.

This result follows easily from the definitions and provides us a useful technique for establishing NP-completeness. From the initial, seminal result concerning SAT, a large number of NP-complete problems has resulted. An initial listing of these problems was collected in [10] and has since been regularly updated in the *Journal of Algorithms*. If any of these NP-complete problems yield to a polynomial time algorithm then they all yield to such an algorithm and P = NP.

We can define an equivalence relation on decision problems by defining Π_1 to be *polynomially equivalent* to Π_2 iff $\Pi_1 \propto \Pi_2$ and $\Pi_2 \propto \Pi_1$. Under this equivalence relation, P and NPC are equivalence classes. P is the 'easiest' such equivalence class in NP, NPC is the 'hardest' such equivalence class in NP. There are, in fact, infinitely many such equivalence classes but most practical problems that have arisen have tended to lie in P or NPC. GRAPH ISOMORPHISM is a possible candidate for being in neither P nor NPC and there are several others.

Another important complexity class is PSPACE, the class of decision problems solvable using workspace of $O(p(n))$ for some polynomial, p, where n is the length of the input string. Clearly P $\subseteq$ PSPACE and NP $\subseteq$ PSPACE but it is widely assumed that P $\neq$ PSPACE. If P were equal to PSPACE, it would follow that P = NP.

Definition 6 A decision problem, Π, is said to be PSPACE-complete iff

(i) $\Pi \in$ PSPACE, and

(ii) for all $\Pi' \in$ NP, $\Pi' \propto \Pi$.

Thus saying that a problem is PSPACE-complete is an even stronger indication that its algorithms will not run in polynomial time than saying it is NP-complete. The first PSPACE-complete problem discovered was QBF [33].

QUANTIFIED BOOLEAN FORMULAE (QBF)
INSTANCE: A well-formed quantified Boolean formulae
$F = (Q_1 x_1)(Q_2 x_2) \ldots (Q_n x_n)E$, where E is a Boolean expression over the variables $x_1, x_2, \ldots, x_n$ and each Q_i is either $\forall$ or $\exists$
QUESTION: Is F true?

From this discovery, more PSPACE-complete problems have been derived but the class has not yet been shown to contain as many important practical problems as NPC.

3 Parallel Processors and Algorithms

There are essentially two types of parallel computers: SIMD (single instruction stream, multiple data stream), and MIMD (multiple instruction stream, multiple data stream). With SIMD machines, a single stream of instructions is executed by multiple processors on different data. Hence at any one moment in time, all processors are executing the same instruction but on different data.

A useful, abstract parallel machine called a P-RAM was proposed in [35]. This models an SIMD machine comprising any number of processes with a shared memory where simultaneous read access is allowed but simultaneous write access is not. This model does not correspond directly to any practical SIMD machine since real world machines do not generally have shared memory. Nevertheless, the P-RAM is the model of a parallel machine used in complexity theory; if significant speed-up on a P-RAM is unlikely then significant speed-up on a real world SIMD machine must also be unlikely.

In an MIMD machine each processor can either have a shared memory accessed via some switching mechanism or each have their own memory and interconnect via message passing. The former type of machine is sometimes called a *multiprocessor* whilst the latter is called a *multicomputer*.

For a fuller discussion of parallel computers and an initial taxonomy, the interested reader is referred to [28].

The design and implementation of parallel algorithms on a wide range of parallel machines has been concerning many computer science researchers over the last few years. Algorithms for the P-RAM have been proposed for a wide range of problems including some of those found in linear algebra [28], combinatorics [8], sorting [27], scheduling [8], and graph theory [4] [7] [20] [21] [23] [31]. There are also many algorithms and strategies for the various practical machines [14] [17] [24] [28] [30].

4 The Parallel Computation Hypothesis

Some of the algorithms for P-RAMs will run in polylogarithmic time provided sufficient processors are available. For example, an efficient sequential algorithm to find the maximum of n unsorted numbers will be $O(n)$ but a data flow technique can produce an $O(\log n)$ parallel algorithm. Similarly, some algorithms cited above are polylogarithmic whilst the sequential versions are $\Omega(p(n))$. Hence, for certain problems there is no doubt that parallelism is offering significant and quite dramatic speed-up. However, there appear to be some problems which require an essentially sequential approach and for which parallelism does not offer such dramatic speed-up. In this section, we describe an attempt by mathematicians working in complexity theory to isolate some class of problems which appears to capture this 'essentially sequential' property.

One important subclass of P is DLOGSPACE. This is the set of all decision problems solvable using a space bound of $\lceil \log_2 n + 1 \rceil$ where n is the length of the input string. DLOGSPACE is unchanged if the space bound is replaced by $c\lceil \log_2 n + 1 \rceil$ for any $c > 0$. DLOGSPACE $\subseteq$ P but it appears to be true that DLOGSPACE $\neq$ P [10]. DLOG^k_SPACE ($k > 1$) is defined analogously to DLOGSPACE except that the space bound is $\lceil \log_2 n+1 \rceil^k$. Then POLYLOGSPACE $= \bigcup_{k=1}^{\infty}$ DLOG^k_SPACE. It is very unlikely that POLYLOGSPACE $\subseteq$ P and it has been proved that P $\neq$ POLYLOGSPACE and that NP $\neq$ POLYLOGSPACE [3].

The *parallel computation hypothesis* asserts that the class of problems solvable in $T(n)^{O(1)}$ time by a P-RAM is equal to the class of problems solvable in $T(n)^{O(1)}$ workspace by a RAM [12]. Therefore, the decision problems in P solvable in polylogarithmic time on a P-RAM are precisely those problems in POLYLOGSPACE $\cap$ P.

Although the hypothesis has not been proved, it has been shown to hold if $T(n) = n^{O(1)}$ and for several other special cases. The polylogarithmic case remains a hypothesis, however. Just as the theory of NP-completeness is based upon the hypothesis that P $\neq$ NP, so it is on the basis of this hypothesis and the assumption that POLYLOGSPACE is unlikely to be a superset of P, that the following definitions are made.

Definition 7 Given two decision problems, $\Pi_1, \Pi_2 \in$ P, a *log-space transformation* $\Pi_1 \propto_{\log} \Pi_2$ is any function, $f : D_{\Pi_1} \to D_{\Pi_2}$ such that

(i) f can be computed by a deterministic Turing machine using space bounded by $\lceil \log_2 n + 1) \rceil$ for all input strings of length, n, and

(ii) $I \in Y_{\Pi_1}$ iff $f(I) \in Y_{\Pi_2}$.

Definition 8 A decision problem, Π, is said to be *log-space complete* for P iff

(i) $\Pi \in P$, and

(ii) for all $\Pi' \in P$, $\Pi' \propto_{\log} \Pi$.

Thus, if Π is log-space complete for P, then $\Pi \in \text{POLYLOGSPACE} \Rightarrow P \subseteq \text{POLYLOGSPACE}$. This suggests that any problem that is log-space complete for P is very unlikely to be in POLYLOGSPACE and hence very unlikely to yield to a polylogarithmic algorithm on a P-RAM.

The computer scientist's interests in parallelism has motivated an increasing research effort into the discovery of problems that are log-space complete for P. Two other complexity classes have also been defined which are of interest in this context. NC (Nick Pippenger's class) contains all problems *solvable in polylogarithmic time on a polynomial number of processors* and SC (Steve Cook's class) [6] contains all problems *solvable in polynomial time and polylogarithmic space on one processor*. Thus $\text{SC} = P \cap \text{POLYLOGSPACE}$ and this possibly equals NC [18].

The first problem shown to be log-space complete for P was identified by Cook:

PATH SYSTEM ACCESSIBILITY (PSA)
INSTANCE: A finite set, X, of nodes, a relation $R \subseteq X \times X \times X$ and
two sets $S, T \subseteq X$ of 'source' and 'terminal' nodes.
QUESTION: Is there an 'accessible' terminal node, where a node $x \in X$ is *accessible* if $x \in S$ or if there exist accessible nodes y, z such that $(x, y, z) \in R$?

To prove this problem is log-space complete for P, a log-space transformation from any $\Pi \in P$ to PSA is required. A similar proof of a seminal problem which is log-space complete for P is given in [19] for unit resolution.

Proposition 2 If Π is log-space complete for P and $\Pi' \in P$ is such that $\Pi \propto_{\log} \Pi'$ then Π' is log-space complete for P.

This result follows easily from the definitions and provides us a practical way to prove a problem log-space complete for P; all we need do is to exhibit a log-space transformation from a problem known to be log-space complete for P.

The number of problems known to be log-space complete for P is growing rapidly. Notable examples are the circuit value problem [11] [13] [26], linear programming [9], maximum flow [13], and a number of problems arising from the use of TSP heuristics [25].

5 Summary

The theory of NP-completeness differentiated between problems which would yield to polynomial time sequential algorithms and those which would (probably) not. The theory developed by mathematicians and theoretical computer scientists has been of enormous value to the more practical computer scientist because this enables the latter to explain why many real world problems are not easily solved. Once a problem has been shown to be NP-complete, it is easier to convince a customer that some approximate solution that can be found quickly and efficiently should be accepted. The vast majority of problems in P yield to algorithms of $O(n^k)$ where k is small, say less than 6.

Complexity theory is not yet providing us with a tool to distinguish between a problem which will yield to an $O(n^k)$ algorithm and another that would require a higher degree polynomial algorithm. This is a possible research area but the performance gulf between exponential time algorithms and (low degree) polynomial time algorithms for even quite moderate sized problems guarantees the success of the current theory and limits the demand for further refinement.

The complexity theory described in Section 4 has been motivated by the practical developments in parallel architectures described in section three. Unfortunately, the gulf between the developed theory and the real world practice is rather large. As explained, the P-RAM does not correspond to any practical machine although it does have some properties of an SIMD machine. Certainly, any problem log-space complete for P will not yield to dramatic speed-up on such machines. It is not clear how these complexity theory results apply to MIMD machines and, even if they do, whether they could possibly be of any practical relevance. This is an important issue and the apparent gulf between theory and practice needs to be closed.

It is important to emphasise that the current theory concentrates on the (very) dramatic speed-up from polynomial to polylogarithmic time. If such speed-up is to be achieved, the machine will require a polynomially large number of processors. In practice, one might be quite pleased to reduce time less dramatically and use less processors. Thus a programmer with an MIMD machine such as a transputer rack (say with 16 transputers) may well be pleased to get moderate speed-ups (say of over ten) and may be even satisfied with less. Such a person could well achieve such moderate speed-up for problems that are log-space complete for P. Linear programming is a case in point; a successful, practical parallel algorithm has been implemented on a transputer rack [2]. Thus linear programming contains some elements of parallelism; it is not a purely sequential problem. Current complexity theory can offer no way of capturing such undramatic parallelism.

References

[1] Aho, A.V. Hopcroft, J.E. and Ullman, J.D., (1974). *The design and analysis of computer algorithms*, Addison-Wesley, (Reading, Mass)

[2] Boffey, T. B. and Hay,R., (1988). 'Implementing Parallel Simplex Algorithms', *Proceedings of CONPAR88*, British Computer Society, 137-146.

[3] Book, R.V., (1976) 'Translation lemmas, polynomial time and $(\log n)^j$-space. *Theor. Comput. Sci.*, 1, 215-226.

[4] Chin, F.Y. Lam, J. and Chen I.-N., (1982) 'Efficient parallel algorithms for some graph problems', *Comm. ACM* 25, No.9, 659-665.

[5] Cook, S.A., (1971). 'The complexity of theorem-proving procedures', *Proc. 3rd. Ann. ACM. Symp. on Theory of Computing*, ACM, (New York) 151-158.

[6] Cook, S.A., (1981). 'Towards a complexity theory of synchronous parallel computation', *Enseign Math (2)* **27**, 99-124.

[7] Dekel, E. Nassimi, D. and Sahni, S. (1981). Parallel matrix and graph algorithms. SIAM J.Comp., Vol.10, pp.657-675.

[8] Dekel, E. and Sahni, S. (1983). Binary trees and parallel scheduling algorithms. IEEE Trans.Comput., C-32, pp.307-315, (1983).

[9] Dobkin, D. Lipton, R.J. and Reiss, S. (1979). Linear programming is log-space hard for P. Inform.Proc.Lett., Vol.8, pp.96-97.

[10] Garey, M.R. and Johnson, D.S. (1979). Computers and intractability: a guide to the theory of NP-completeness, W.H. Freeman, San Francisco.

[11] Goldschlager, L.M. (1977). "The monotone and planar circuit value problems are log-space complete for P", SIGACT News, 9.2, pp.25-29.

[12] Goldschlager, L.M. (1982). A universal connection pattern for parallel computers. Journal ACM, Vol. 29, pp.1073-1086.

[13] Goldschlager, L.M. Shaw, R.A. and Staples, J. (1982). The maximum flow problem is log space complete for P. Theor.Comp.Sci., Vol.21, pp.105-111.

[14] Hockney, R.W. and Jesshope, C.R. Parallel computers: architectures, programming and algorithms, Hilger, Bristol.

[15] Hopcroft, J.E. and Ullman, J.D. (1969). Formal languages and their relation to automata, Addison-Wesley.

[16] Horowitz, E. and Sahni, S. (1978). Fundamentals of Computer Algorithms, Computer Science Press,Rockville,Maryland.

[17] Hwang, K. and Briggs, F.A. (1984). Computer Architecture and Parallel Processing, McGraw-Hill, New York.

[18] Johnson, D.S. (1983). The NP-completeness column: an ongoing guide (7th edn).J. Algorithms, Vol.4, pp.189-203.

[19] Jones, N.D. and Laaser, W.T. (1976). Complete problems for deterministic polynomial time.Theoretical Comput. Sci., Vol.3, pp.105-117.

[20] Karchmer, M. and Naor, J. (1988). A fast parallel algorithm to color a graph with D colors, J. Algorithms, Vol.9, pp.83-91.

[21] Karloff, H.J. and Shmoys, D.B. (1987). Efficient parallel algorithms for edge coloring problems. J. Algorithms, Vol.8, pp.39-52.

[22] Karp, R.M. (1972). Reducibility among combinatorial problems. In Complexity of Computer Computations (eds. R.E. Miller and J.W. Thatcher), pp.85-104.

[23] Karp, R.M. Upfal, E. and Wigderson, A. (1986). Constructing a perfect matching is in random NC. *Combinatorica*, Vol.6, pp.35-48.

[24] Kindervater, G.A.P. and Lenstra, J.K. (1986). An introduction to parallelism in combinatorial optimization. Disc. App. Math., Vol.14, pp.135-156.

[25] Kindervater, G.A.P. and Lenstra, J.K. (1986). The parallel complexity of TSP heuristics. Report OS-R8609, Centre for Mathematics and Computer Science, Amsterdam.

[26] Ladner, R.E. (1975). The circuit value problem is log space complete for P. SIGACT News, Vol.7, No.1, pp.18-20.

[27] Muller, D.E. and Preparata, F.P. (1975). "Bounds of Complexities of Networks for Sorting and Switching", JACM, Vol.22, pp.195-201.

[28] Quinn, M. J. (1987). Designing efficient algorithms for parallel computers.. McGraw Hill, New York.

[29] Rayward-Smith, V.J. (1986). A first course in computability, Blackwell Scientific.

[30] Rayward-Smith, V.J. McKeown, G.P. and Burton, F.W. (1988). The general problem solving algorithm and its implementation. New Generation Computing, Vol.6, pp.41-66.

[31] Savage, C. and Ja' Ja', J. (1981). Fast efficient parallel algorithms for some graph problems. SIAM J. Comput., Vol.10, pp.628-691.

[32] Shepherdson, J.C. and Sturgis, H.E. (1963). Computability of recursive functions. Journal ACM, Vol.10, pp.217-255.

[33] Stockmeyer, L.J. and Meyer, A.R. (1973). Word problems requiring exponential time. Proc. 5th. Ann. ACM Symp. on Theory of Computing, ACM, New York, pp.1-9.

[34] Turing, A.M. (1936). On computable numbers with an application to the Entscheidungs problem. Proceedings of the London Mathematical Society, (Series 2), Vol.42, pp.230-265.

[35] Wyllie, J. C., (1979).The complexity of parallel computations. Ph.D. Dissertation, Dept. of Computer Science, Cornell University, Ithaca, NY.

The Mathematics of Complexity in Computing and Software Engineering

N. E. Fenton

Centre for Software Reliability, City University
Northampton Square, London EC1V OHB

Abstract

For programming-in-the-small, the notion of *complexity* is normally synonymous with classical algorithmic complexity, i.e. algorithmic time and space efficiency and computational complexity. There are precise mathematical models used for the measurement of such complexity. Moreover we argue that in this case measurement holds the key to controlling complexity of the 'systems' (i.e. small programs) produced. For programming-in-the-large, the notion of *complexity* is synonymous with cognitive notions associated with the difficulty of constructing and understanding large systems comprising many interrelated components. Although mathematical models and methods exist which supposedly help control this notion of complexity, it appears that the necessity for measurement in this case has been overlooked. This paper investigates why this is so and what can be done about it.

1 Introduction

The notion of *complexity* in computing has been around for many years. Rather than attempting a comprehensive definition in this context, I shall look at what it's 'default' meaning has been over the last few decades. I will argue that this default meaning has changed fundamentally as a result of the trend from *programming-in-the-small* to *programming-in-the-large*. Despite this there are crucial similarities between the meanings, but the important lessons learned about early notions of complexity are being largely ignored for the current notion.

In the early days of digital computing, technology was so constrained that the kinds of software systems which could be constructed effectively were rather small in comparison with those of today. That is not to say that it was not difficult to construct them. On the contrary, but all the difficulties with such construction were confined to the programming task. In short the problems to be solved were generally very specific and well-defined, whereas the task of producing algorithms and coding these into machine readable form was the *complex* part, due to memory and programming language limitations. Thus computing was largely synonymous with algorithm construction and implementation, and consequently complexity in computing was synonymous with the *execution* of the underlying algorithms. Specifically, complexity was associated with time and space efficiency of algorithms, rather than any cognitive notions which were considered irrelevant for this programming-in-the small. Following on naturally from this work, the theory of computational complexity, in the early seventies and subsequently, was concerned with the underlying complexity of algorithmic problems (as opposed to specific algorithms) and this topic has given rise to an enormous wealth of interesting mathematics [11,13]. For ease of discussion let us refer to all work on complexity of algorithms and algorithmic problems as *small-scale complexity*. Even today *complexity* has this default meaning to some computer scientists.

The advent of third generation computers and high level programming languages in the sixties drastically changed the scope of tasks which were now more readily solvable by computers, and also greatly decreased the difficulty of coding algorithms into machine readable form. Suddenly it became possible to construct very large programs which might even be used by people who were not themselves computer experts!

Thus, the advent of a new discipline – Software Engineering – as distinct from Computer Science was justified by the observation that the construction of software was in general no longer a one-person or even small-team activity. Emphasis on computer science research based on the theory of algorithms was inadequate for the kinds of problems associated with large scale software construction which were more like those of civil engineering than any other engineering discipline. Topics such as requirements capture, documentation control and structure, modularisation, formal specification, automated development, and configuration control had to be considered. In short, the drastic change could be attributed to the increase in the 'complexity' of the software and systems constructed, where now the default meaning of complexity was a cognitive one involving size and structure and the associated difficulty of human understanding of these due to scale of the systems and their interactions with the environment. In short complexity was synonymous with the cognitive notions associated with *programming-in-the-large* (although there has also been some interest in the cognitive complexity of small programs or modules).

No longer were increasingly more efficient algorithms important to the success of software projects; what mattered most was the ability to control and understand the complexity of the evolving system, where for example 'complexity' could be manifest at the requirements/specification level in terms of size and relationships between required components. It would be fair to say that in addition to a mushrooming of pragmatic research on tools and methods, there has been an enormous wealth of mathematical work (see for example [2]) which has arisen out of this new notion of complexity (which for ease of discussion we shall refer to as *large-scale complexity*). The bulk of this work has been concentrated in the area of mathematical notations and theories for specification, abstraction and program and system development.

There is a major common theme underlying practical research in both small-scale and large-scale complexity. This is that such research has been directed at *controlling* the (relevant notion of) complexity of the resulting software. For small-scale complexity this meant choosing the most efficient algorithm, and for large-scale complexity this meant using an appropriate specification and design technique based on approaches like top-down design, data abstraction, modularisation, theory construction, and structured programming. However, although the *theme* of research is similar, there is a fundamental underlying principle present for small-scale complexity which is generally absent for large-scale complexity. This principle is MEASUREMENT: Thus, algorithm refinement controlled small-scale complexity because new algorithms could be shown to be superior in a quantifiable manner. This will be properly articulated in the context of measurement theory in this paper. In the case of research in large-scale complexity, the popular attempts to define or establish measurable criteria have largely been misdirected, with scant attention paid to the obligations of scientific measurement. A rare early example [23] of a rigorous approach to the topic which potentially laid the necessary theoretical foundations appears to have been completely ignored by the software community. The result of all this is that there are few objective means of judging the effectiveness of methods which supposedly control complexity, nor of the complexity of the resulting systems. Until this omission is more widely recognised much of the promising new (mathematically based) work will have nothing like the same kind of lasting value as the research on small-scale complexity.

Using the (mathematical) theory of measurement it will be shown that:

- the theories of computational and algorithmic complexity are instances of classical measurement in extended symbol systems
- the limited theoretical work on measuring large-scale complexity is largely incompatible with classical measurement (and hence of little value)
- the way forward to measuring large-scale complexity, must be based on recognising the infeasability of measuring 'cognitive' complexity and concentrating on observable system attributes which are believed to contribute to such complexity, and to approach the measurement of these by appealing to measurement theory.

In short it will be argued that measurement is demonstrably important for both understanding and controlling complexity in computing and software engineering, and that there is even some nice mathematics to support it.

2 Some Measurement theory

Due to shortage of space it will not be possible to do more here than give a flavour for measurement theory. An excellent, thorough, and highly relevant account of the topic is presented in [22], which also contains a clear exposition of all the mathematical prerequisites (sets, functions, relations, logic, probability). The reader is also referred to [9]. Strictly speaking we present here the formal *representation theory of measurement*, which consists of

1. an empirical relational system corresponding to a quality;
2. a number relational system;
3. a representation condition;
4. a uniqueness condition.

An equally valuable mathematical approach to measurment is given in [19].

2.1 Quality as an empirical relation system

Let us assume that for a given class of objects C we have identified some quality attribute Q which is possessed by each member of C.

Let us additionally suppose that the quality attribute Q induces a set $\mathcal{R}$ of empirical relations $R_1, R_2, \ldots, R_n$ on C, i.e. our understanding of Q leads to the observation that these relations hold.

In Figure 1 our class of real world objects C is a set of human beings, and the attribute Q possessed by these is *height*. Note that such an attribute exists in advance of its measurement. Thus in advance of measurement we observe (establish empirically) that the attribute imposes a number of relations on the class of objects. For example, consider the relations R_1, R_2, R_3, R_4 defined on human beings:

$R_1 \subset C \times C$ given by $(x,y) \in R_1$ if 'x is Taller than y'.
                      e.g. (Hermann Munster, Superman)$\in R_1$.

$R_2 \subset C$ given by $x \in R_2$ if 'x is tall'. e.g. Hermann Munster $\in R_2$,
                      and also probably Superman $\in R_2$, whereas
                      Peter Pan $\notin R_2$.

$R_3 \subset C \times C$ given by $(x,y) \in R_3$ if 'x is much taller than y'.
                      e.g. (Hermann Munster, Peter Pan)$\in R_3$,
                      but (Hermann Munster, Superman) $\notin R_3$.

$R_4 \subset C \times C \times C$ given by (x,y,z) 'x is higher than y when sitting on z's
                      shoulders'. e.g. (Peter Pan, Hermann Munster,
                      Superman) $\in R_4$.

Note that the relation R_1 'higher than' is defined on pairs of humans, the relation R_2 is (although defined on single humans, and the relation R_4 is defined on triples of humans. There will generally be some dispute about details of the relations and there may be a need to set down certain standards and definitions (like whether or not we include hair height, or set criteria for posture etc.).

It is the relations which characterise the quality attribute. Thus in general, given

$$\mathcal{R} = <R_1, \ldots, R_n>$$

the quality attribute Q is represented by an empirical relation system

$$\mathcal{L} = <C, \mathcal{R}>$$

2.2 Numerical relational system

In order to have measurement of an attribute we need to have a mapping of objects possessing the attribute into a 'number system', for example the real numbers $\mathbb{R}$.

More generally we shall need a *numerical relational system*, which consists of a set together with one or more relations on that set. Thus let N represent a set and let

$$\mathcal{P} = <P_1, P_2, \ldots, P_n>$$

be a set of relations defined on N. Then the numerical relation system $\mathcal{N}$ is the pair:

$$\mathcal{N} = <N, \mathcal{P}>$$

For example we may have $N = \mathbb{R}$ (the set of real numbers) and we may have the relations:

$P_1 \subset \mathbb{R} \times \mathbb{R}$	given by	'$(x, y) \in P_1$ if $x > y$.
$P_2 \subset \mathbb{R}$	given by	'$x \in P_2$ if $x > 70$.
$P_3 \subset \mathbb{R} \times \mathbb{R}$	given by	'$(x, y) \in P_4$ if $x > y + 15$.
$P_4 \subset \mathbb{R} \times \mathbb{R} \times \mathbb{R}$	given by	'$(x, y, z) \in P_3$ if $0.7x + 0.6z > y$.

3 The Representation Condition

In order to have a measure M for our attribute we need to identify a numerical relational system $\mathcal{N} = <N, \mathcal{P}>$ whose relations $\mathcal{P}$ 'correspond' to the empirical relations $\mathcal{R}$ under a mapping $M : C \longrightarrow N$.

It is worth noting here that in many texts on measurement there is an assumption that only the set $\mathbb{R}$ of real numbers (as in the example) is acceptable for N. Examples of the use of the complex numbers in electrical engineering should dispel the notion that $\mathbb{R}$ is the only useful numerical relational system for measurement. Moreover in [8] the theory of measurement is shown to be easily extended for general numerical relational systems, and in the crucial example in this paper of measuring algorithmic efficiency, we show that we are only able to achieve true measurement when we drop the restriction to $\mathbb{R}$.

Thus M maps relations (in $\mathcal{R}$) onto appropriate relations (in $\mathcal{P}$), in addition to assigning specific values in N to each object in C possessing the attribute Q. This must be done in such a way that all the empirical relations are preserved in the numerical relational system (as shown in the example). This is called a *homomorphism*. This extends to a completely general and formal definition of measurement. Thus, if Q is an attribute defined on the class of objects C, and if $\mathcal{R} = <R_1, \ldots, R_n>$ is the set of relations in C defined by Q and $\mathcal{P} = <P_1, \ldots, P_n>$ is a set of relations in some numerical relational system $\mathcal{N}$, then $M : <C, \mathcal{R}> \longrightarrow <N, \mathcal{P}>$ is a *measure for* Q iff

$M(R_i) = P_i,$ and

$R_i(x_1, \ldots, x_{k_i})$ if and only if $P_i(M(x_1), \ldots, M(x_{k_i}))$

The second condition is called the *Representation Condition*. This determines whether or not we have genuinely captured the attribute in question. Figure 1 illustrates this by showing how the relations R_1 and R_4 are mapped homomorphically into P_1 and P_4 respectively.

The representation condition requires that measurement be the establishment of a correspondence between objects in C (or more appropriately of a suitable model of C) and numbers in such a way that the relations induced by Q on C imply and are implied by the relations between their images in the number set.

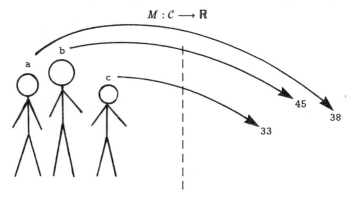

$(x, y) \in R_1$: "x higher than y" $(n, m) \in P_1$: "$n > m$"

$(x, y, z) \in R_4$: "x higher than y when sitting on z's shoulders" $(n, m, p) \in P_4$: "$0.7n + 0.6p > m$"

Figure 1. The Representation Condition for Measurement applies to R_1 and R_4: $R_1(x, y)$ iff $P_1(M(x), M(y))$ and $R_4(x, y, z)$ iff $P_4(M(x), M(y), M(z))$. Here $M(a) = 38$, $M(b) = 45$, $M(c) = 33$ so , for example, $R_1(b, a)$ since $P_1(45, 38)$, and $R_4(c, b, a)$ since $P_4(33, 45, 38)$, i.e. $.7 \times 33 + .6 \times 38 = 45.9 > 45$.

Thus the representation condition may be regarded as the formal definition of what we mean by a measure describing or characterising an attribute.

3.1 Uniqueness and Scales

Note that the homomorphism given in the representation condition is not in general unique. In the above example the numbers may be assigned according to centimetres or inches among many choices of *scale*. In fact, in the above example, if M is a measure for height then so is any scalar multiple αM. We say that αM is a *rescaling* of M.

Just which rescalings are allowed depends on the properties of the relational system $< \mathcal{C}, \mathcal{R} >$.

Scale types can be characterised by the *admissible rescalings* (or *transformations*) as summarised in Figure 2.

Thus if the relational system is invariant under all 1-1 mappings then the scale type of the measure is *nominal*. Similarly if it is invariant under all scalar multiplications (as for height in the previous example) then the scale type is *ratio* etc.

Admissible Transformations	Scale Types	Examples
$M' = F(M)$ (F 1-1 mapping)	Nominal	Labels
$M' = F(M)$ (F monotonic increasing ie $M(x) \geq M(y)$ $\Rightarrow M'(x) \geq M'(y)$)	Ordinal	Preference
$M' = \alpha M + \beta$ ($\alpha > 0$)	Interval	Time, Temperature (Fahrenheit, Centigrade)
$M' = \alpha M$ ($\alpha > 0$)	Ratio	Time interval, Length, Temperature (Absolute)
$M' = M$	Absolute	Counting

Figure 2. *Scales of measurement*

4 Algorithmic complexity seen as measurement

Having presented the classical definitions of measurement theory, it is natural to consider whether algorithmic complexity measurement is truly measurement in that sense.

First of all let us distinguish between the several types of 'measurement' being considered for algorithmic efficiency. We shall restrict the discussion to time efficiency (although it applies equally to space efficiency).

4.1 Absolute Time Efficiency

In this case we measure the actual speed of an algorithm running in a particular environment.

Example: For specific input X algorithm A runs in 23.6 nanoseconds on on machine M

In this case there is no 'new' attribute of algorithms being measured – it just captures the physical notion of speed. This is not particularly useful for measuring algorithmic efficiency, since the measure must not only be computed for specific input values and sizes but will be dependent on the implementation of the algorithm and the machine environment. Thus comparisons between different algorithms are not particularly enlightening this way.

4.2 Time efficiency in terms of primitive operations

Here and henceforth, we need to make assumptions about the kind of primitive machine operations performed; in most cases we identify a single type of operation of interest (e.g. comparison in the case of sorting algorithms). This is a reasonable assumption, since after all, even for high level programming languages, a particular set of primitive instructions is always fixed.

Example: For specific input X algorithm A requires 358 operation executions

Although no longer machine dependent (modulo the same primitive operations), it is still input specific, and again it fails to capture any intuitive attribute associated with comparative efficiency of algorithms. It actually captures the notion of counting so there is nothing new from the measurement viewpoint.

4.3 Time efficiency in terms of primitive operations as function of input size

Example: For input size n algorithm A requires at most $4n^2 + log(n)$ operations

Here we assume either worst-case, average-case, or best-case efficiency (usually the former). We also generally assume that different inputs are characterised by a single size parameter, say n and finally we express the measurement of efficiency as a function $f(n)$ where $f(n)$ is the number of operations needed to solve the problem for input size n.

In this case we have a 'measure' of efficiency which appears to be neither machine nor input dependent. In what sense is it measurement? We have to ask which relation(s) is being captured in which numerical relational system.

Here the numerical relational system's underlying set must be the set F of functions of n. Unfortunately, there is no clearly defined relation in F which corresponds (in the sense of the representation condition) to any intuitive relation on algorithms. For example, although there is a partial order over F in which $n < 2n$, an algorithm of time efficiency $2n$ will run faster than an algorithm of time efficiency n if we run it on a machine whose primitive operations execute over twice as fast. Moreover, if we execute the algorithms on the same machine, although the n algorithm is always faster, for large input n the *rate* at which it is faster is constant. This contrasts with the difference between an n and an n^2 algorithm, yet both pairs $(n, 2n)$ and (n, n^2) lie in the same relation. Worse still consider a $100n$ algorithm compared with an n^2 algorithm. The former is only faster for inputs $n > 100$. We appear to have a poorly defined notion of efficiency if this case.

4.4 Time efficiency (worst case) in terms of Big O notation

Example: For input size n algorithm A requires at most $O(n^2)$ operations or algorithm is $O(n^2)$ in worst case

Recall that we have an equivalence relation *asymptotically dominates* defined on functions from $\mathbb{N}$ to $\mathbb{R}$, such that g asymptotically dominates f if there exist $k \geq 0$ and $m \geq 0$ such that $|f(n)| \geq m|g(n)|$ for all $n \geq k$. The set of all functions which are asymptotically dominated by a given function g is denoted by $O(g)$. If $f \in O(g)$ then f is said to be $O(g)$.

Do we have measurement in a true sense now? I argue that we do. There is a well-understood attribute of efficiency which determines a relation *asymptotically (order of magnitude) greater*. This intuitive relation is made precise by the mapping into the numerical relational system $\mathcal{N}$ whose underlying set's elements are the sets $O(g)$ for functions g, and whose corresponding 'order' relation is given by normal set inclusion. Note that we get round all the problems encountered in the previous case; here both the n and $2n$ algorithms are mapped to $O(n)$ – they are equivalent since (as we previously argued informally) they both formally asymptotically dominate eachother. On the other hand the n algorithm is 'superior' to the n^2 algorithm since $O(n) \subset O(n^2)$, and this is quite right for this view since no matter how much we speed up the execution time of the primitive operations for a machine running the n^2 algorithm, for sufficiently large input it will still always run slower than the n algorithm. This is the sense in which Harel [13] asserts that the Big O notation is *robust* – truly machine independent. Specifically what he meant was that it was *true measurement* since it preserved the representation condition for an intuitively understood attribute. Also note that we have no problem here in asserting that the $100n$ algorithm, which is $O(n)$, is superior to the n^2 algorithm.

Finally it is also worth noting that the Big O notation also preserves other interesting relations derived from efficiency on the class of algorithms. For example we have an intuitive understanding of whether or not an algorithm is *feasible*. This unary relation is preserved under the defined mapping with respect to the relation P in N given by $O(g) \in P$ iff g is *bounded from above* (see [13]) by a polynomial function of n.

By similar arguments to those above, it can be seen that the hierarchy of complexity classes (see [13] for a good introduction to this topic) $\mathcal{P}$, $\mathcal{NP}$ $\mathcal{NP}$ – *complete* etc. together with the *conjectured* relations between classes constitute a conjectured measurement system for problem complexity. Alternatively, for those problems for which the minimal asymptotic bound is theoretically known, we can assert that a measure of the problem complexity is the corresponding Big O value.

5 Controlling large-scale complexity

Thus *measurement* in its strictest sense was central to work concerned with controlling small-scale complexity. I shall now consider the case of large-scale complexity. As far as the bulk of research is concerned, the attitude is quite different. First of all there is an assumption that such complexity can not really be measured at all. We shall return to this point. However, people believe that although complexity cannot (and need not) be measured, it can be controlled by imposing certain types of structural disciplines during the process of creating large systems. In fact almost all major developments in software engineering – which is after all concerned primarily with improving quality by controlling complexity of large systems – have focused on imposing (or assuming) specific (structural) attributes on software products (documents) throughout the life cycle. Examples are *structured programming, modular programming, software reuse, object oriented programming, structured design* methods, data abstraction, verification methods, static analysis, formal specification methods.

Thus for example, in [17], Jones, comments on his VDM development method (which is both popular and highly representative of developments in specification and design techniques):

> ....it will be best to consider the development method as providing a structure in which a design can be documented.

Although the notion of software quality is rarely broken down into identifiable and measurable attributes, advocates of the above methods assume (without recourse to measurement) that their use leads to greater *quality* software products. There is often an almost *axiomatic* assertion about this.

Thus for example, Brooks in his famous book [1] bestowes the virtues of Wirth's philosophy on top-down design thus:

> A good top-down design avoids bugs in several ways. First, the clarity of structure and representation makes the precise statement of requirements and functions of the modules easier. Second the partitioning and independence of modules avoids system bugs. Third the suppression of detail makes flaws in the structure more apparent. Fourth, the design can be tested at each of its refinement steps, so testing can start earlier and focus on the proper level of detail at each step

In [16] Howden relates quality to modularity:

> High-quality designs consist of modules that have a high degree of functional cohesion and a low degree of inter-module data coupling.

For specification languages, Zave [24] declares that one of only two means to achieve 'understandability' and 'modifiability' is by

> providing means by which complexity can be decomposed

To support reuse, we have Belady and Leavenworth [10]:

> An important aspect of reducing complexity is to fight the "part number explosion" by designing around reusable and perhaps standard program parts

Interested readers will no doubt have come across hundreds of quotes in a similar vein to those above. These were just quickly picked out from references close to hand. I am not criticising their content at all (that in itself would require resorting to measurement) – I am merely using them to support my hypothesis.

Let us assume that we construct a software product using every appropriate state of the art specification, design and coding technique. We produce a series of documents (terminating let us say with the object code) each of which is well-structured according to specific structuring rules, makes full-use of modularity, data-abstraction and reused components, has low intra-module coupling and high cohesiveness and is generally presented in a highly structured manner. In addition the software has been subjected successfully to a number of structural tests with high proportion of test strategy coverage.

We may even have correctness proofs for certain critical components and algorithms. Given all of these (internal) product attributes we may find that the final system is highly unreliable, totally unusable, unmaintainable, non-portable and untestable, ie the 'user-view' quality attributes are very low. This is possible because all software engineering development techniques provide only the *framework* for producing 'good' software by encouraging (or even forcing) the use of particular structuring mechanisms; none of these methods guarantees high levels of satisfaction of the user-view quality attributes since none can guarantee the necessary intellectual level required to apply the methods properly to the problem in hand.

However, if I were a software manager given the choice between two software systems built to fulfil the same requirements, the first of which had product documents with the desirable internal attributes described above, and the second had documents which did not possess these properties (or worse still some of the documents were missing), then I would certainly have greater confidence that the first system exhibited satisfactory levels for the user-view quality attributes than the second. We have to accept the argument that good structure, modularity, abstraction, reuse etc. are the attributes which lead to improved (user-view) quality although there are no explicit validated theories establishing all of these links; one way or another modern software engineering development techniques do little other than provide mechanisms for improving just these structural attributes.

Then we need to be able to measure the extent to which these attributes are present in our products. We would need to do this for purposes of quality assurance and prediction, and for comparing and evaluating processes and methods, and also for ultimately validating the relationship between the internal and user-view attributes. From the point of view of measurement theory this is not an unrealistic expectation. If we have sufficiently well defined notions of, for example, *modularity* then it seems reasonable to be able to establish empirical relations for example which establish a partial ordering of documents with respect to their level of modularity, and then subsequent measures which preserve these relations.

6 Measuring Complexity?

What has all the above to do with complexity? It appears to me that if it is the internal structural attributes which ultimately determine the external quality of a system, then the notion of large-scale complexity is actually some derived measure of the measures of the various structural attributes. Moreover, contrary to some current views on software quality modelling and measurement [12,18] where it is asserted that there is no relevance in measuring the internal structural attributes if we are interested in the external view of quality, it would appear to me that the former are the attributes that we should always be concerned about measuring.

Software measurement (which is actually a collection of widely diverse activities) is generally a disrespected area of research. However it is only recently [4,6,14,21,25] that people have argued about the reasons for this – namely that much of this work suffers because there has been a lack of attention to the obligations demanded of measurement theory. Notably there has often been no attempt to define the specific attributes being measured. Nowhere is this more true than in the so-called measurement of software complexity. Well-known metrics (the software engineering term for measures) such as those of McCabe [20] and Kafura and Henry [15] are claimed to be measures of "complexity" in the most general cognitive sense. For example it is claimed that they are "indicators" of such diverse notions as *comprehensibility, correctness, maintainability, testability*, and *ease of implementation*. In fact McCabe's metric is nothing more than a count of the number of linearly independent paths through a program's control flowgraph; since this turns out to be equal to the number of decisions in the program (plus 1), it is a valid measure of a specific program attribute (decisions!), but this does not make it a measure of any of the other attributes mentioned above. In order to show that it is a *predictor* or part of a derived measure of some other attribute we need to establish a *theory* via experimentation in the classical sense. This has never been done to the best of my knowledge.

The Kafura-Henry metric is actually a function defined in terms of the flows between modules of a system. It is apparently attractive in the sense that it can be derived at the design stage for software. In fact rather than measuring any of the attributes listed above, it appears to measure some non-intuitive hybrid notion of intra-module information and control flow.

Is it all doom and gloom then in this area? The answer according to my earlier arguments must be no, but we must be a little less ambitious in our aims for software measurement. We have to concentrate on measuring the partially understood structural attributes. There would have been nothing wrong if Kafura and Henry had explicitly declared that there was an attribute of, say *information flow connectivity* apparent from the module design phase. They could have gone on to determine the kind of relations on software designs which this imposed and then sought a measurement mapping (defined on their own model perhaps of designs) to preserve these. This is exactly the approach which we have pursued relatively successfully for the attribute of *structuredness* [3,5,7]. Specifically we had a special flowgraph model and attempted to characterise different views of structuredness and the relations imposed on flowgraphs via axioms. As an example we observed [6] that many of the so-called complexity metrics of the literature were implicitly attempting to characterise a notion of *pure global control flow structuredness*. This is never explicitly stated, but it can be inferred by comments like "Metric A does not take account of the extra complexity associated with nesting and therefore wrongly rates program x equal to y. Therefore I propose metric B which correctly yields a higher value for program x". Our argument is that if pure global control flow is sufficiently well-understood then there should be a concensus about the (partial) order it imposes on the class of all structures, and once this relation is established we can then attempt to capture it by finding a measurement mapping which satisfies the representation condition. This approach has also been recognised in [21,25].

We are currently involved in research which is attempting to extend this approach to the measurement of specific attributes such as (various views of) modularity, internal reuse, and connectivity. Moreover, we believe that by defining appropriate abstract models which capture the different structural notions, it should be possible to perform such measurement on early life-cycle phase products (such as for example formal specifications).

7 Conclusions

I have argued that researchers in software engineering have to a large extent failed to learn from the experiences of their computer science predecessors, in recognising the central role that measurement (with its underlying mathematical obligations) should play in their discipline. Industrial take-up of many of the current research techniques (which themselves often have interesting mathematical and logical foundations) which have been developed to help control complexity will continue to be slow or non-existent unless researchers are prepared to evaluate both the methods and the products which arise from their use. Measurement theory may hold the key to effective evaluation.

Acknowledgements

The time spent with Richard Bache and Hans-Ludwig Hausen at GMD (Bonn) in 1988, has certainly influenced some of the ideas presented here. The comments of Pat Hall and Robin Whitty have also improved the text. I would also like to thank Jeffrey Johnson and the other members of the IMA conference committee for their support.

References

[1] Brooks FP, 'The mythical man-month', Addison Wesley, 1975.

[2] Cohen B, Harwood WT, Jackson MI, The specification of complex systems, Addison Wesley, 1986.

[3] Fenton NE, 'The structural complexity of flowgraphs', in Graph Theory and its applications to Algorithms and Computer Sci (publ. J.Wiley & Sons 273-282, 1985.

[4] Fenton NE, 'Software Measurement', CSR Newsletter 7, 35-62, 1988.

[5] Fenton NE, 'Software Measurement and analysis: a case study in collaborative research' Proceedings of 22nd HICSS (Vol 2) (Ed. Shriver), IEEE, 95-104, 1989.

[6] Fenton NE, Kaposi AA, 'Metrics and software structure', J. Information & Software Tech., Butterworth, 301-320, 1987.

[7] Fenton NE, Whitty RW, 'Axiomatic approach to software metrication through program decomposition', Computer J. 29(4), 329-339, 1986.

[8] Finkelstein L, 'Representation by symbol systems as an extension of the concept of measurement', Kybernetes, Vol 4 215-223, 1975.

[9] Finkelstein L, 'A review of the fundamental concepts of measurement', Measurement Vol 2(1) 25-34, 1984.

[10] Freeman H, Lewis II PM (Eds), Software Engineering, Academic Press, 1980.

[11] Garey MR, Johnson DS, Computers and intractability, W.H. Freeman Co., 1979.

[12] Gilb T, Principles of Software Engineering Management, Addison Wesley, 1987.

[13] Harel D, 'Algorithmics', Addison Wesley, 1987.

[14] Hausen H-L, 'Yet another modelling of software quality and productivity', in Measurement for Software Quality and Control (ed. Littlewood), 1988.

[15] Henry S, Kafura D, 'Software structure metrics based on information flow', IEEE Trans Soft Enf SE-7 5, 1981.

[16] Howden WE, Functional Program Testing and Analysis, McGraw-Hill, 1987.

[17] Jones CB, Systematic program development, in 'Software Specification Techniques', ed Gehani N, McGettrick, Addison Wesley, 1986.

[18] Kitchenham B, 'Towards a constructive quality model', IEE Software Eng. J Vol 2(4), 105-113, 1987.

[19] Kyburg HE, 'Theory and Measurement', Cambridge University Press, 1984.

[20] McCabe TJ, 'A complexity measure', IEEE Trans Software Eng SE2,308-320, 1976.

[21] Melton AC, Bieman JM, Baker Al, Gustafson DA, 'Mathematical perspective of software measures research', to appear *IEEE Trans Soft. Eng*, 1989.

[22] Roberts FS, 'Measurement Theory with applications to decision making, utility, and the social sciences', Addison Wesley, 1979.

[23] Van Emden MH, An analysis of complexity, *Mathematical Centre Tracts 35* Mathematical Centre, 49 2e Boerhaavestraat, Amsterdam, 1971.

[24] Zave P, 'An operational approach to requirements specification for embedded systems', IEEE Trans. Software Eng, Vol SE-8 (3), 250-269, 1982.

[25] Zuse, H., Bollman, P., 'Software Metrics: using measurement theory to describe the properties and scales of static software complexity metrics', Technische Universität Berlin, 1989.

The Mathematics of Calibration

K. L. Tse
Nijenrode, The Netherlands School of Business
Straatweg 25, 3621 BG Breukelen, The Netherlands

R. W. Whitty
Department of Mathematics, Goldsmiths' College
University of London, New Cross, London, SE14 6NW

1 Introduction

In the construction of computer systems, software engineering attempts to minimise human error while maximising human creativity. To achieve this goal, software engineers have consciously set themselves the task of applying the traditional engineering strategy of using sophisticated tools based on applied mathematics. In analogy with the electrical and mechanical engineers they try to derive system models, to apply analytical techniques to these models in order to predict system behaviour, and to certify the safe and satisfactory behaviour of the system itself. Indeed, the analogy is inescapable because computers and their software are often embedded in electrical of mechanical systems. But electrical engineers have over a century of applied experience in using classical continuous mathematics for modelling and analysis; modern mechanical engineering has been in existence for at least twice that period. The software engineer, in contrast, has had scarcely fifty years to come to grips with some relatively untried areas of mathematics: set theory, logic, graph theory, probability and statistical theory and modern algebra. Small wonder if sometimes the mathematics is adapted or reconstructed to suit the application.

At the present time, software engineering is faced with sharp contrast. The traditional scientific method, rooted in mathematical and statistical concepts, offers no short-term solutions to the human creativity and human fallibility issues in computer systems construction. Meanwhile, advances in hardware technology offer the use of hundreds of megabytes of computer memory which can be addressed in nanoseconds; artificial intelligence has generated correspondingly sophisticated methods of knowledge representation and manipulation. We believe that this imbalance is leading to an implicit rejection of traditional scientific thinking and the emergence of a new engineering mathematics based on the brute force incorporation of huge quantities of software product and production information into otherwise vacuous mathematical models. We call this the mathematics of calibration.

We shall illustrate the mathematics of calibration by a discussion of the techniques of cost estimation in software construction. This illustration is chosen because it is possible to make a direct comparison with the microeconomic approach to the study of production which adheres to the traditional scientific method. This comparison is the central topic of the paper; it leads to some conclusions which need to be appreciated by software engineers concerned in cost management.

2   Software engineering

We begin by recalling some of the activities of the software engineer. These activities are often summarised in terms of the *software life cycle*:

Specification By modelling and analysing the user requirements and the profile of the users application domain, a precise definition of a software system is derived.

Design The specification is analysed in terms of available construction technology (materials/tools/methods).

Implementation Using the technology as planned in the design, a working system is constructed. Components of the system are tested to ensure that they conform to the specification of their roles in the system.

Testing Field trials are conducted before the system is put into general use to ensure that the system matches the user's requirements and the user profile.

General Use The system goes into service. Over the period of its working life it is constantly upgraded or repaired.

The name 'life-cycle' derives from the idea that each of the above phases may lead to a re-iteration of the preceding phases, for instance, all upgrades during general use should in theory start with a proper specification.

The life cycle activities are not necessarily adhered to chronologically, or at all, in practice. However, each activity is of great importance in its own right and has evolved a body of theory to support it, based on appropriate mathematical techniques. Thus specifications can be written down using predicate logic, set theory or category theory; construction can be supported by applying graph theory to flowcharts; testing appeals to statistical and probability theory and so on. It is generally agreed that the mathematical support available for any of the life cycle activities should not be relied upon in isolation to guarantee best possible results. A conjunction of predicates used to specify the functional behaviour of a software system cannot in practice adequately capture a user's requirements; current reliability models do not support realistic certification of software for safety purposes. In every aspect of software development there is inadequate understanding about what are the critical issues and how they can be reasoned about mathematically.

It is probably true to say that mathematics has so far been much more successfully applied to software *products* than to the *process* by which these products are developed. This is partly because this process is one of *construction* rather than *production*. There are no production lines in software engineering and no raw

The Mathematics of Calibration   259

materials; operating systems or process control systems emerge from an initial engineering design more in the manner of bridges or ships than of bricks or motor cars. Nevertheless, there are is an important point of contact between production and construction engineering, they both have to solve problems in the prediction and controlling of resources. This is a major area of interest for software engineers, particularly in the design and implementation stages of software construction.

Closely related to this area of software engineering is the theory of production and production modelling evolved by economists. A comparison between the two approaches is revealing in that economists have taken a traditional scientific approach to production modelling, which is described in the following section. This comparison reveals that software engineers are adopting an approach which relies more on exploiting mathematical models as quantitative decision-making tools than on using models to explore properties of the production process.

3 Microeconomic Production Modelling

Economists begin by observing the production phenomenon and on the basis of their knowledge, reasoning and logic, a theory is built to explain the phenomenon. For the theory to be of any value, it must be consistent in the sense that it must be logical and be derived from a set of assumptions. These initial assumptions are called axioms and the conclusions are 'theorems' of the theory. The acceptance of the theory depends on how it accords with the facts. The robustness of the theory can be tested by relaxing some of the underlying assumptions.

In the case of the economic theory of production which relates production factor inputs and outputs, this theory is summarised in the empirical literature in terms of a production function, which is a mathematical expression which represents the maximum output obtainable from every feasible combination of inputs:

$$Y = f(L, K) \qquad (3.1)$$

where

$Y =$ units of output

$L =$ units of labour input

$K =$ units of capital input

The production function is not a device for calculating a prediction or requirement given various known values; it summarises, in a convenient mathematical form, a hypothetical relationship between the inputs and outputs of a production process; it is a 'stylised' representation of the transformation of inputs into outputs. The aim of this is to examine the different underlying asumptions built into the economic theory of production. Different functional forms of the production function are used in the empirical literature, reflecting the different underlying assumptions built into the theory. For example, the presence or absence of scale economics or homotheticity of factor inputs. Thus the empirical testing of a particular production function also involves the testing of these assumptions. A particular functional form will usually involve some constant coefficients, whose values have specific economic interpretations. Changing the values of the coefficients has implications for the interpretation of the production function.

As an example, a popular form of production function in empirical literature is the Cobb-Douglas production function which has the form:

$$Y = kL^\alpha \cdot K^\beta \tag{3.2}$$

where k, α and β are constants.

Implicit in the economic interpretation of the Cobb-Douglas equation is the assumption that the production function is linear and homogeneous ($\beta = 1 - \alpha$), yielding a constant return to a proportionate increase in the use of all inputs. As an example of the manner in which economists exploit this production model, it can be deduced that the above formulation (Equation 3.2) shows that if labour is paid a wage equal to its marginal product ($\partial Y/\partial L$), this production function will yield a share of wages relative to total output which is fixed and independent of the values of Y, L and k. In fact the ratio between total wage income and total output must, in these circumstances, be exactly equal to α, in the Cobb-Douglas production function. The same applies to capital income. This is essentially a well-known theorem of Euler and says that if a production function involves constant returns to scale then the sum of the marginal products will actually add up to the total product.

However, despite the ability of the Cobb-Douglas equation to represent mathematically certain basic economic concepts, there is nothing universal about this model. According to Baumol (1965),

> It is to be observed that there is no *a priori* reason for accepting the validity of the Cobb-Douglas production function. It is merely an empirical hypothesis which has been proposed to explain an empirical observation. But that is good scientific procedure which always builds and tentatively accepts theoretical constructs only because their implications accord with (explain) observed phenomena. It is to be added that, at least so far, the statistical evidence does not appear to contradict the Cobb-Douglas hypothesis.

The Cobb-Douglas and other forms of the production function, are models derived for the purpose of studying the relationship between factor inputs and outputs. Limitations imposed by research methods such as problems associated with input measurement, aggregation across heterogeneous inputs and production activities, mean that few economists are prepared to make predictions or recommendations on the basis of their models. To illustrate, in a recent study on production efficiency in farming in Tennessee, the Bagi and Huang (1983) concluded that

> we should not expect the measured inefficiencies to give an indication of how much potential exists for raising output from existing inputs.

The Mathematics of Calibration   261

On the other hand, if agricultural experts were to be sent to Tennessee to help farmers produce more, one might feel confident in using the measured inefficiencies as a guide to which farm to send them to first

Hence, in economics, a clear distinction is made between *positive* and *normative* approaches. The former aims primarily to *describe* an economic phenomenon, while the latter aims to *prescribe* solutions. Thus Layard and Walters (1978, p. 277) observe that (our italics):

> the economists' theory of production is often criticised on the grounds that it does not help businessmen to make decisions. The criticism has some force: the theory was devised to predict how businessmen *would* behave rather than to help them to decide *how* to behave.

4 Cost estimation

In common with the economists' approach to production, software engineers are interested in the relationship between factor inputs and outputs in their attempt to predict and control manpower requirements and duration in software construction. Models similar in structure to the production function (Equation 3.1) have been developed. The models, which are known as *algorithmic cost estimation models* ideally aim to allow for the setting of detailed project milestones, the breakdown of the production process into activities and the generation of cost profiles for each activity.

An example of a cost estimation model is one developed by Putnam (1978). The model relates factor inputs and output and has the functional form:

$$S_s = C_k \cdot K^{1/3} \cdot t_d^{4/3} \qquad (4.1)$$

Where S_s is Product size, C_k is a state-of-technology constant, K is the labour (man years) for the *whole* life cycle, and t_d is development time (years).

It has been argued that the Putnam model is theoretically based to the extent that it was derived, in conjunction with other empirically observed relationships, on a well-known engineering development phenomenon as summarised by the Rayleigh curve. But for the Putnam model to provide estimates of cost, a transformation process is necessary. For instance, Kitchenham and Taylor (1984) derived from Equation 4.1 the following:

$$E = a \cdot C_k' \cdot C' \cdot B \cdot S_s^{1.286} \qquad (4.2)$$

where E is Effort in Man Months, C_k', C' and B are constants depending on the type of project and a is a conversion factor.

We derive the functional form Equation 4.2 in order to obtain an explicit formula for calculating development cost from a given projection of Product size (S_s). This is the general form of the algorithmic cost estimation model, summarised in Equation 4.3:

$$\text{Cost} = f(\text{Product size}, \text{Bias factors}). \qquad (4.3)$$

Here, *Cost* is usually expressed in terms of effort, i.e. labour input and/or elapsed time (duration of production process); *Product size* is normally measured in terms of code size and *Bias factors* are multipliers for factors, other than product size, which are considered to have an impact on cost: personnel, project, product and technology attributes. The manner and extent to which the bias factors are incorporated into the cost estimation equation vary from model to model.

The key difference between the production function and the cost estimation model is indicated by the fact that, whereas the functional form in Equation 3.1 has output as the dependent variable, Equation 4.3 has input as the dependent variable and output (measured by Product size) as an independent variable. Since the Bias factors capture further factor inputs, we can contrast the economist's production function form and the software engineer's cost estimation form as, respectively:

$$output = f(input, input, \ldots) \qquad (4.4)$$

and

$$input = g(output, input, \ldots). \qquad (4.5)$$

For example, Putnam's original Equation 4.1 takes the first form whereas the transformed Equation 4.2 takes the second form, which is precisely the cost estimation form in Equation 4.3.

Moreover, in the case of the cost estimation models, the functional relationships are derived *not* to support an underlying theory, but purely as empirical observations which may support management decision making. The coefficients and exponents in a cost estimation model have no valid theoretical interpretation, they are merely values observed as giving the closest relationship between the dependent and independent variables.

The mathematical manipulation which transforms Equation 4.1 into Equation 4.2 is a valid method of arriving at a cost estimation equation because (and only because) the derivation of the Putnam's model from the Rayleigh curve phenomenon is not relevant to this mathematical manipulation. The only issue is whether the manipulations preserve the empirical basis of Equation 4.1. Therefore, concerning the manipulations, no connection between the mathematical form of the equations and their underlying theoretical interpretations is assumed.

To take our argument further we examine a well-known and representative cost estimation model, COCOMO (Constructive Cost Model) developed by B.W. Boehm (Boehm 1981). Information about the model is fully available and of late different variants of the model as well as software tools to support these models have been developed (Cowderoy and Jenkins, 1986).

One of the basic equations for COCOMO provides the following specific form of the functional relationship shown in Equation 4.3:

$$\text{Effort} = b \cdot (\text{Product size})^c \cdot (\text{Bias factors}). \qquad (4.6)$$

Product size is measured in terms of thousands of Delivered Source Instructions (kDSI), b and c are constants and their values depend on the development environment. For example, in an 'organic' environment, which reflects relatively small software teams developing software in highly familiar in-house environment, Boehm proposed the values for b and c in Equation 4.6 to be 3.2 and 1.05 respectively, independent of the size of the product to be produced and bias factors.

In COCOMO, there are fifteen bias factors known as *cost drivers*, an example (of a personnel bias factor) being 'analyst capability'. Since the bias factors in Equation 4.3 are assumed to have a multiplicative effect, this gives the following form for Equation 4.6:

$$\text{Effort} = b \cdot (\text{Product size})^c \cdot \text{factor}_1 \cdot \text{factor}_2 \cdots \text{factor}_{15}. \qquad (4.7)$$

Their values were generated by Boehm from a set of 63 completed projects at the Thompson Ramo Woolridge (TRW). If, for example, the project environment is 'organic' and the project under consideration is ranked as having 'high product complexity' and 'very high usage of software tools', then the Equation 4.7 is adjusted by the 2 bias factors with values of 1.65 and 0.83 respectively:

$$\text{Effort} = 3.2(kDSI)^{1.05} \cdot 1.65 \cdot 0.83. \qquad (4.8)$$

All other cost drivers are assumed to have *nominal* values of 1.00.

In addition to the $Effort$ equation, COCOMO also provides a specific form for the functional relationship in Equation 4.3 in which *Cost* is represented as project duration:

$$\text{Elapsed time} = d \cdot \text{Effort}^f. \qquad (4.9)$$

Again d and f are constants and their values depend on the project environment mode.

It can readily be seen that the assignment of appropriate values to the cost drivers forms a significant part of the COCOMO estimation process. But the values given in (Boehm, 1981) for the cost drivers, the actual functional forms of the basic equation as well as all the coefficients in Equations 4.6 and 4.9 were derived empirically in the TRW environment. This environment concentrates on very large scale software development in the aerospace industry. On this basis the model developed by Boehm may not be directly applicable to other production environments. This can be either because the functional relationships are not a proper reflection of a different specific production process or because the given values assigned to the cost drivers are no longer appropriate in the different environment, or a combination of both. Two solutions are apparent: to abandon the original model and re-model the new environment, or to adjust the values given to the cost drivers. The latter approach is known as calibration.

5 Calibration

The aim of the calibration process is to ensure that a cost estimation model is fine-tuned to fit a particular production environment. This is done not by altering the basic functional form of the relationship in Equation 4.3 for a specific model, but by changing the values of the coefficients, exponents and, in particular, the cost drivers. New sets of values for the cost drivers are derived from past projects in such a way as to ensure that the new set of values for the cost drivers give a reasonable *ex-post* estimate of cost for the past projects.

The underlying assumption of the calibration approach is that once the cost estimation model has been fine tuned to the past history of an environment, future project costs in that environment will be estimated with reasonable accuracy. Ideally, calibration should be a process of continuous adaptation to take account of ongoing changes to the environment. So far, experimental validation of the approach has concentrated on demonstrating that particular cost estimation models do have the capability to be accurately calibrated for past project costing.

For COCOMO the original studies were carried out by Boehm (1981). Using a database of 63 completed projects at TRW he showed that, using the most detailed version of COCOMO, 70% of the project costs were estimated to within 20% of actual costs. More recently, Kemerer (1987) using data from fifteen completed projects, found that, using COCOMO without calibration, the estimated effort (man months) required overshot the actual effort by 600%. The result after calibration proved to be significantly better. Similarly, Miyazaki and Mori (1985) made an extensive evaluation of COCOMO using data from 33 completed projects and found that, in the absence of calibration, the model gave a marked over estimation of costs and duration, the average deviation from actual being 166%. They then calibrated the COCOMO model by eliminating a number of cost drivers which were considered not relevant in their production environment and deriving new values for the remaining cost drivers on the basis of their old project data. The calibrated model produced much better estimates with an average deviation of about 17%.

Calibration is a general technique. It may be observed that the manipulations which gave the cost estimation form of Putnam's model (Equation 4.2) can be immediately justified if calibration can be applied to this resulting equation so that it gives reasonably accurate estimates of costs for a specific project environment. In fact, calibration has become widely accepted as the preferred method of dealing with the effects on cost estimation models of changes in production environments.

There are compelling reasons for preferring calibration to remodelling. Creating a new model requires a thorough understanding of the underlying relationships between factor inputs and outputs in the production environments, if for no other reason than to appreciate why the original model fails to work satisfactorily. This is effectively adopting the economists' approach to production which, as discussed above, must be regarded as positive or descriptive rather than normative or prescriptive. Conversely, the calibration process does not involve changing the functional relationship between cost, product size and bias factors, and therefore the internal consistency, the rationale, the theory behind the basic model remains unchanged and can be disregarded.

If no underlying theory is assumed for a cost estimation model then none of the values assigned to the coefficients and bias factors in Equation 4.3 have any interpretation. Therefore no restrictions are placed on how these values are reassigned and a great range of mathematical and statistical techniques may be applied without regard to whether they are appropriate to the data sets involved.

Calibration can be supported by a range of powerful techniques from knowledge engineering, such as knowledge elicitation and expert systems technology. Most of the recently developed variants of cost estimation models rely to some extent on these techniques (Cowderoy and Jenkins, 1986, Cuelenaere et al, 1987).

Calibration for a new environment depends on how data has been collected characterising the environment. Mass availability of cheap storage, together with the introduction of project database tools such as Integrated Project Support Environments (IPSEs), mean there is now a possibility of deriving values for cost drivers based on huge volumes of past project measurements.

6 Discussion

Although the calibration approach applies to mathematical models and employs statistical and mathematical techniques, it is not, in the traditional sense, a mathematical, scientific method. This new approach to modelling can be regarded as making an over-simplistic model work by brute force, in contrast with the more traditional approach of re-examining the underlying relationships of the model. Calibration involves the abandoning of the idea that a model should be a stylised representation of the real world.

Taken to its extreme *any* equation relating inputs and outputs might serve as a basis for calibration and cost estimation. Indeed, a number of well-known models are proprietary; the users of such a model may have no knowledge whatsoever of its functional form and must put their faith entirely in calibration.

It can be argued that, the developers of cost models are encouraging the emergence of a new kind of mathematics which is based not on accepted ideas of elegance, rigour or universality but on short-term applicability and the ability to collect and process vast amounts of structured data. This is not to say that calibration is 'second class' science or relies upon 'second class' mathematics. It is a new approach and it is to be expected that the mathematical and statistical techniques involved will be applied in new ways and in conjunction with new sophisticated disciplines such as knowledge engineering.

Judging by its popularity in software cost estimation, calibration is being accepted wholeheartedly by software engineers. It is clear that they see it as outperforming traditional approaches as far as their requirements are concerned. This cannot be repudiated other than on philosophical grounds, but it is important that practitioners of calibration should fully realise that they are making a clear distinction between an theoretical understanding of *why* the production process behaves as it does and a practical knowledge of *how* this behaviour effects their costs. This practical knowledge is only as reliable as the data on which calibration has been preformed. There seems to be no reason to hope that the reliability of a calibration can be quantified without resorting back to some theoretical understanding of the production process.

The implications of radically new approaches to scientific problems have to be discussed and understood as widely as possible. It seems inevitable that the calibration should not confined to software engineering. It is symptomatic of the scale and complexity of most undertakings in modern engineering. As an example, the Glasgow firm of Babtie Shaw and Morton has developed a cost model called Bridget for predicting the resources used in building highway structures such as motorway bridges. The model is based on historical data from existing structures and focuses on a list of key items in the engineer's 'Bill of Quantities', which act as cost drivers for the construction. The model is currently in use in Scotland on a number of projects, including the M74. If the model were to be used in a different country then the techniques of calibration would be required to transfer the model successfully. Indeed, even within Scotland, the database for Bridget is constantly being updated (Meall, 1988).

References

Bagi and Huang (1983). Estimating production technical efficiency for individual farms in Tennessee. *Canadian Journal of Agricultural Economics*, vol. 31, 249–256

Baumol, W.J. (1965) *Economic theory and operations analysis. (Second edition)*, Prentice Hall, Englewood Cliffs, N.J.

Boehm, B. W. (1981). *Software engineering economics*, Prentice Hall, Englewood Cliffs, N.J.

Boehm, B. W. (1988). Understanding and controlling software costs. *IEEE Trans. Softw. Eng.*, vol. SE-14, no. 10, 1462–1477.

Cowderoy, A. and Jenkins, J. O. (1986). Survey of cost modelling. *ESPRIT Project 938, Report IMPC-WP5-A-002*, Imperial College London, August 1986.

Cuelenaere, A. M. E., van Genuchten, M. J. I. M. and Heemstra, F. J. (1987). Calibrating a software cost model: why and how. *Information and Software Technology*, vol. 29, no. 10, 558–567.

Kemerer, C. F. (1987) An empirical validation of software cost estimation models. *Comm. ACM*, vol. 30, no. 5.

Kitchenham, B. A. and Taylor, N. R. (1984). Software cost models. *ICL Technical Journal*, May 1984, 73–102.

Layard, P. R. G. and Walters, A. A. (1978). *Microeconomic theory*, McGraw Hill, New York.

Meall, L. (1988). The significant cost of building bridges. *Accountancy*, June 1988, p. 135.

Miyazaki, Y. and Mori, K. (1985). COCOMO evaluation and tailoring. *Proc. 8th Int. Conf. on Softw. Eng.*, IEEE.

Putnam, L. H. (1978). A general empirical solution to the macro software sizing and estimating problem. *IEEE Trans. Softw. Eng.*, SE-4, no. 4, 345–361.

PART IV

Is There a Mathematical Revolution Inspired by Computing?

Computing and Foundations

J. M. E. Hyland

Department of Pure Mathematics, University of Cambridge
16 Mill Lane, Cambridge, CB2 1SB

Abstract

The types and functions which are used in computing are very different from traditional mathematical sets and functions; and yet in practice they are treated as much the same. Doing so in a systematic way raises foundational issues, as it is not clear why the traditional mathematical objects should be given privileged ontological or epistemological status in any such treatment. This paper sketches the mathematical contexts (based on categories known as toposes) in which types and functions in functional programming languages can be treated on a par with ordinary mathematical sets and functions. It argues that the existence of such a perspective undermines the commonly accepted view of foundations based on the primacy of that notion of set analysed in modern set theory.

1 Introduction

1.1 Mathematical Logic

How can there be any connection between the engineering science of computing and the esoteric world of the foundations of mathematics? The suggestion seems like a conscious paradox. The link is provided by mathematical logic. Mathematical or formal logic is concerned with the precise mathematical properties of syntax both in itself and in relation to appropriate semantics. The historical roots of the discipline lie in philosophical enquiry, in particular in questions relating to the philosophy of mathematics. However as computing also requires a precise formal syntax, mathematical logic is in principle and (as it turns out) in practice one of the theoretical bases for the new engineering science.

Now the informed reader may well consider that mathematical logic provides a rather superficial connection between computing and foundations. There is after all nothing unusual in areas of pure mathematics later becoming applicable to a new science. Certainly, if there were nothing more to the story, then we would not be justified in talking of the 'mathematical revolution inspired by computing' in relation to the foundations of mathematics. In this paper I try to describe some of the ways in which conceptual problems raised by computing are rousing us from our dogmatic slumbers about foundations.

1.2 The Foundations of Mathematics

One can distinguish two aspects to the foundations of mathematics: questions of ontology (of what there is or can be) and matters of conceptual organisation (theories of natural kinds). Most mathematicians have lost interest in what they regard as ontological questions. To the extent that they consider the matter, they are generally satisfied with some form of set theory as the foundation for mathematics. However even when the cumulative hierarchy of sets is presented in the best possible light, it must seem philosophically problematic. Its main justification is pragmatic: it does the job. (But what job?) On the other hand mathematicians continue to develop the conceptual foundations of their subject without making any fuss about it. The creation of powerful new notions, which serve both to unify strands of mathematical thought and to solve outstanding problems, is part of the natural activity of the mathematician.

Now there is a considerable tension between (these attitudes to) ontological and conceptual foundations. What seem natural mathematical constructions are 'implemented' in current set theory in an apparently arbitrary fashion. For example the fundamental notion of an ordered pair is standardly 'implemented' by Kuratowski's definition:

$$(x, y) = \{\{x\}, \{x, y\}\}.$$

But there is nothing canonical about this definition. On the other hand there seems to be something behind the idea of conceptual or epistemological priority. Some mathematical notions can naturally be explained in terms of others, but not vice-versa. But what has this to do with ontological priority?

This tension between ontological and conceptual foundations suggests that at bottom the issues involved in them may be two sides of the same coin. The idea that some components of our conceptual equipment are ontologically primary is problematic: but also we feel embarrassed if what is for us conceptually primary is taken in any serious sense to be ontologically secondary.

Now at a trivial level any mathematical discipline must have an effect on the conceptual apparatus of mathematics: it introduces its own basic concepts. But theoretical computer science goes further than that. It can claim an effect on conceptual *foundations* because it challenges the accepted fundamental notions of set and function. This is most clear in the problems associated with the rational design of programming languages, some of which are described below.

1.3 Functions in Computer Science

Many branches of pure and applied mathematics share the feature that some mathematical notion of a function is central; but the notions realised in programming languages are peculiarly problematic. One such notion, which seems to be fundamental, is encapsulated in the pure lambda calculus. This is the calculus which underlies LISP and other modern functional programming languages. In accounting for it, one has to account for quite arbitrary definitions by recursion; and hence, in some sense, for the existence of fixed points for arbitrary functions. This is quite at odds with the classical set theoretic conception of a function. In section 2, I explain a modern view of models of the lambda calculus, and describe the mathematics that is needed to make sense of it.

1.4 Types in Computer Science

The practice of programming generally requires one to distinguish between different *types* of entity: between the data types of Booleans and Lists for example. Similarly in mathematics one distinguishes, for example, real numbers from continuous functions. It is of the essence of type theory to make these distinctions; this results in a more rigidly structured universe than the set-theoretic one. For example, there is no primitive meaning to be attached to the intersection of two types; and one has to provide a function (coercion) to map a natural number to the corresponding real number. However, in this paper I will not stick to any firm distinction between sets and types: the distinction has no effect on the semantics which I discuss in Section 3.

Both in programming and in mathematics, types can seem more important conceptually than sets. However there are problematic issues connected with types in programming languages, in particular with such full blown polymorphic types as occur in the calculus of constructions (this calculus forms the basis for the LEGO proof system described in this volume by Burstall [5]). The most important of these issues are those of the modularity and genericity of programs.

1.5 Modularity and Genericity

Modularity

Large pieces of code must be written by many people collaboratively. Thus one wants to be able to write programs in small pieces, each of which does something identifiable, and then slot the components or modules together to form larger programs; and then one wants to iterate the process. This old philosophical idea, that the meaning of the parts should determine the meaning of the whole, is also the ideal of structured programming. At its simplest it suggests writing programs which define extensional operators (without side effects). The most developed form of this is functional programming, where modularity is most effectively controlled by explicit type systems.

Genericity

The idea is to avoid unnecessary work. One should exploit similarity of structure between routines, by writing very general routines which may be used over and over again in different contexts. A traditional example is that of sorting lists. In principle any of the basic sorting algorithms acts on data consisting of a list of elements (of given type) together with a (decidable) total order (on the type). So they act generically on ordered types. Systematising this idea is the task of polymorphic type systems (a sketch of one such system is given in this volume by Burstall [5]).

The ideas of modularity and genericity are clearly related: in practical terms each makes the other more useful. However there is also a tension: crude modularity suggests making as many distinctions between kinds of code as possible; on the other hand genericity suggests identifying pieces of code in so far as they do essentially the same thing. For this reason, the type systems which organise programming languages in which it is feasible to write general purpose programs in a modular fashion raise difficult technical questions.

1.6 Overview

The rational design of programming languages is now an area of very intense research, but the basic issues are quite accessible. In this paper I attempt to make them clear by focusing on two novel forms of abstract mathematics which have resulted from the needs of computing. These are

1. the problem of describing in a civilised fashion what is a model of Church's lambda calculus;

2. the problem of giving an account of polymorphic functions.

Section 2 is devoted to the pure lambda calculus and section 3 to typed versions. In each case I have tried to indicate how the ideas bring pressure to bear on traditional views of foundations.

Though I have attempted to keep references to category theory to a minimum, this paper makes propaganda for a view of foundations informed by that subject. Category theory is an essential organising principle in modern computer science, and the effect of computing on foundations is best seen in the light of that experience.

2 The Pure Theory of Functions

2.1 The General Notion of a Function

What is a function? With the questionable benefit of hindsight we can see that this question played a significant role in the development of modern mathematics. We associate with Dirichlet the example of the function of a real variable taking the value 1 on rational and 0 on irrational numbers. It seems but a short step from this to the general idea of a function being determined by its graph: that is to the modern set-theoretic notion of function. Probably this is an entirely superficial history of ideas; the example predates serious set theory. However it seems clear that mathematicians in the past did not work with the idea of a function sanctioned by current ideology. We can well imagine that they expected functions to be defined by formulae. (One might liken this to the notion of propositional function as it appears in *Principia Mathematica*.)

In teaching, we are inclined to present the various kinds of function which arise in branches of mathematics as parasitic on the set-theoretic notion of a function as a graph. In analysis we have functions, continuous functions, differentiable functions, analytic functions. In algebra, we have homomorphisms between different kinds of algebraic structure. There is nothing wrong with this set-theoretically based mathematics. Indeed I teach it with pleasure. However we should not suppose that the set-theoretic notion is the only possible basic notion of function. Where we see it under strain (e.g. rational mappings in algebraic geometry, random sequences in probability) we should detect the need for other foundations.

2.2 Computable Functions

Since the advent of computing, we all think that we know what computable functions are: functions which 'in principle' can be computed on a digital computer. (Of course 'in principle' covers a multitude of sins. The computer must be possessed of unlimited memory and be capable of running for an unlimited time.) Unfortunately our intuitive understanding does not readily give rise to a useful theory of computable functions. Classical recursion theory [18] treats the subject, but its aims are limited; in particular attention is largely restricted to the data type of the Natural Numbers. Furthermore most generalised recursion theory equally fails to address issues which are significant to the practice of computing. Among these issues are those of *Intentionality*, *Non-termination*, *Fixed points* and *Effectivity*. (I am unable to provide a comprehensive treatment of these, but I list them to give an impression of the range of the conceptual problems in just one aspect of theoretical computer science.) It seems best therefore to treat computable functions via a theory of functions stripped to its bare bones, that is, via the pure lambda calculus.

2.3 The Syntax of the Lambda Calculus

This formal system was constructed around 1930 by Alonzo Church. He intended it to be a foundation for mathematics based on a universal theory of functions. Church's original system incorporated a system of logic at the same level as the functions, and turned out to be inconsistent [11]. The fundamental observation is that functions in the lambda calculus have fixed points; a fixed point for negation is a proposition equivalent to its own negation, and so is a contradiction. However the part of the theory which deals only with function application and abstraction is (in a suitable sense) consistent. This theory is called the *pure lambda calculus*; it encapsulates a pure theory of functions.

The lambda calculus is a theory of *terms*. We suppose that we are given a (countably infinite) set of variables $x, y, z, \ldots$. The set of terms of the pure lambda calculus is then defined recursively by the following clauses:

(Base clause)   *a variable is a term;*
(Application)   *if s and t are terms then (st) is a term;*
(Abstraction)   *if r is a term and x a variable then $(\lambda x.r)$ is a term.*

In the application clause, we think of s as a function, t as its argument (input) and then (st) is the value of s at t (the output). In the abstraction clause, we imagine that for each possible input x, we have a corresponding output $r(x)$, so that

$$x \longrightarrow r(x)$$

is a function from inputs to outputs; then $(\lambda x.r)$ denotes this function-as-object. It is a feature of this theory that functions of many arguments can be reduced to functions of one. For example

$$(x, y) \longrightarrow f(x, y)$$

reduces to

$$x \longrightarrow (y \longrightarrow f(x, y))$$

and so is represented by the term

$$(\lambda x.(\lambda y.f)).$$

This motivates the standard bracketing convention in the lambda calculus:

$$st_1 t_2 ... t_n \text{ stands for } (...((st_1)t_2)...t_n).$$

(Implicitly here s is a function of n arguments so that we think of $st_1 t_2 ... t_n$ as $s(t_1, t_2, ..., t_n)$.) There is a corresponding convention for iterated abstractions:

$$\lambda x_1 x_2 ... x_n.r \text{ stands for } (\lambda x_1 (\lambda x_2 (...(\lambda x_n.r)...))).$$

(We think of $\lambda x_1 x_2 ... x_n.r$ as an n-argument function abstracted from r.) We use these conventions and also drop brackets where they do not add anything in the rest of this paper.

2.4 Computing with the Lambda Calculus

The main computation rule for the lambda calculus is the following rule of β-equality

$$(\lambda x.s)t = s[t/x].$$

Here $s[t/x]$ is the result of substituting term t for all free occurrences of x in s, with stipulations which prevent 'dynamic binding' of the variables free in t. So λ acts as a variable binding operator. (There is also a rule of η-equality, which makes for cleaner semantics; but it is not computationally essential, so I will not discuss it.) The computation process which results is called β-reduction. A term u reduces in one step to a term v just when v is the result of replacing a subterm of form $(\lambda x.s)t$ by the corresponding one of form $s[t/x]$ in the term u. We write $u \geq v$ when v is obtained from u by a sequence of one step reductions; in particular $(\lambda x.s)t \geq s[t/x]$.

We illustrate the computation process using the traditional Church numerals. We define inductively terms $f^n(x)$ of the lambda calculus for n a natural number by

$$f^0(x) = x$$
$$f^{n+1}(x) = f(f^n(x)).$$

We can associate with each natural number n the term

$$\lambda f.\lambda x.f^n(x)$$

which encapsulates the notion of n-fold iteration. This term is called the Church numeral for n. We write $\hat{n}$ for the Church numeral $\lambda f.\lambda x.f^n(x)$.

One can compute with Church numerals as codes. We give a formal statement of this result. (The notion of partial recursive function is explained in the classic text book by Rogers [18], and in many more recent books on logic. Simpson's paper in this volume also discusses recursive functions [21]).

Theorem 2.1 *For any partial recursive function f there is a term t_f of the lambda calculus such that*

$$t_f \hat{n} \text{ reduces to } \hat{m} \text{ if and only if } f(n) = m.$$

And conversely, if t is a term of the lambda calculus, then the function f defined by

$$f(n) = m \text{ if and only if } t\hat{n} \text{ reduces to } \hat{m}$$

is partial recursive.

This was essentially first proved (but for a definition of recursive function in terms of primitive recursion and minimalization) by Kleene, but the most natural proof procedure is that given by Turing [22]. By way of an example, we write down one of the simplest codes for an arithmetical function in the lambda calculus, and perform a computation with it. We define

$$mult = \lambda abf.a(bf),$$

and compute $mult\hat{2}\hat{3}$ as follows. (Here we adopt the convention that a reduction is denoted by $\geq$.)

$$\begin{aligned}
(mult\hat{2})\hat{3} &= ((\lambda abf.a(bf))\hat{2})\hat{3} \\
&\geq (\lambda bf.\hat{2}(bf))\hat{3} \\
&\geq \lambda f.\hat{2}(\hat{3}f) \\
&= \lambda f.\hat{2}((\lambda h.\lambda z.h^3(z))f) \\
&\geq \lambda f.\hat{2}(\lambda z.f^3(z)) \\
&= \lambda f.(\lambda g.\lambda x.g^2(x))(\lambda z.f^3(z)) \\
&\geq \lambda f.\lambda x.(\lambda z.f^3(z))^2(x)) \\
&= \lambda f.\lambda x.(\lambda z.f^3(z))((\lambda z.f^3(z))(x))
\end{aligned}$$

$$\begin{aligned}
&\geq \lambda f.\lambda x.(\lambda z.f^3(z))(f^3(x))\\
&\geq \lambda f.\lambda x.f^3(f^3(x))\\
&= \lambda f.\lambda x.f^6(x)\\
&= \hat{6}.
\end{aligned}$$

(I do not expect the reader to be very impressed by this computation.)

2.5 Models for the Lambda Calculus

The good answer to the question 'what is a model of the lambda calculus?' is suggested by our basic intuition: the lambda calculus is a theory of *all* functions. Thus everything must be both a potential argument (input), a potential value (output) and a function from arguments to values; what is more all functions are assumed to occur. Thus a model for the lambda calculus should be a set D which is equal to the set D^D of all functions from D to D. (In fact to model β-equality alone, a retraction from D to D^D will do.) Unfortunately unless the set D has just one element D^D can never be even a retract of D; the cardinality of D^D is too great (this is essentially Cantor's Theorem). Thus there are no non-trivial models of the lambda calculus in the intuitive sense.

For this reason, a bad answer to the question 'what is a model for the lambda calculus?' is used in much of the literature. Typically this amounts to the following: a set D and subset $F \subseteq D^D$ together with maps

$$ap : D \times D \longrightarrow D \qquad \text{and} \qquad rep : F \longrightarrow D$$

such that the equations of the lambda calculus are satisfied when ap is used to model application and rep to model lambda abstraction. Saying this in any precise form is clumsy and conceptually unilluminating; for we have lost the basic intuition of a theory of all functions. There is a discussion of the various primitive definitions of a model for the lambda calculus in Chapter 5 of Barendregt's book[1].

These conceptual problems are a partial explanation for the fact that during the early history of the lambda calculus the focus of interest was on the purely syntactic properties of the computation rules. Of course the pioneers understood the problems, and implicitly responded to them with the thought that the lambda calculus must be a theory of a countable world of intensional functions. This is a very interesting idea, which in principle should have had a profound effect on foundational questions. However the lack of a traditional semantics for the lambda calculus meant that few took the idea seriously.

What counts as serious semantics for some syntax is not an absolute mathematical question; it is very much a matter of tacit agreement by the mathematical community. Any answer reflects the view of the community as to the natural kinds of structures available in mathematics. Even our picture of the relation between syntax and semantics changes: there is no longer such a clear distinction between them. Indeed for the lambda calculus the computation rule of β-reduction provides what computer scientists (unfortunately) call an 'operational semantics'.

The first mathematical model for the lambda calculus was discovered by Dana Scott [19]. (For more on the background see 2.2 of Johnson's paper in this volume [9]). Scott's approach was typical of modern conceptual pure mathematics; he first found an appropriate category in which to work. In these categories continuity is used as an analogue of effectivity, and this keeps control of the structure of function spaces. Other kinds of structure have since been used for this purpose, and most of the 'categories of domains' that provide 'denotational semantics' for functional programming languages contain objects D with categorical function space D^D isomorphic to D.

2.6 Topos Theoretic Models

There is a sense in which the semantics initiated by Scott is unsatisfactory. We have to be conscious of types as 'sets with structure'. The necessary structure is not itself represented in the programming language, but rather (at best) is a reflection of some intuition about how computations are carried out. This conflicts with our original intuition of types-as-sets.

The right response seems to be to adopt the more flexible notion of set inherent in the 'universes of constructive mathematics' called toposes. A category consists of 'objects' and 'maps' (or 'morphisms' or 'arrows') from one object to another; there are 'identity maps' for a notion of 'composition' with obvious axioms but no further structure. (Mac Lane's book [14] remains the best introduction to category theory for the mathematically educated.) Intuitively a topos is a category equipped with such structure as to make it an abstract category of sets and functions. Formally a topos may be defined to be a category equipped with (i) all finite limits, and (ii) power objects. One can think of the objects of a topos as (constructive) sets and the morphisms (or maps or arrows) as (constructive) functions between sets. The epithet constructive is in order as the internal logic of the topos is intuitionistic logic; the lattices of subobjects of an object may be Heyting algebras. The finite limits provide some finitary constructions on sets, while the power objects provide a full power set (set of all subsets) with associated membership relation. This gives a very rich essentially set-theoretic or (perhaps better) type theoretic structure. The standard reference to topos theory is still Johnstone's book [10], but see Bell's book [2] for a good account of the logician's perception of a topos as a world of constructive mathematics (model of type theory or local set theory).

Now, if we are prepared to take an undogmatic approach to foundations, we can readily recapture our basic intuitions about the lambda calculus. For the cardinality problems associated with Cantor's Theorem do not bite in constructive mathematics. There it is perfectly consistent that there be sets D which are equal to (or better, isomorphic to) the set D^D. In such models we have the technical and conceptual advantage of being able to argue (albeit constructively) as if we were dealing with arbitrary sets and functions. A description of the simplest kind of topos, in which one can find objects (that is constructive types) D with D^D isomorphic to D is given in [20]. (They are 'toposes of presheaves': they capture a very primitive notion of 'variable set'.)

Now topos theory does more than make our mathematical models run more smoothly. For once we can conceive of a world of sets in which there exist non-trivial models of the the pure lambda calculus, we come naturally to regard it as a defect of the classical universe of sets that it contains no such model. Considerations of this kind may lead to quite radical forms of relativism. (See the preface of the book by Lambek & Scott [13] for a hint of such a position.) But we need not go that far. Once we have an interest in alternative formal systems which can serve as a foundation for mathematics, the absolute status of the classical set-theoretic foundations must come into question.

3 The Theory of Types

3.1 Typed Programming Languages

The value of types in a programming language is that they provide a basic guide to the programmer who needs to structure complex programs. In particular in many typed languages it is possible to detect syntax errors at compile time, that is effectively as the program is being constructed. The usual analogy is with dimensional analysis in physics. We can write

$$\begin{aligned} x &\in L \\ v &\in LT^{-1} \\ m &\in M \\ F &\in MLT^{-2} \end{aligned} \tag{3.1}$$

to signify that x is a length, v a velocity, m a mass and F a force. Then we can deduce that

$$\frac{1}{2}mv^2 \in ML^2T^{-2}$$

and that

$$Fx \in ML^2T^{-2}$$

so that

$$\frac{1}{2}mv^2 = Fx$$

makes sense, while

$$\frac{1}{2}mv^2 = F$$

does not.

3.2 The Simple Typed Lambda Calculus

For our purposes we only need consider the type structure of this theory, and not the associated computation rules which give it meaning. (These are in fact just the rules of the pure lambda calculus restricted to typed terms.)

We start with some basic types and their associated constants and functions. For the simplest kind of programming, these might be the types *Nat* and *Bool* of Natural Numbers and Booleans, together with some basic constants and functions and with definition by cases (*if...then..., else...*). But the details are not important. The collection of all *Simple Types* is then generated by the single rule:

if A and B are types then so is $(A \to B)$.

The idea is that $(A \to B)$ is the type of all functions from elements of type A to elements of type B. Corresponding to this are the rules

($\to$-elimination) if $s \in (A \to B)$, and $t \in A$ then $(st) \in B$;

($\to$-introduction) if $r \in B$ given $x \in A$, then $\lambda x.r \in (A \to B)$.

Now we are forced to distinguish $f \in (Nat \to Bool)$ from $g \in (Nat \to Nat)$ and can apply neither to $t \in Bool$.

The reader may find it useful to check that in this calculus each of the Church numerals of 2.4 can be given the type

$$(A \to A) \to (A \to A)$$

independently of what A may be. This is a hint of genericity which cannot be explicitly handled by the simple typed lambda calculus.

The mathematical structure needed to model this calculus with full $\beta\eta$-equality is that of a cartesian closed category. The connection is explained in full detail in the book by Lambek and Scott [13]. There is a plentiful supply of cartesian closed categories, including the familiar category of sets. Thus the semantics of the simple typed lambda calculus is unproblematic. (Of course, it is not so straightforward to find models useful in computer science.)

3.3 The Syntax of the Second Order Lambda Calculus

It is altogether more problematic to describe the semantics of a system in which the genericity of functions in the simple typed lambda calculus is made explicit. Here is a brief sketch of the simplest such extension of the simple typed lambda calculus described above.

To the rules for *Simple Types* we add type variables $X, Y, Z, ...,$ and a further rule of type formation:

if A is a type and X is a type variable, then $\Pi X.A$ is a type

This gives us the collection of *Second Order Types*. The idea of the new rule of type formation is as follows. Call types with no free type variables constant types, and suppose for simplicity that A has at most the variable X free. Then for each constant type B there is a (constant) type $A[B/X]$ where B has been substituted for X in A. Then $\Pi X.A$ is the type of all (choice) functions from the collection of all constant types B which pick an element of the corresponding type $A[B/X]$. (So of course X is bound in $\Pi X.A$).

We simultaneously add further operations on terms:

(2nd order application) if s is a term and B a type then (sB) is a term;
(2nd order abstraction) if r is a term and X a type variable then $(\lambda X.r)$ is a term.

Then there are the typing rules:

(Π-elimination) if $s \in (\Pi X.A)$, and B is a type then $(sB) \in A[B/X]$;
(Π-introduction) if $r \in A$ with X a type variable, then $\lambda X.r \in (\Pi X.A)$.

The intended meaning of the term-forming operations should be clear enough from the typing rules. The meaning of second order application is that $s \in (\Pi X.A)$ denotes a function from types X to elements of $A(X)$, and then $(sB) \in A[B/X]$ denotes its value at B. To understand second order abstraction, we imagine that as X varies over types, r takes values in the corresponding types $A(X)$; then $\lambda X.r \in (\Pi X.A)$ denotes the corresponding function-as-object. (The new notions of abstraction over types and of application of terms to types give rise to new computational rules of $\beta\eta$-equality, but we do not go into these here.)

The extended system of types and terms which we have just described is the *second order lambda calculus*. It was first considered by Girard in the course of proof theoretic investigations [6] and rediscovered by Reynolds in the context of computer science [16].

3.4 Computations in the Second Order Lambda Calculus

In the second order lambda calculus we explicitly have generic types. For example, for every natural number n, we have

$$\lambda X.\hat{n} \in \Pi X.(X \to X) \to (X \to X),$$

using the notation of 2.4. In fact we can regard $\Pi X.(X \to X) \to (X \to X)$ as an implementation of the natural numbers, and so write $\mathbb{N}$ for this type. And we write $\bar{n}$ for $\lambda X.\hat{n}$.

Theorem 3.1 *Up to $\beta\eta$-equality the only terms of type $\mathbb{N}$ are of the form $\bar{n}$.*

This result says in effect that there are no non-standard natural numbers. The first explicit statement and proof that I know is rather late [3]. However the great expressive power of the second order lambda calculus was known to Girard [6] which includes (amongst many other things) a characterisation of the computations which can be coded as terms of type ($N \to N$).

Theorem 3.2 *For any function f, provably recursive in analysis, there is a term $t_f \in (N \to N)$ of the second order lambda calculus such that*

$$t_f \bar{n} \text{ reduces to } \bar{m} \text{ if and only if } f(n) = m.$$

And conversely, if $t \in (N \to N)$ is a term of the second order lambda calculus, then the function f defined by

$$f(n) = m \text{ if and only if } t\bar{n} \text{ reduces to } \bar{m}$$

is provably recursive in analysis.

Here, as usual in logic, 'analysis' refers to any formalisation of second order arithmetic with full comprehension. The provably recursive functions of analysis form a large class of functions containing all those total recursive functions which seem likely to arise in practice (and many many more).

3.5 Models of the Second Order Lambda Calculus

The question of what is a good notion of model for the second order lambda calculus is quite different in detail from the corresponding question for the pure lambda calculus, which we considered in section 2. However it is similar in impact and causes the same kind of heartache. There is an answer in accord with our intuitions: essentially we must have a collection of sets and (all) functions closed under sufficiently large products. (For the natural interpretation of $(\Pi X.A)$ is as the product of all sets $A(X)$.) But there turn out to be straightforward cardinality problems with this idea within the world of classical set theory. (A more general problem with models of the second order lambda calculus emerges from the analysis in a paper by Reynold [17].) As with the pure lambda calculus, the result is that many people work with a conceptually unilluminating answer. (The details are too awful to be worth sketching here; they can be found in the paper by Bruce & Meyer [4].) But again this is a situation in which constructive mathematics comes to the rescue. As first suggested by Eugenio Moggi, sufficiently complete collections of sets and functions do exist in suitable constructive universes. (For a general construction of such toposes see Pitts' paper [15].)

One particular kind of constructive mathematics (based on 'realizability') in which we can model the second order lambda calculus is very attractive. One of the least appealing features of traditional category theory (as developed in Mac Lane [14] for example) is the need for size restrictions (the solution set condition) in the fundamental Adjoint Functor Theorems. But these are not necessary for small categories. The problem for traditional category theory is that the only small complete categories are preordered sets. However, there are toposes containing very rich small complete categories and for these, the Adjoint Functor (and related) Theorems can be exploited in the very simple form appropriate to small categories. I give a sketch of this perspective is in [8].

Overall, the effect of the constructive view of models of the second order lambda calculus is much the same as that of the constructive view of the pure lambda calculus. We are driven emotionally from any view which gives the less conceptually rich world of classical set theory any primary status.

4 Conclusions

4.1 Universes of Constructive Mathematics

In this paper I have described how the ideas in a particular area of computer science make non-standard constructive worlds of mathematics seem very attractive. I am myself struck by how closely some of these mirror views about foundations sketched long ago by Kolmogorov in 1932 [12].

It is worth emphasising that the way we usually present toposes presupposes a definite classical set theory. However this is in no way essential and cannot be used to give any primacy to classical set theory. (Equally, I believe there are no strong arguments for any other kind of foundations.)

Quite generally, the concepts of classical set theory are inappropriate as organising principles for much modern mathematics and dramatically so for computer science. The basic concepts of category theory are very flexible and prove more satisfactory in many instances. As rightly stressed in 2.3 of Johnson's paper in this volume [9] much category theory is essentially computational and this makes it particularly appropriate to the conceptual demands made by computer science.

It is true that category theorists (in particular topos theorists) have for the last twenty years been engaged in activities that are antipathetic to traditional foundations. However since in most cases these activities have involved the exploitation of nothing more subtle than the possibility of doing algebra in a topos, the effect on our thinking about foundations has been small. The honourable exception is the study of Synthetic Differential Geometry initiated by Lawvere, where spaces of functions are exploited in a natural and elegant fashion. Regrettably this subject remains a minority interest. By comparison, the radical conceptual demands made by the concrete practice of computing have given the process of rethinking foundational questions a quite definite focus.

The emerging view does not make set theory redundant; it may remain a crucial component of the foundations of mathematics. But its place in the scheme of things looks radically different. In particular, the idea that there should be some definite foundation for mathematics in the traditional sense looks less secure.

4.2 Other Aspects of a Revolution?

I should emphasise that the story that I tell here is both partial and far from over. In the first place I have had to omit serious discussion of a number of areas of mathematical logic which have recently been transformed by questions arising in computer science. (In particular I regret not being able to discuss Girard's linear logic, which has recently challenged our view of what logic is. The curious reader might like to compare Girard's paper [7] with traditional books on logic.) And secondly we are in the midst of very exciting times in the development of an abstract mathematical view of logic. Mathematicians may still refer to category theory as abstract nonsense, but the IT revolution has transformed this abstract nonsense into a serious form of applicable mathematics. In so doing, it has revitalized logic and foundations. The old complacent security is gone. Does all this deserve to be called a revolution in the foundations of mathematics? If not maybe it is something altogether more politically desirable: a radical reform.

References

[1] Barendregt, H. P., (1981). *The Lambda Calculus, its Syntax and Semantics.* North-Holland, Amsterdam.

[2] Bell, J. L. (1988). *Toposes and Local Set Theories.* Clarendon Press, Oxford.

[3] Böhm, C. and Berarducci, A. (1985). 'Automatic synthesis of Typed Λ-Programs on Term Algebras' *Theoretical Computer Science* 39, 135-154.

[4] Bruce, K. B. and Meyer, A. R. (1984). 'The semantics of second order polymorphic lambda calculus.' In: G. Kahn et al (eds), Semantics of Data Types, *Lecture Notes in Computer Science 173*, (Springer-Verlag, Berlin), 131-144.

[5] Burstall, R., (1991) 'Computer assisted proof for mathematics: an introduction using the LEGO Proof System', in *The Mathematical Revolution Inspired by Computing*, J.H. Johnson & M.J. Loomes (eds), Oxford University Press, (Oxford)

[6] Girard, J.-Y. (1972). 'Interprétation Fonctionelle et Elimination des Coupures dans l'Arithmétique d'Ordre Supérieur.' Thése de Doctorat d'Etat, (Paris).

[7] Girard, J.-Y. (1987). Linear Logic. *Theoretical Computer Science* 50, 1-102.

[8] Hyland, J. M. E. (1988). 'A small complete category.' *Annals of Pure and Applied Logic*, 40, 135-165.

[9] Johnson, J. H., (1991) 'An introduction to the mathematical revolution inspired by computing', in *The Mathematical Revolution Inspired by Computing*, J.H. Johnson & M.J. Loomes (eds), Oxford University Press, (Oxford)

[10] Johnstone, P. T. (1977). *Topos Theory* Academic Press, (London).

[11] Kleene, S. C. and Rosser, J. B. (1935). 'The inconsistency of certain formal logics'. *Annals of Mathematics* (2) **36**, 630-636.

[12] Kolmogorov, A. N. (1932). 'Zur Deutung der intuitionistischen Logik'. *Mathematische Zeitschrift* **35**, 58-65.

[13] Lambek, J. and Scott, P. J. (1986). *Introduction to higher order categorical logic.* Cambridge University Press, (Cambridge)

[14] Mac Lane, S. (1971). *Categories for the Working Mathematician.* Springer-Verlag, (Berlin)

[15] Pitts, A. M. (1987). 'Polymorphism is Set Theoretic, Constructively'. In: D. H. Pitt et al (eds), *Category Theory and Computer Science, Lecture Notes in Computer Science 283,* Springer-Verlag, (Berlin), 12-39.

[16] Reynolds, J. C. (1974). 'Towards a Theory of Type Structure.' In: Programming Symposium, *Lecture Notes in Computer Science 19,* Springer-Verlag, (Berlin), 408-425.

[17] Reynolds, J. C. (1984). 'Polymorphism is not set-theoretic.' In: G. Kahn et al (eds), Semantics of Data Types, *Lecture Notes in Computer Science 173,* Springer-Verlag, (Berlin), 145-156.

[18] Rogers, H. (1967). *Theory of Recursive Functions and Effective Operations.* McGraw Hill, (New York)

[19] Scott, D. S. (1972). 'Continuous Lattices.' In: F. W. Lawvere (ed), *Toposes, Algebraic Geometry and Logic, Lecture Notes in Mathematics 274,* Springer-Verlag, (Berlin), 97-136.

[20] Scott, D. S. (1980). 'Relating theories of the lambda calculus.' In: J. R. Hindley & J. P. Seldin (eds), *To H. B. Curry: essays on combinatory logic, lambda calculus and formalism* (Academic Press, London), 403-450.

[21] Simpson, D. (1991) 'A Euclidean Basis for Computation', in *The Mathematical Revolution Inspired by Computing,* J.H. Johnson & M.J. Loomes (eds), Oxford University Press, (Oxford)

[22] Turing, A. M. (1937). 'Computability and λ-definability.' *Journal of Symbolic Logic,* **2**, 153-163.

The Mathematical Revolution Inspired by Computing.   J H Johnson & M J Loomes (eds)
©1991   The Institute of Mathematics and its Applications.   Oxford University Press

The Development and Use of Variables in Mathematics and Computer Science

Meurig Beynon & Steve Russ

Department of Computer Science, University of Warwick, CV4 7AL

Abstract

There is a wide variety of uses of variables in mathematics which we cope with in practice through conventions and tacit assumptions. Experience with computers has made us articulate, criticise and develop these assumptions much more carefully. Historically the term 'variable quantity' was introduced in the context of describing and calculating changing quantities which corresponded to phenomena in the observable world (e.g. the velocity, or fluxion, of a moving body). The evolution of the concept has divorced it from these 'roots of reference' and required us to establish the formal apparatus of interpretation and valuation. While the changes considered are highly structured this may be satisfactory, but computing power invites us to cope with change in vastly more complex, unstructured situations such as in simulation of 'real world' processes. We relate this challenge to the distinctive differences in the use of variables in mathematics and practical computing, and we develop a general framework in which all uses of variables can be described in a unified way.

1 A general framework for variables

1.1 State and the notion of primitive variable

From about 3000 BC until about 10 years ago people occupied with engineering, navigation, commerce, science or mathematics depended to a great extent on the construction and use of large tables of data. These tables formed a vital part of their capacity to organise, explain and predict changes in the world. They ranged from Babylonian tablets recording astronomical data alongside data about the weather, the frequency of earthquakes and prices of wheat, through later tables for such matters as high-tides, mortality and compound interest, up to the 4-figure logarithm tables used by generations of school-children until the late 1970's. Such systematic association of different values with the same family of symbols illustrates the use of a primitive concept of variable that will be significant in the general framework to be proposed. Such primitive variables are typically used for describing *state*.

There are two fundamental elements involved in our notion of a primitive variable: a *symbol* and the capacity for that symbol to be associated with a *value*. A variable (of any description) is not the *name* of a value, like π, since it is characteristic that different values may be associated with the same variable (for example, choosing different points on a parabola or computing successive values of a counter in a loop). It is often convenient and powerful not to associate any value with a variable but only to manipulate it in ways appropriate to its type. The capacity to be associated with different values or no value is reflected in the picture of a variable (common in computing) in which we associate with a symbol a unique 'location' which may, or may not, 'contain' a value. A *primitive variable* serves this function of a 'marker' for a pigeon-hole. This expresses the idea of 'capacity to be associated with a value', while also matching the way our minds seem to deal with different values being associated with concepts. For example, if x metres is the height of a projectile, when we make a program declaration, 'x : real;' there is a physical memory location associated with x; and when we say 'as x increases, at some time it will be greater than ... ', we mean by 'it', not the symbol x but the value associated with the height of the projectile and so with x.

In the past, our methods for recording complex state information were very laborious, and of limited scope. A major contribution of mathematics was to provide means of simplifying the representation of state. By identifying relationships between variables, it became possible to reduce the amount of information to be recorded in order to characterise the *state of a system*. In this way, modern mathematics has a central interest in systems whose behaviour can be understood in terms of states in which the possible values of variables form a class: $\{v_1, v_2, v_3, \ldots | P(v_1, v_2, v_3, \ldots)\}$, where P is a system of relationships that might include explicit definitions of one variable in terms of others, or an equational constraint governing several variables.

In mathematics an isolated primitive variable has only a minor role. The applications for which there are no relationships between variables to be recorded, either because we are unable to perceive them or because they don't exist, have traditionally been divorced from mathematics. The nearest equivalent to the mathematical equation or functional relationship may be the methodologies by which we seek to systematise cataloguing, and the theories by which we organise species. The advent of the computer has revolutionised our power to represent state. By physically constructing arrays of millions of Boolean variables that can be freely assigned, we can in principle conveniently record many more states than by any other means. Through our dependence upon computer records, the whole economy of advanced nations has come to rely upon such power. A potent side-effect of practical computing has been to liberate the potentially anarchic primitive variable upon which procedural programming at every level of abstraction depends, and to challenge mathematics to discipline its use.

1.2 The uses of variables and the notion of context

In modern practice a mathematical variable is used in a wide variety of ways. It is used as an indeterminate, as an arbitrary constant or as an unknown value to be found. It is used to express relationships and functions, as a parameter to a class of problems or objects, as a dummy (bound) variable in the context of operator symbols like summation and integration, and it is used in the context of limit operations. Comparing the notion of a primitive variable with the apparently more complex mathematical variable raises the issue of whether we should be speaking of different kinds of variables as well as different uses of variables. The question will become even more pertinent when we include variables as used in spreadsheets and variables over compound data structures in both computing and mathematics.

We shall use the term *valuation* to mean a relation between a set of primitive variables and a set of values of appropriate types. Then a *context* may be defined as a set of primitive variables together with a means for providing their valuations. An example of a valuation is the state of a computer system or experimental environment in which the set of values have been (or at least could be) 'observed' simultaneously. In a different kind of context some or all of the values in any valuation may be related to one another (e.g. the co-ordinates of points on a parabola). Contexts and their valuations can be viewed in many ways and form a convenient unifying framework for discussing different uses of variables. When using procedural variables and directly assigning valuations we may be interested in the transitions from one valuation (state) to another as a result of a program statement. A mathematical use of variables in writing equations or other relationships will lead us to consider the set of all possible valuations implicitly defined by the equations. To understand the use of a variable in a problem or theory it is necessary to know how valuations can be generated in the context for that variable. This might, for example, be a matter of function evaluation, of assignment of some variables and observation of others, or by the updating of a spreadsheet.

It is the main claim of this paper that all the uses of variables we find in mathematics and computing can be described in terms of this general framework of a context. Given a set of variables which are being used together in some way (for the solution of a problem, or within a procedure, for example), the relevant context will contain a corresponding set of primitive variables ('corresponding' in the sense of using the same symbols). But whereas the use of the original variables might be rather complicated, the use of the corresponding primitive variables is simple—serving only to mark the association of a symbol with a value. The distinctive, possible uses of the original variables relate to the ways in which valuations of the context arise (i.e. valuations of the primitive variables in the context). On the above issue of 'kinds versus uses' we therefore propose that there is only one kind of variable—the primitive variable—but numerous more or less sophisticated uses.

By way of illustration, the following references to a variable x : $x := y + 34$ (procedural program), $\{(x,y)|x = y + 34\}$ (problem-solving), $x = y + 34$ (spreadsheet) have very different semantics, and assume different characteristics of the variable. We can nonetheless think of these diverse constructions as representing particular assertions about the value, or actions to be performed upon the value associated with x. For instance, in the context of a particular valuation of a set of primitive variables in which x occurs, $x := y + 34$ denotes the action of assigning to x the value of y plus 34. That is to say, a procedural variable is associated with transition when valuations are interpreted as states. In contrast, $\{(x,y)|x = y + 34\}$ is the set of possible valuations of (x,y) such that the value of x equals the value of y plus 34.

Finally the spreadsheet formula $x = y + 34$ expresses the fact that in any particular valuation the value of x is the result of adding 34 to the value of y, and that x will be updated so that this relationship still pertains if y is subsequently redefined.

The most common use of variables consists of the expression of relationships and their manipulation within the appropriate language system (e.g. in number theory, logic, plane geometry, functional programming); this usually depends only on the type of the value, not on any particular valuation. With some uses (e.g. in cost-analysis, weather-forecasting, procedural programming) where there are few relationships known or simple enough to be written down, it is the changing valuations that matter most. Yet other uses of variables (e.g. in spreadsheets) involve some characteristic combination of these requirements. All of these uses can be described in terms of a context and its valuations and the remainder of the paper is largely devoted to illustrating and elaborating this claim.

2 Mathematical Variables and Problem-Solving

2.1 Historical development

It was not until the second half of the 16th century that developments occurred in the use of symbolic notations that could really merit description as the beginnings of algebra. François Viète (1540–1603) was among the first clearly to be aware of the significance of symbols that generalise numbers or magnitudes. He introduced the use of a vowel for a quantity assumed to be unknown or undetermined and consonants for quantities that are assumed to be given. The idea of 'given' quantities being represented by a letter is the beginning of the modern role of variables as 'placeholders' for values. Thus Boyer remarks, "Here we find for the first time in algebra a clear-cut distinction between the important concept of a parameter and the idea of an unknown quantity" [11]. In 1636 Fermat (who had studied Viète) was writing equations like :

$$A \quad in \quad E \quad aeq \quad Z \quad pl.$$

The Latin 'in' means 'times' and 'pl' is an abbreviation for 'planus'.) He showed that the corresponding locus is a hyperbola (cf. $xy = c^2$)[13]. Viète and Fermat (at this time) still regarded the magnitudes associated with letters in a geometrical way so that a product was an area etc.

Descartes, in *La Geometrie* [12], introduced a radically new vision of mathematics in which he regarded all quantities to be of the same one-dimensional kind (so capable of taking the same range of values). Even when a solution to an equation was obtained geometrically Descartes used Viète's conventions to distinguish coefficients and variables, and to express the solution as a general algebraic solution to a whole class of equations. Throughout the work of Descartes and Fermat there is evidence of a major movement from the primarily geometric mode of thought which was widespread at the beginning of the 17th century to an algebraic mode of thought which was becoming dominant in the outlook of most of the leading mathematicians by the end of that century. The increasing use and understanding of variables over this period was a major factor in making possible the much greater generality and algorithmic power of the algebraic approach.

The rapid progress of analytical geometry led to many authors by the mid-17th century dealing with higher order curves (cubics etc) and transcendental (i.e. non-algebraic) curves such as the cissoid, cycloid and quadratrix. Newton studied *La Geometrie* in 1664 and mastered the algebraic methods there thoroughly although this in no way affected the fact that his outlook was always primarily geometric. He also had no hesitation about the incorporation of motion into geometry and mathematical methods. The independent development of the calculus by Newton in 1666 [16] and by Leibniz in 1684 [15] illustrated very well the fact that profound ideas and methods could be expressed in both geometric and algebraic patterns of thought. This has significance for our theme in that, "Newton considered variables as changing in time. He applied concepts of motion; the variables were considered as flowing quantities .... . In modern terminology one might say that Newton considered all variables as functions of time .... . Leibniz considered variables as ranging over sequences of infinitely close values. In his calculus there is little use of concepts of motion." [2]. Newton in fact seems to have taken velocity as a primitive concept. It is likely to have been the prevalence of such kinematic ideas which promoted the use of the term 'variable quantity' (and hence 'variable'). For example, here are some samples of Newton's language suggesting that he has in mind an abstraction from the variable quantities which we can observe:

" I consider mathematical quantities in this place not as consisting of very small parts; but as described by a continued motion. ......Therefore considering that quantities, which increase in equal times, and by increasing are generated, become greater or less according to the greater or less velocity with which they increase and are generated; ...... calling these velocities of the motions or increments fluxions, and the generated quantities fluents. ....Let a, b, c,d, &c. be determinate and invariable quantities, and let any equation be proposed involving the flowing quantities z, y, x, &c. as $x^3 - xy^2 + a^2z - b^3 = 0$. ...." [17]

While there was no analytic concept of function (for Newton a function was identified with a curve), the notion of variable was, in at least some of its uses, playing the role of a function [10]. For example, the use of variables to express a dependency relation between the abscissae and ordinates of points on a curve was in many cases expressing a functional relationship. The 'fluents' used by Newton can usually be thought of (in modern terms) as functions of time. So the history of the notion of a mathematical variable is inextricably linked with the history of the function concept [22]. Not surprisingly the association of time and motion with the idea of a variable quantity and a function led throughout the 18th century to physical intuitions of dynamic behaviour obscuring the possibility of an arithmetic concept of the continuity of a function. The same sort of problems arose with the convergence of infinite series which the operational freedom of algebraic symbolisms made available long before they were understood. (Already in 1666 Newton was using infinite series to express areas under various curves [18].) It was not until 1817 that Bolzano published a paper ([9],[19]) in which the modern concepts of convergence and continuity are made clear and used for the first time. An essential ingredient of Bolzano's work is the clear understanding of the need for two distinct variables to define each of these concepts and the significance of arbitrary choice for the way values are assigned to one of them—the 'ϵ' of usual modern forms—further details appear in [20]. At this stage the dynamic concept of variable as a near-relation of varying quantities has been clearly replaced by a variable for which the changes in valuation arise in a precise context and in a specifiable manner. The mathematical variable had by now 'come of age' and was being used fluently as a placeholder and vehicle for expressing relationships of a quite abstract nature.

2.2 Problem-solving and the 'such-that' construction

Historically, variables have been closely associated with problems of finding unknowns from givens. The study of such problems has been so influential in the development of mathematics that it is tempting to regard all mathematics as a problem-solving activity. Computer science has in some respects reinforced this view. A large body of theoretical computer science is concerned with the solution of algorithmic problems that take the form of finding unknowns from givens. Declarative approaches to programming can be seen as an attempt to cast all computation into this paradigm.

A typical use of variables in formulating a mathematical problem requires a logical framework established by naming function and predicate symbols defined over a domain, or a sorted family of domains. The constraints of the problem to be solved are expressed by means of a system S of well-formed formulae containing free variables $v_1, v_2, v_3, \ldots$. On fixing an interpretation of the domains, and the function and predicate symbols, the system S implicitly defines a context (set of valuations) which is the set of sorted values that can be substituted for the free variables so as to satisfy all the formulae in S: $S(v_1, v_2, v_3, \ldots) \equiv \{v_1, v_2, v_3, \ldots | \forall P \in S\,.\,P(v_1, v_2, v_3, \ldots)\}$. A set of valuations generated implicitly in this way is what we mean by the 'such-that' construction and it serves to represent a very general class of problems. It is important to note that the use of variables as a means of representing all possible valuations does not in itself distinguish one valuation from another, nor does it describe any relationship between valuations. In this context, the class $S(v_1, v_2, v_3, \ldots)$ is being used to distinguish two kinds of valuation: $viz.$ those that satisfy the system of formulae S, and those that do not. In describing a problem in this fashion, some of the free variables used in specifying the problem take the form of parameters to be given for each instance of the problem, and the rest are the variables to be instantiated appropriately.

3 Variables in computer science

3.1 Problem-solving versus modelling and simulation

The computer is a physical system with a superhuman capacity to represent state. Each memory location corresponds to a primitive variable, and each assignment of values to memory locations defines a valuation, or state. The set comprising all such valuations is one possible context within which a computation may be viewed as taking place. The primitive operations that the processor can perform are the means by which transitions occur, and a program essentially comprises a prescribed sequence of such transitions.

There are two principal ways in which computers can be used. We can exploit their power to perform routine computation fast and accurately when following a particular recipe for the solution of a given problem. We can also use a computer to record state information. These uses correspond to two quite different ways in which contexts (in the technical sense of Section 1) are typically specified. In problem-solving, a system of relations between variables is used to define a set of valuations implicitly. In modelling physical systems with which we interact, observed or specified system states are represented by sets of valuations of primitive variables. The former is associated with mathematical variables as they occur in the 'such that' construction, and describes a context declaratively. The latter is associated with the assignment to primitive variables, and determines a context procedurally. In a *declarative program*, there are explicit relationships between variables, but only implicit valuations. In a *procedural program*, the valuations are explicit, and the relationships between variable values implicit.

The potential conflict between the declarative formulation of a problem, and the procedural nature of its computer solution is evident. A most striking aspect of problem-solving by computer is the manner in which computation has to be expressed using variables whose values can be freely assigned, and can only be constrained implicitly by imposing an order upon the procedures that manipulate them. The relationship between these valuations within the machine and problem-solving at the highest level of abstraction may be very obscure. For example, there is no reason to suppose that a valuation of the variables significant in the problem should be encoded in any intermediate state of the computation. The principal difficulty in describing computer solutions to problems can be seen as identifying the states through which the computation must pass in the construction process. The need to unify a procedural and a declarative use of variables has led to the development of methods of describing and manipulating values that we can interpret as novel uses of variables within the framework introduced above. For example, there is spreadsheet programming and definition-based programming as described below.

3.2 Procedural Versus Declarative Programming

The procedural use of variables serves as a direct way to capture state. Procedural methods reflect the manner in which state is manipulated in a conventional computer, but their adoption has provoked controversy. Some extreme criticism is made of the concept of variables that are given different valuations at different times within the same program. For the advocates of pure declarative programming: "variables always denote the same quantity, provided we remain in the same context" [8]. This concept of *referential transparency* seems to involve a restrictive interpretation of the term *context* (*viz.* that supplied by a specific fixed valuation), rather than a legitimate interpretation of the usual concept of variable. After all, the semantics of variables is determined not by a single chosen valuation, but by a set of possible valuations.

Procedural methods are characterised by the explicit manipulation of valuations. The simplest procedural approach to problem-solving can be viewed as manipulating the valuations of the characteristic variables of the problem directly, as when enumerating the possible set of valuations in solving a combinatorial problem such as the 8 Queens, or computing integers P and Q such that $Pp + Qq = gcd(p, q)$ by the extension of Euclid's greatest common divisor algorithm. In such programs we can think of the characteristic variables of the problem as corresponding abstractly to literal machine locations that are set to initial values and attain appropriate values on termination.

Declarative programming methods have evolved in response to the difficulties of representing relationships in a procedural framework, especially in connection with problem-solving. The most direct way to support mathematical problem solving on a computer is to develop a programming system that permits the specification of relationships between variables and as far as possible automates the extraction of satisfying valuations. As explained in detail by Kowalski [14], logic programming aims to automate problem-solving in just this way.

An alternative declarative approach is adopted by functional programming. Its underlying philosophy is that all computation is equivalent to the evaluation of an expression over a rich and sophisticated algebra, and that program development entails the formulation and manipulation of such expressions. In pure functional programming, explicit functional relationships between variables are established by a script that determines the environment for all variable evaluation.

Pure functional programming systems most effectively eliminate state. In a referentially transparent framework, there is no scope for using valuations for state information; supplying alternative parameters has no side-effect upon the environment. Though variables representing higher-order functions can express very subtle parametric systems, the limitations of a purely functional approach where state information is concerned stem from adherence to a single environment. Several methods of capturing state have been proposed (cf. Backus [1]). Some involve the encapsulation of history in values. As studied in data flow models of computation, this involves variables that are in certain respects analogous to Newton's flowing variables (cf. Lucid [21]).

3.3 A Definition-Based Approach To Programming

We shall briefly consider a novel approach to programming which combines characteristics of both procedural and declarative styles and that has been under development at the University of Warwick over several years. This work was originally concerned with interaction between the user and computer, but has developed in many different directions. A central theme in this research is the representation of state as it appears to an agent participating in a computation through a use of variables that integrates both declarative and procedural characteristics. An extended exposition is beyond the scope of this paper; we shall merely identify the main points relevant to this paper. For more details see [3], [4],[5],[6].

The central abstraction in our approach ('definitive = definition-based programming') is variable definition. As a simple example, the integer variable a may be defined by the formula $a = b + c$. This is to be interpreted as saying that the value of a is specified as the sum of the values of the variables b and c. Such definitions of variables resemble those of a functional programming script; but definitive programming differs from functional programming in that redefinition of a variable is a legitimate action within the programming paradigm. If for instance the value of the variable b is incremented, the value of a is incremented by the same amount. As in functional programming, some procedural action is regarded as invisible, *viz.* the mechanisms by which the value of a variable is guaranteed to be consistent with its definition at all times. This automatic updating of variables according to their definitions is the essential principle underlying the spreadsheet.

The use of variables in definitive programming has many points of contact with the other paradigms discussed above. Of particular interest is the way in which a definitive program can effectively 'edit its own script'.

From the perspective of this paper, this integrates the manipulation of relationships that apply to the value of a variable and changing valuations. This feature is especially important in applications such as design, where the relationships between variables must be fluid and easily modified within the programming system [4].

4 Conclusion

We have argued that the many uses of variables can be described in terms of an underlying concept of primitive variable that serves to establish an association between symbol and value. This association has to be interpreted with reference to a context defined by a set of valuations of a family of primitive variables. Such contexts are familiar as models of physical systems that we observe and manipulate. Contexts provide the cognitive basis for a variety of abstractions, and their description and manipulation corresponds closely to different uses of variables. Each method of using variables deals with the set of valuations of a set of primitive variables in a characteristic way. In some cases, it is more natural to think of each valuation as a state, and to consider possible transitions between one state and another. In other cases, the entire set of valuations—rather than the individual valuations—has primary significance. Reference to values through their symbolic variable names is required in whatever manner we choose to address the set of valuations. The semantics of the variable in each case is derived from the nature of the assertions being made about values, or the actions being performed upon them.

We have seen how the use of variables in mathematics illustrates many different ways in which contexts are described and manipulated. In mathematical argument, or in the process of constructing a solution to an algorithmic problem, we may focus attention upon a particular valuation, or construct sequences of valuations. In modern computing, the importance of state-transition models of computation has revived interest in the explicit manipulation of valuations. The need to develop programming paradigms that are well-suited to problem-solving, and to modelling and simulation, has forced us to re-examine the relationship between state-based and relational views of primitive variables in a context, and led us to seek a more satisfactory synthesis of the procedural and declarative computational perspectives they respectively represent. In particular, it appears that more powerful methods of dealing with sets of valuations as states and state transitions are required, and that it is necessary to consider when, and by what agents, state changes are enabled. We have proposed the application of 'definitive methods', a generalisation of the principle underlying the use of variables in a spreadsheet, as a way of addressing some of these problems. The interested reader is referred to [7] for a fuller version of this paper.

References

[1] Backus, J., 'Can programming be liberated from the Von Neumann style?' *Communications of the ACM*, **21**,8 (August), 613–641,1978.

[2] Baron, M.E., and Bos, H.J.M., *History of Mathematics : Origins and Development of the Calculus 3* , Open University Unit AM289 C3, page 55, The Open University, Milton Keynes, MK7 6AA, 1974.

[3] Beynon, W.M., *Definitive principles for interactive graphics*, NATO ASI Series F: 140, 1083–1097, 1988.

[4] Beynon, W.M., 'Definitive programming as a framework for design', *Preliminary Proceedings Third Eurographics Workshop on Intelligent CAD Systems*, CWI, 325–335, 1989.

[5] Beynon, W.M., Norris, M.T. and Slade, M.D., 'Definitions for modelling and simulating concurrent systems', Proceedings IASTED Conference ASM 1988, Acta Press, 94–98, 1988.

[6] Beynon, W.M., Slade, M. and Yung, Y.W., 'Parallel Computation in Definitive Models', CONPAR'88. BCS Workshop Series, Cambridge University Press, 359–366, 1989.

[7] Beynon, W.M. and Russ, S.B., 'Variables in Mathematics and Computer Science', Research Report 141, Department of Computer Science, University of Warwick, 1989.

[8] Bird, R.and Wadler, P., *Introduction to Functional Programming*, Prentice-Hall, page 4, 1988.

[9] Bolzano, B., *Rein analytischer Beweis des Lehrsatzes....* , Prague, 1817.

[10] Bos, H.J.M., 'Differentials, Higher-Order Differentials and the Derivative in the Leibnizian Calculus', *Archive for History of Exact Sciences*, 14, Sec.1., 1974

[11] Boyer, C., *A History of Mathematics*, Wiley, page 335, 1968.

[12] Descartes, R., *La Geometrie*, 1637; tr. Smith, D.E. and Latham, M.L., *The Geometry of Rene Descartes*, Dover, 1954.

[13] Fermat, P., *Isagoge ad locos planos et solidos*, 1679. [written 1636].

[14] Kowalski, R., *Logic for Problem Solving*, North-Holland, 1979.

[15] Leibniz, G., 'Nova methodus pro maximus et minimis ...... ', *Acta Eruditorum*, 1684.

[16] Newton, I., *De Analysi per Aequationes Numero Terminorum Infinitas*, 1669.

[17] Newton, I., 'Tractatus de quadratura curvarum' in *Opticks* , 1704 , tr. Struik , D.J., *A Sourcebook in Mathematics, 1200-1800*, Harvard University Press, p.303., 1969.

[18] Newton, I., 'Epistola Posterior', 1676, in Fauvel, J. and Gray, J. (eds) *The History of Mathematics : A Reader*, Macmillan and Open University, page 404, 1987.

[19] Russ, S.B., 'A Translation of Bolzano's Paper on the Intermediate Value Theorem', *Historia Mathematica* 7, 156–185, 1980.

[20] Russ, S.B., 'The Mathematical Works of Bernard Bolzano published between 1804 and 1817', Open University, 1980, unpublished Ph.D thesis, Ch.4.

[21] Wadge, W., Ashcroft, E.A., *Lucid - the dataflow programming language*, Academic Press, 1985.

[22] Youshkevitch, A.P., 'The Concept of Function up to the Middle of the 19th century', *Archive for History of Exact Sciences*, 16, 1976.

The End of the Defensive Era of Mathematics

C. Ormell

School of Education, University of East Anglia

Norwich NR4 7TJ

Abstract

It is argued that until recent times the applicability of mathematics has been poorly understood, not only by the educated public at large, but also by mathematics users and mathematicians themselves. Successful applications of mathematics have been widely recognised *after* the event, but have rarely been foreseen. This, it is claimed, is the fundamental reason why mathematicians throughout history have treated successful applications as an 'extra' or 'bonus' on top of the business of doing mathematics. The root reason for this state of affairs, it is claimed, is that successful applicability relies on computability and in the pre-computer era computability was an hit or miss affair depending on whether the power of conceptual simplification could be brought to bear on a problem. In the past the educated public and mathematics users tended to over-estimate the power mathematics could offer to solve numeric, geometric and logical problems. This put a pressure on applied mathematicians to obtain answers, which they were often unable readily to deliver. It produced a ***defensive*** attitude among mathematicians towards would-be users of mathematics. The arrival of copious computing power together with a cogent account of the final *purpose* of an application, enables us at last to throw off this defensive attitude and the associated pure-foremost conception of mathematics. This is a subtle but historic change in the relationship between mathematics and the rest of society. It implies a new role for the mathematician in society and a new form of mathematical education to prepare him or her for this new approach.

1   The Defensive Era

It is necessary to begin by showing in some detail how mathematics has been stuck in a defensive rut, before going on to demonstrate this 'Defensive Era' in mathematics coming to an end. By a *defensive* stance I mean the kind of attitude exhibited by a person who behaves *defensively* under questioning in ordinary conversation. He or she does not *initiate* lines of inquiry, but merely responds, somewhat reluctantly, to them. This reluctant response says it all: there is no attempt to inhabit the universe of discourse of the other, or positively to *help to clarify* the matters under discussion. Instead, what shows through is a resistance to such involvement, a lack of appetite for mutual enlightenment, a grudging compliance with the conventions of conversation, but no more.

Some have argued in the past that the potency of mathematics in terms of its applications was commonly over-rated, so that any slight defensiveness on the part of mathematicians was simply the natural defensiveness of someone being asked an inappropriate question stemming from a major misapprehension, e.g. a person wrongly assumed to be very rich, being asked to contribute to a prestige fund. From this point of view mathematicians will, of course, *appear* a little 'defensive' when assailed by questions presuming that mathematics has a greater applicative potency than it actually has.

Today less is heard of this argument: because the sheer extent of the applicability of mathematics + computing has become public knowledge. For example, John Bowers in his *Invitation to Mathematics* [1], shows his preference by the statement that "...mathematics is best studied as pure mathematics..." having previously conceded that: " ...applied mathematics is all the mathematics that has been applied. At present this covers more than half of mathematics...". Bowers 'knows' that mathematics is now recognised as being a powerful applicative discipline, but he does not *show* that he fully realises the significance of this fact: namely that it alters the balance of meaning in mathematics and makes an approach via applicability more intelligible for many readers.

The essence of the reason for the 'defensiveness' of pre-computer mathematics is, I think, to be found in the fact that mathematicians, insofar as they contributed to other areas of society via the illuminative effect of applications, could ever only answer a *fraction* of the questions put to them. The classic example of a question put to mathematicians in the pre-computer age which they could not answer was the 'Three Body Problem'. The question was easy to conceptualise and not difficult to formulate mathematically: but apparently impossible to solve! [1]

It was always extremely easy to stump the resources of pre- computer mathematics with questions drawn from the frontiers of technology and science.

More generally, the fact that mathematics *took* its principal challenge-problems from physics, astronomy and engineering for more than two thousand years implies that mathematics was not really setting its own agenda. People came to the mathematician with extremely difficult mathematical problems. Hardly surprisingly he was often unable to solve them; though the issues were, of course, borne in mind as important problems needing eventual resolution.

After the emergence of 'Modern Mathematics', mathematicians discovered that they could create their own agenda, *provided that it was a pure mathematical agenda*. Those who seized this opportunity created a new stance *vis-a-vis* the non-mathematical public's request for mathematical answers, namely, one of defiance. So, if we date the emergence of 'Modern Mathematics' to around 1830, there was a wholly 'defensive' era from the beginnings of the subject up to about 160 years ago, and since then there has been a *mixed* era in which some mathematicians were

---

[1] The problem of predicting what would happen, in the general case, to three moving, gravitationally mutually attracting bodies. Other problems in the same case include the general pursuit path problem and the travelling salesman problem. A Twentieth Century example from Physics is the failure of pre-computer mathematics to solve the equations of Wave Mechanics for any but the simplest atomic systems.

'defensive' and others were, in effect, 'defiant' of society, because they "washed their hands of applications"! [2] Of course not all those who worked in so-called *Modern Mathematics* did so from an autonomous agenda: some parts of this area are simply modern examples of applicable mathematics, and when pursued with applicability in mind are in the same case as the older applicable topics such as calculus.

When one says that there was a long defensive era of mathematics and that even today some mathematicians are essentially 'defensive' in their attitude to the question-askers, one is not saying that these mathematicians were not confident of their subject, proud of their skills, etc.

A similar distinction can be drawn in sport, where some of the most formidable tennis players have stayed, defensively, on the baseline, and some of the most successful football clubs have shown great restraint in their initiation of attacks.

The full significance of the 'defensive' era in mathematics can only be brought out by considering its opposite. Many people have so deeply internalised the attitudes of defensive mathematics that they cannot see them as anything other than normality, commonsense, etc. My thesis is that the defensive era has now ended. When we have looked at the new *non-defensive* mathematics we shall return to the defensive era to consider some of its less obvious implications, and the role of computing in ending it.

It should be said at this stage that the mere arrival of computing power need not necessarily have resulted in a change from the 'defensive' stance. An applicable mathematician with a supercomputer may still operate defensively, not getting involved in the thinking from which his or her problems have arisen.

2 The New applicability of Mathematics

Around 1960 the idea of *mathematical modelling* suddenly came into circulation in mathematical circles. It implied a new attitude towards the applicability of mathematics quite distinct from the style of *applied mathematics* which had previously been accepted. It implied (a) that mathematics could be used to describe chunks of reality; systems, situations... not merely bits of a problem. It implied, (b) that the *description* achieved might be at various levels of accuracy; the platonist premise that mathematics automatically gave one the 'inside story' about physical reality disappeared. It implied, (c) that the attitude of the mathematician could be akin to play; that one would look at or discuss the situation in a relaxed way, not necessarily converging straightaway onto a particular parameter or answer.

Since 1960 most mathematicians who are concerned with applying mathematics to the real world have conducted this 'applying' through the vehicle of mathematical modelling. But there has not been much explicit philosophising on, or reflection

---

[2] For example, G. H. Hardy, J Dieudonne: "The 'real' mathematics of the 'real' mathematicians, the mathematics of Fermat and Euler and Gauss and Abel and Reimann, is almost wholly useless ... " [4]page 58. "So one may say that in principle modern mathematics, for the most part, does not have any utilitarian aim, and that it constitutes an intellectual discipline, the 'utility' of which is nil." [3]

about, the profound key-shift which this change entails. The lack of an explicit new account of the applicability of mathematics has meant that the changeover to 'modelling' mathematics has been a fairly muddy and confused exercise [3]. To some the new emphases were scarcely more than slogans, and they simply carried on as before, taking care to talk about what they were doing in the new terms. The arrival of the computer gave them additional power (to arrive at answers) but, apart from that, nothing fundamental had changed.

In my opinion this is a mistake. The idea of mathematical modelling contains within itself the embryo of a 'new applicability' for mathematics. This new applicability may be stated baldly as the thesis that mathematics can be most usefully and splendidly applied to the *possibilities* of practical life and theoretical science. Mathematics provides society with a sort of 'illuminative headlamp' capable of looking ahead to pick-out the main implications of such 'possibilities'. Of course we can only use mathematics to 'discuss' those 'implications' or 'possibilities' which are essentially deterministic or genuinely probabilistic.

This is a much larger role for applicable mathematics in society than was envisaged by anyone prior to 1960, though Charles Sanders Peirce stated something very like this view at the end of the Nineteenth Century [14].

The 'possibilities' which, it now appears, are the central target of applicable mathematics, are not, however, easily defined, categorised or delimited. A more enlightened public regime will, almost by definition, concern itself with far more active possibilities than a 'narrow' one. Rigid, hidebound societies do not look at many 'possibilities' of action or thought before deciding what to do: they simply follow precedent. An innovative, imaginative, dynamic society will look at large numbers of possibilities, scanning them for the most attractive and worthwhile variations. So one striking consequence of the new view of the applicability of mathematics is that the scope for serious applicable mathematics in a given society is very largely a function of that society's dynamism and enlightenment.

Thus the *natural target* for applications of mathematics is huge, in the sense that there is a vast, potentially limitless area of currently relevant possibilities out there waiting to be modelled. It is also a constantly renewing 'target', because the possibilities which are relevant to progress today are not those of yesteryear. Tomorrow a shift of perspective will ensure that yet further relevant possibilities come into view. We are not talking about fantasies, whims or mere 'logical possibilities', but genuine, realistic *possible directions* of progress in science, technology and development. If politics is the *art* of the possible, then applicable mathematics is the *science* of the possible; and 'the possible' means similar things in the two doctrines, that is, things which need to be considered because they could happen in the reasonably near future.

The previous view of the applicability of mathematics might be summed-up by saying that mathematics was used as a *tool* to "solve problems" in technology and

---

[3] The present author gave an explicit account in various publications [7,8,9] but it made little impact on the prevailing opinion. Only in more recent papers [11,12,13] have these views become more widely known.

science, and as a "powerful means for communicating" technological and scientific information. This is the point of view on which the Cockcroft Report is based [2].

But such a formula tells us nothing about the character of these 'problems', or of the essential aim facilitated by their 'solution'. The hackneyed phrase about mathematics being a "powerful means for communicating" seems to mean little more than that scientists and technologists use mathematical equations to express their results.

The real content of this view of the applicability of mathematics may be summed up as follows:

(i) Scientists and technologists bring their problems to be solved by the mathematician,

(ii) These problems possess no recognisable common characteristics,

(iii) The reasons why scientists and technologists want to solve these problems are outside the province of mathematics.

Of these items (i) tells us that applicable mathematics has been defensive: applicable mathematicians did not seek out new problems for themselves, but simply waited for the technologist/scientist to bring them along. (ii) and (iii) tell us that the applicable mathematician made virtually no effort to understand the world views of the scientist and/or technologist, the background of the problems presented or the purposes which a 'solution' might achieve.

The new view of the applicability of mathematics asserts that mathematics is a possibility-simulating activity [4] and that the purpose of such simulation is to find the fact-implications of a possible new theory and the main value-implications of a new gadget, machine, system or development. Thus the *vacuum* represented by (i) and (ii) in the old view of applicability has been replaced by substance. This, in turn, implies that we now know, in outline terms at least, what we are at when we model the real world. The enormous potential target, together with this awareness of what we are 'at'; means that we can now begin to search-out our own problems, to set our own mathematical-applicative agenda.

This, I submit, is the *End of the Defensive Era in Mathematics*. Mathematicians may still be brought problems by the technologist and scientist and may still fail to solve some of them. However the presence of the computer and the range of new computer-mathematical strategies gives us a better chance to succeed than in the past [9]. We may 'solve' these problems, but not to the level of accuracy currently perceived to be necessary by the scientist or technologist. But we need not be 'defensive' about it: we can involve ourselves in the background from which these problems arose. We may suggest alternative formulations or alternative problems, or even alternatives to the hypotheses/programmes from which the 'given' problem

---

[4] The argument that the new positivity of applicable mathematics only succeeds in putting more pressure on computer centre operations should not, in the author's opinion, be accepted. Mathematicians should solve their own problems, by devising more powerful algorithms, not expect computer scientists to do it for them.

arose in the first place. Nor need we be spending all, or perhaps even most, of our time on these 'given' problems. We shall be working actively on areas of possibility known to be matters of live concern both in science and technology. In these areas we shall be setting our own agenda, formulating both outline hypotheses and programmes and actively searching-out their major implications.

We are in the position of a tennis player who abandons mere baseline defence and begins to dominate the net: of a football team which stops packing its own penalty area and begins to launch daring attacks on its opponent's goal.

It is interesting to note, too, that the revolution of thinking in *pure* mathematics which occurred around 1830 arose from the realisation that pure mathematicians could legitimately concern themselves with formal possibility, with non-arithmetic 'numbers', non-numerical algebras, non- Euclidean geometries. These were *possible* pure mathematical structures, and their properties were of genuine interest to mathematicians. This realisation liberated pure mathematics and led directly to a pyrotechnic display of new ideas in modern mathematics which began with groups, complex numbers and hyperbolic geometry, but soon led to Boolean algebras, matrices, elliptic geometry, rings, fields... etc. continuing right up to the present day.

The liberation of applicable mathematics also owes its origin to the realisation that applicable mathematics can legitimately concern itself with *possibilities* of science and technological development. We may call these possibilities "big ifs", because they are, in effect, alternative scenarios in science and technology. Mathematics' contribution is at the very cutting edge of science and technology; identifying the observable implications of a new theory (thus enabling us to decide whether it permits surprising or previously unenvisaged predictions of empirical fact) and the features of a new system or gadget which impinge on people (thus enabling us to decide whether it is likely to be wanted by a client, a consituency, or a market).

3   The Dimensions of the Change

What is different about the new view of the applicability of mathematics? How has it come about? What has been the role of computing in this changeover?

On one level the change of viewpoint is all about taking care in making practical and theoretical decisions. In the domain of the practical the difference is that we no longer decide on non-mathematical grounds to do something, and then look at the details of that, already adopted, course of action. In a highly competitive market it is essential to create products which optimally efficiently achieve their aims. Of course 'optimally' is not a fully defined concept in such circumstances, but an aspiration to make distinct leaps-forward in performance. This exerts a pressure on businesses to scan the horizon for *possible* designs.

At the same time the arrival of 'innovation-based' economies in the leading industrial countries means that the scope for variations and new possibilities of design has never been so wide. This combination of competitive need and amazingly open technical facility points businesses firmly in the direction of *projective* mathematical modelling; that is, using mathematical modelling to 'discuss' the main user- and producer- implications of new, possible designs. This is the practical face of the 'new applicability'.

We may describe such modelling as IF-if modelling, because it centres around the "big ifs" of possible designs and then consists in teasing out their implications.

All mathematical modelling consists in the pursuit of "ifs" in a deterministic or genuinely probabilistic system, but projective modelling consists in the pursuit of such "ifs" within a system which is only a gleam in someone's eye and is if fact itself only an "IF".

In science, too, similar factors apply. The amount of scientific research now being undertaken makes science a far more competitive enterprise than it previously was. And the era of what Kuhn calls *Normal Science* [5] has now definitely ended: instead of a period dominated by a single paradigm in science, we now have a multiplicity of paradigms, which can be used as required, e.g. NASA has always used *Newtonian* mechanics in the navigational systems on its space probes. So there is not just *one* kind of explanation of a new phenomenon in science, which everyone in principle accepts: in some areas of science there are currently as many as twenty live, viable hypotheses jostling to give the best (i.e. most accurate overall) account of the facts. Here too, mathematical modelling needs to be directed, open mindedly, at all the live hypotheses. Each hypothesis is a theoretical "big if" and the modelling here too is of the IF-if variety.[5] This is the theoretical face of the new applicability.

How different is projective modelling from non-projective modelling? A typical example of non-projective modelling occurred in the first major use of Operational Research during the Second World War: to find the optimum timing to set the detonators on mines dropped from aircraft onto fast submerging U-boats. The German Navy Command got the false impression that a new weapon was being used against their submarines! If a new weapon had been introduced, the earliest stages of its design *would* have consisted of projective modelling. But the modelling actually done was ordinary "if" modelling, looking at the most likely depth-probability distribution for U-boats which were trying to disappear at maximum speed.[6]

Clearly projective modelling was done in science as far back as Aristarchus[7] and in technology as far back as Archimedes, who looked at the stability in water of parabolic hulls for ships. But it is only since around 1960 that the computing power needed to do projective modelling widely has been available. We assume, of course, that anyone doing mathematical modelling since 1960 has made substantial use of

---

[5] of course scientists have used If-if modelling to try to work out the observable implications of their hypotheses throughout history. It is the scale of the activity which has increased so markedly in recent times.

[6] This brings out the weight of the difference between a "big" and a "little" if. If the object is simply to reset the existing detonators the "if" in question is a "little if". But introducing a new kind of detonator, e.g. a magnetic one (which is probably what the U-boat Command imagined) would involve a "big IF". In this case, though, its "bigness" is relative. There is actually a continuum of cases in which more and more intervention/rebuilding/redesign is envisaged. It is a matter of linguistic convenience at which point we start talking about a "big" if.

[7] For example, Aristarchus' method for calculating the relative distances of the Sun and Moon from the Earth.

the computer.

The main contribution of the computer, though, is probably via the conceptualisation of the new applicability. Before 1960 the applicability of mathematics was tied to the *computability* of answers[8], and this depended in a totally *ad hoc*, arbitrary way on manipulative tricks which managed to harness deeper level abstractions, like calculus of differences and complex partial fractions leading to otherwise intractable integrations. It is the removal of this idiosyncratic constraint on applicability which has enabled us to formulate the account of the 'new applicability' given above.

So this particular revolution in mathematics has not been so much *inspired* by computing as *enabled* by the simplification resulting from the removal of the arbitrary computability condition. We can now see a common theme in applications of mathematics to the real world, as opposed to the vacuum represented by (ii) and (iii) in the earlier view. Projective modelling enables us to "zoom in" and focus on the observability and usability implications of what are initially mere fuzzy "good ideas". It is really only since around 1960 that people have had the stomach to take on the hugely increased scope for modelling which this implies. The computing power available has tipped the balance, making the new interpretation of the applicability of mathematics into a habitable position.

4 The Challenge of IF-if Modelling

So far many applicable mathematicians have behaved rather like an animal which has not yet realised that the bars of its cage have been removed. Let's face it, the new role of positive applicable mathematics, where the mathematician seeks out modelling opportunities, is very different from the old. It requires new interests, new skills, a new willingness to become involved in practical and scientific progress, and centrally a thorough understanding of the human, non- mathematical purposes of modelling. The main purpose of modelling when used in "planning to do things, planning to build things"[9] is to disperse the fog of possibilities surrounding these operations and this fog-dispersal is wanted, not so much for its own sake, as for the sake of the peace of mind one gets from it. Fog can harbour nasty surprises, even golden opportunities which, if not taken, become bitter disappointments.

These principles apply even on the simplest levels, e.g. checking one's change to see that one has not been cheated. In general the use of mathematical modelling in "planning to do things, planning to build things" is that it generates confidence and motivation. One can concentrate on a task much more wholeheartedly than before when modelling reveals that what one is doing will not run into a crevasse on the one hand or be revealed as going the long way round the mountain on the other.

---

[8] G.H.Hardy pointed this out [4]page 74 "...pure mathematics is on the whole distinctly more useful than applied. For what is useful above all is technique, and mathematical technique is taught mainly through pure mathematics;"

[9] These are two items in a twelve level Taxonomy of purposes of application proposed by the author [10].

IF-if modelling, then, is a way of life which few applicable mathematicians are actually living at present. Some may query whether a 'mathematician' should be expected to become so involved in considerations of technical/scientific (or even social/commercial) progress and the human wants to which it relates. The 'new applicable mathematician' needs to be both a generalist and involved in the concerns of mankind. It is like a stonemason stepping up to become architect of a medieval cathedral. Although the two worlds of stonemason and architect are closely related, the *feeling* of the two roles is completely different. If the historic opportunity of 'new applicable mathematics' is to be fully seized, we shall all need to re-think our careers, and, crucially, we shall have to re-think the *education* of the up and coming generations. IF-if type modelling, IF-if type problems and illustrations tailored to the needs of the classroom (i.e. interesting to the youthful mind, simplified, solvable) will be needed on an impressive scale if we are to prepare youngsters properly for the challenges ahead.[10]

Although there may be few fully fledged positive applicable mathematicians operating at present, there has been a subtle crossing of the line into IF-if or 'projective' modelling. It may not yet be dominant in the work of individuals, but it is *there*, if we take the scene as a whole into account. It was used, we know, in planning the cleaning up of the River Thames [6] pages 91-88, the NASA Moon (and other) Missions, feasibility studies for a possible Severn Barrage, the Roskill Commission's assessment of possible sites for London's third airport. The ground has shifted under our feet, almost without our noticing it. We are, indeed coming to *assume* that any new project like the Channel Tunnel, the mapping of the human genotype, Hotol, the Arctic House at San Diego Zoo will be thoroughly "discussed" mathematically at the planning and pre-planning stages. There is no actual barrier preventing the development of projective modelling (for the reasons explained in footnote 6). With the gradual development of new facilitating techniques like background information databases, powerful computer graphics, and non-linear optimisation methods (e.g. hillclimbing), the barrier will slowly become a thing of the past.

One final caveat. *Positive applicability*, when it eventually emerges as a recognised role for individuals, will still have to operate for many years in a predominantly maths phobic world. Being positive should not imply being brash, aggressive or insensitive. In a democratic, free, market-led society every development needs to be 'sold' to its appropriate constituency. Twenty- five years ago this Spring I spent two weeks visiting Dr Houlden's National Coal Board Operational Research Unit on the South Bank of the Thames. Dr Houlden's philosophy was very positive, but he never tried to *force* any mathematical analysis onto a colliery manager or mining research director. His unit waited to be *asked* to make a contribution. This policy required a lot of patience, but gradually it began to pay off, and in the end the unit had to adopt a highly selective approach to requests for help; namely, that

---

[10] The author has been engaged on producing quantities of such material since the mid-Sixties. Mathematics Applicable (1969-78) produced a ten book series. Since 1978 the MAG (Maths Applicable Group) has produced quite a lot of new materials on these lines. [MAG, School of Education, UEA, Norwich NR4 7TJ.]

only if the likely economic benefit would be ten times the cost of the operational research effort, would they accept a commission. It is an object lesson, both in the effectiveness of the work done and in the diplomacy displayed.

References

[1] Bowers, J. *An Invitation to Mathematics*, Blackwell (Oxford), 1988.

[2] Cockcroft, Sir W. (ed) *Mathematics Counts*, H.M.S.O. (London), 1982.

[3] Dieudonne, J. in *Philosophia Mathematica* 1, 2, 1964.

[4] Hardy, G. H. *A Mathematician's Apology*, Cambridge University Press (Cambridge), 1940.

[5] Kuhn, T. *The Structure of Scientific Revolutions*, (Chicago), 1962.

[6] Lighthill, J. (ed) *The Newer Uses of Mathematics*, Penguin Books (London), 1978.

[7] Ormell, C. P., Ch.2 in *Mathematics through Geometry*, Pergamon (London), 1964.

[8] Ormell, C. P., 'Mathematics Science of Possibility' *Int. J. Math. Educ. Sci. Tech.*, 1972.

[9] Ormell, C. P., *Mathematics Changes Gear*, Heinemann Eductional Books (London), 1975, *The Applicability of Mathematics*, MAG (EDU, University of East Anglia), 1982.

[10] Ormell, C. P., 'Maths with Bite', *Education*, 3-13, 1983.

[11] Ormell, C. P., 'Breaking Out the Big Ifs' *Times Higher Ed. Supplement*, 1989a.

[12] Ormell, C. P., 'Maths Devoid of Myths', *Times Higher Ed. Supplement*, 1989b.

[13] Ormell, C. P., 'A Modelling View of Mathematics' a paper given at the 4th International Conference on the Teaching of Math. Modelling & Applications, Rosklide, Denmark, 1989c.

[14] Peirce, C. S. 'The Essence of Mathematics' in *The World of Mathematics*, Simon and Schuster, New York, 1956.

Revolution, Evolution or Renaissance?

D. J. Cooke

Department of Computer Studies

Loughborough University of Technology, Leics, LE11 3TU

Abstract

The first computers were constructed to assist mathematicians in 'crunching numbers'. In those days it was difficult to distinguish the disciplines of Computing and Numerical Analysis. Today's computer systems are much more sophisticated and programs should be engineered. Rigour is the guiding maxim: programs cannot be exhaustively tested; so algebra and logic are important in describing system requirements and checking that a system conforms with those requirements. This paper takes a brief look at some of the mathematics associated with programs and programming languages. As in any living science, Computing is in a continuing state of change; seeking rationalisation and simplification in an attempt to push back the frontiers which hinder our understanding. But progress is slow; despite the pressure to deliver the goods 'yesterday', mature techniques in computing technology do not appear 'on demand'. In good programming, as in good mathematics, the concepts that survive are disarmingly simple. They are based on the fundamental ideas that have been intrinsic to undergraduate pure mathematics for many years, they are not really new, they are merely interpreted in an 'unreal' context.

1 Preliminaries

The important words in the title of this paper are 'evolution', suggesting *gradual* development, and 'renaissance', inferring a *revival*, of mathematics. 'Revolution' is either concerned with *turning full circle*, thus getting back to where we started, or with a *radical change* which is thorough, drastic, harsh, and rapid. We will argue that the first interpretation is most appropriate for mathematics.

Before going further it must be stressed that Computing is interpreted as in Computing Science and *not* Computer Science i.e. the activity itself, not the fruits of such an activity, and certainly not the equipment used.

While it is true that recent advances in Computing and Computer Science have meant that certain, hitherto impractical, calculations are now viable; computability theory still holds, we can't now do anything we couldn't do before, we can just do some things faster.

It is also important to recognise that computers are mathematical machines - their theories and models are exact - not approximations as found in theories and models in the natural sciences. We do need a notion of approximation, but one in which 'better' means 'more' (more information, more *consistent* information), not 'more accurate' *per se*. So there are bound to be differences but these are outweighed by the similarities.

Software Engineering is the part of Computing Science that addresses, amongst other things, the specification and implementation of programs and programming languages. (Of course, the specification of a language can be viewed as a specification for a translator for that language and since a translator is itself a program, this is really all the same - although different specification techniques are appropriate.) In this paper discussion is restricted to the mathematics associated with Software Engineering; that is, to the mathematics inherent in any computer system, no matter what it is used for. To non-Computing Scientists (ambiguity intended) this might prove enlightening and hopefully educative, though the presentation can in no sense be considered comprehensive.

The development and use of computer systems essentially reduces to three phases: these relate to the client, the implementer, and the machine:

- the client says what is to be done, (specification)
- the programmer says how it should be done, (implementation)
- and the machine actually does the 'computation' (execution).

We don't do the calculations, but we need to know how to do them. The resulting description of what to do is exactly a program (and nothing else). Moving from a specification to a program implies an increase in precision/formality, the process is called *refinement*.

Mathematics is important here in two ways, firstly in describing what is required of a system (*not* saying how it is to be achieved, *not* giving a recipe, *not* giving an algorithm or writing a program), and secondly in using this description either to verify that a program satisfies the requirements given in the specification, or to derive a program from the requirements by algebraic transformation techniques.

In discussing the interface between computing and mathematics much is often made of the dubious distinction between continuous and discrete mathematics. Although this is sometimes convenient it can also be misleading. A better comparison is the discrepancy between school calculus and analytical calculus as practised by pure mathematicians in tertiary academic establishments. The latter demands great attention to detail and the same is true of specifying and writing programs.

To illustrate consider the derivative $\frac{d}{dx}(\sqrt{x^2 - 1} + \sqrt{1 - x^2})$. How many mathematicians or engineers would blast away applying the differentiation rules without considering whether they were valid or sensible? Of course the exercise is a futile waste of time.

Similarly one might be tempted to plunge headlong into solving the equation: $f(x) = 0$ by the iterative scheme $x_{n+1} = x_n - f(x_n)/f'(x_n)$ or $x_{n+1} = F(x_n)$. At the solution, $x = F(x)$, x is a fixed point of F. This is the familiar Newton-Raphson 'formula', which under suitable conditions can be used to progress from an initial guess at the answer to an approximation of a root of f to within the desired degree of accuracy. To guarantee that this works requires application of the

Contraction Mapping theorem – F being the mapping in question - and the Mean Value theorem. In practical terms, F must be defined on some closed interval I of $\mathbb{R}$, it must be differentiable on I, $|F(x)| \leq C < 1$ for all $x \in I$ and the initial guess, x_0, must be within I. In computing we must tread carefully, just like proper mathematicians - not mathematical handle-turners. The Software Engineer must, either by self-imposed discipline or with the aid of a suitable computer system, work within well-defined sets of rules.

This paper presumes 'some' exposure to programming. The 'standard' undergraduate mathematics degree topics which form the basis of 'formal' Software Engineering are:

* ⋆ sets: subsets, powersets, Cartesian products etc.
* ⋆ relations: equivalence, partial orders, (least) upper bounds, etc.
* ⋆ functions: monotonic (increasing), continuity, etc.
* ⋆ algebraic structures: common examples, rule following
* ⋆ proofs: direct, contra-positive, disproof by counter example, simple induction
* ⋆ "fixed point characterisation of solutions to equations."

The equation $f(x) = 0$ above has a solution y if $y = F(y)$, under appropriate conditions we can compute y to any required accuracy. This notion reoccurs in many contexts in computing as we shall see later.

We illustrate some of the ways in which mathematics is used in Software Engineering – avoiding the need to digress into areas which require extra 'non-mathematical' background. The three topics addressed are the specifications of programs, algebraic structures, and languages.

2 Program Specifications

Several different varieties of specification system have emerged in recent years. Here we use VDM-like specifications [12] which are based on relations: a type line to identify the source and target i.e. the *signature*, a *pre-condition* to characterise the domain, and a *post-condition* to identify the underlying relation (any implementation should deliver a unique output for an input value that satisfies the pre-condition, and is therefore a function; but several different functions may be acceptable and hence the specification generally yields a relation).

Here we meet a (trivial) instance of a situation in which software engineering requires a higher degree of precision than found in most mathematics. As with many areas of mathematics the terminology is non-standard but the distinctions are important. We follow the notation of [17] and name the sets associated with the function F as shown in Figure 1.

In computing it is crucial to distinguish between the source and the domain since an attempt to calculate the result of function applied to a data value not in its domain would fail or never terminate. This is of obvious concern in computing but is no great problem when performing an evaluation by hand.

Figure 1. The sets associated with the function F

In the following examples $\triangleq$ means 'is defined to be', and $\wedge$ means 'and'.

example1
type: $\mathbb{R} \to \mathbb{R}$
pre-example1(x) $\triangleq$ true
post-example1(x, y) $\triangleq$ $(x = x^2 - z) \wedge (2*y = z + 3)$

This defines a function from the reals to the reals. There is no extra domain constraint, so the domain equals the source equals the reals. The underlying relation is given by the set $\{(x, y)| \exists z, (x = x^2 - z) \wedge (2*y = z+3)\}$ although the introduction of z is usually implicit, after all, since $x = x^2 + z$ then z has to be $x^2 - x$.

Manipulating the post-condition to get y as the subject, using the field axioms applicable to $\mathbb{R}$, we get $y = (z+3)/2$, $z = x^2 - x$, and $y = (x^2 - x + 3)/2$. This is a simple example and we can extract the 'obviously correct' single statement program:

$y := (x^2 - x + 3)/2$

which indicates that the value of the expression $(x^2 - x + 3)/2$ is passed to the 'variable' y. Also in this example the pre-condition was true regardless of the input value x. In such cases it is common simply to leave out this part of the specification. A familiar instance when a non-trivial restriction is required now follows.

example2
type: $\mathbb{R} \to \mathbb{R}$
pre-example2(x) $\triangleq$ $x \geq 0$
post-example2(x, y) $\triangleq$ $(y^2 = x) \wedge (y \geq 0)$

This is simply a definition of what in Fortran would be SQRT, but there is absolutely no indication of how to compute it – and there shouldn't be. Here, as already noted, the domain is not the same as the source; the $x \geq 0$ condition must hold in order that the implicit calculation can take place. Notice also that this condition cannot be evaluated unless the source (here $\mathbb{R}$) is a data type on which appropriate operators (here $\geq$) are defined.

Transformations, be they straightforward, as in example 1, or more involved as in recursion removal, equate to using theorems – under the appropriate conditions. (The post-condition is a relation which may be formulated in a recursive fashion. Removal of recursion to achieve more conventional iterative programs is non-trivial; see Appendix D of [18]).

The logical connectives $\wedge, \vee, \neg$ (*and, or, not*) used in specifications are commonly known from Boolean algebra used in logic design. Specifications can also employ conditional clauses, quantifiers and recursion (rather than explicit iteration). The implication connective may also be used in specifications but, more importantly, it is required in formal proofs. Explicitly, for example 2, a routine called SQRT would need to satisfy the theorem

$$(\forall x \in \mathbb{R})(\text{pre-example2}(x) \Rightarrow \text{post-example2}(x, \text{SQRT}(x)))$$

i.e. $(\forall x \in \mathbb{R})((x \geq 0) \Rightarrow ((\text{SQRT}(x))^2 = x) \wedge ((\text{SQRT}(x)) \geq 0)))$. Of course, although this *specification* is perfectly acceptable these conditions are almost impossible to achieve. A more sensible and realistic condition in the specification might be: $(|y^2 - x| \leq 10^{-7})$.

3 Specification of Algebraic Structures

An algebraic structure is a set together with operations defined on that set. These operations could be defined via functions as in the previous section, however we cannot carry on indefinitely defining each layer of operators in terms of more elementary ones, we must stop somewhere.

To overcome this difficulty an alternative, algebraic, style of definition can be used. To illustrate we give two examples, one familiar in mathematics, one in computing. What follows is essentially the declaration of a group in the language CLEAR. (Depending on where you sit this is either a specification language or a programming language. The closely related language OBJ is described in [8]).

ops	(i.e. operations)	This part of the specification gives the syntax (though this may be 'abstract') and type checking requirements. Here * is a diadic infix operator that operates on two values from the set 'elt', I is a specific constant within elt, and the prime (') is a postfix monadic operator.
	elt * elt $\to$ elt	
	I $\to$ elt	
	elt' $\to$ elt	

eqns (equations, axioms)
$(a * b) * c = a * (b * c)$
$I * a = a$
$a * I = a$
$a' * a = I$
$a * a' = I$

This part indicates what can be done within the system being defined. 'Eqns' are 'the rules of the game'. We don't explicitly define * and ' but say how they behave. The signatures and axioms effectively (but *implicitly*) specify all we need to know about a group.

An example of the same approach applied to an algebraic structure that could be considered to be an abstract data type is the set of all finite lists of integers. This presupposes the existence of a definition of the integers - but that is another problem; one not addressed here. What we give is an 'enrichment' of the set of types Integer and Boolean and makes no pretence at being 'complete'. Here we presume that the types 'integer' (on which the equality test, $=_i$, is defined) and 'Boolean' already exist; we are extending the set of objects (by adding lists) and operations to manipulate them.

ops

<>	→ list	the empty list
list ⊙ list	→ list	concatenation
<int>	→ list	construction of single element list
head(list)	→ int	see 'eqns' below
tail(list)	→ list	see 'eqns' below
list $=_\ell$ list	→ bool	equality between lists

eqns

$$(\ell_1 \odot \ell_2) \odot \ell_3 = \ell_1 \odot (\ell_2 \odot \ell_3)$$
$$<> \odot \ell_1 = \ell_1$$
$$\ell_1 \odot <> = \ell_1$$
$$\text{head}(<i>) = i$$
$$\text{tail}(<i>) = <>$$
$$<\text{head}(\ell)> \odot \text{tail}(\ell) = \ell$$
$$(<> =_\ell <>) = true$$
$$(\ell_1 =_\ell \ell_2) = (\ell_2 =_\ell \ell_1)$$
$$(<i_1> \odot \ell_1) =_\ell (<i_2> \odot \ell_2) = (i_1 =_i i_2) \wedge (\ell_1 =_\ell \ell_2)$$
$$(<> =_\ell <i> \odot \ell) = false$$

Referring to the signatures above, ℓ_1, ℓ_2 etc. must be lists and i_1, i_2 etc. must be integers.

err eqns

$$\text{head}(<>) = error$$
$$\text{tail}(<>) = error$$

Notice that in this example (i) we have 3 different kinds of equality, we define $=_\ell$ but presume the existence of $=_i$, and we have the general equality in the 'eqns' themselves – which simply state which replacements can be carried out; (ii) there is a need to generate 'false' result, so we must go further than conventional axioms where one is only interested in knowing what is 'true', and (iii) we need to catch errors *within* the specification system; it shouldn't give rise to illegal expressions. Specifications of this form can be used as the basis of verification procedures in which the proofs are performed by term rewriting. Alternatively, implementations of the list operations can be verified directly by substitution into the 'eqns' which must then all reduce to 'true'. In this case the '=' of the specifications must, in each case, be replaced by the equality operator of the appropriate type, e.g. $(\ell_1 \odot \ell_2) \odot \ell_3 =_\ell \ell_1 \odot (\ell_2 \odot \ell_3)$, $\text{head}(<i>) =_i i$, etc.

It is not necessary for the set of equations to be minimal but the set should be consistent. All evaluation is done by rewriting sub-expressions by equivalent ones (as in the equations). The equations might be 'directed' so as to form so-called reduction rules which aim at securing termination of the rewriting sequence and yield 'an answer'. No elementary texts on this subject are readily available but for an insight the reader may consult [11]. Every time we (mathematicians) simplify an algebraic expression we replace a sub-expression with another having the same value, usually with the intention of eventually evaluating the entire expression. Systems like CLEAR and OBJ merely represent the computing equivalent of this process but again more detail is needed to ensure only legal replacements can take place.

4 Language Specifications

In discussing programming languages we first must give meanings to (parts of) programs, the meanings are denoted by functions. Even before addressing this topic we need to be more particular about some commonly used – but not well understood – mathematical terms.

4.1 Lambda Notation

What are constants and variables? (This is a good old mathematics question – with no easy answer! [3]). In most programming languages identifiers are constants and so are their values, but the 'ownership' function changes as the execution (of a procedural program) proceeds.

Again computing (or computing-related mathematics) comes to the rescue. Consider the notion of 'place holders'; often referred to as 'variables'. Suppose $f(x) \triangleq x^2 - 2$. This can be rewritten as $f \triangleq \lambda x.(x^2 - 2)$ which means, '...substitute for x in $(x^2 - 2)$. We can now perform a function evaluation thus, 'substitute 3 for x' gives:

$$f(3) = (\lambda x.(x^2 - 2))(3) \quad = \quad 3^2 - 2 \quad = 7.$$

Here, using a notation that predates computing, Church [4] separates the function from its argument and makes it possible to talk about f rather than $f(x)$; we want to be able to say things about a program *without* direct reference to its data. Notice also that the use of λ-notation helps clarify the distinction between the related but distinct notions of formal parameters (used in the definition of a function) and actual parameters (used in a specific evaluation) as follows.

In '$f = \lambda x.(x^2 - 2)$', x is the formal parameter of f; when the function is called using $f(3)$, the 3 is an actual parameter. This is a natural distinction for programmers (not that they all fully appreciate it) but the idea is alien to most mathematicians who 'just do it' without fully realising the mechanics of what they are doing. Of course some mathematicians know, and care, about such matters. The 'λx' effectively binds x in the subsequent expression; any identifier in the expression which is not bound is said to be *free*. Notice also the following substitutions

$$(\lambda x.(x^2 - 2))(y) = (y^2 - 2)$$

$$(\lambda n.(\lambda x.(x^2 - 2)))(2) = (\lambda x.(x^2 - 2)) = f$$

$$(\lambda m.(\lambda n.(\lambda x.(x^n - m))))(2))(2) = f$$

$$(\lambda g.(g(y+3)))(f) = f(y+3) = (\lambda x.(x^2 - 2))(y+3) = ((y+3)^2 - 2), \text{ etc..}$$

So functions can be treated as perfectly respectable 'first class' objects which can be used as inputs to (higher order) functions, and delivered as results of function evaluations. See also [10].

Other common occurences of so-called variables are really constants whose value is not yet obvious. Given $x^2 - 4x + 4 = 0$ we get $(x-2)^2 = 0$, which yields $x - 2 = 0$, which yields $x = 2$. These 'variables' are of no particular interest in computing.

Any non-trivial program involves repetition; it is not merely a linear sequence of statements that are executed one after the other. Lambda notation, as illustrated so far, relates only to straight forward evaluation of expressions. Inclusion of certain combinators, notably the fixed point combinator, Y, enables us to deal with recursive definitions of iterative processes – and much more. Combinators are (higher order) operators that can be defined without using bound variables but we need extensive discussion of λ-calculus to show how this fits (see [9][15]).

The classical fixed point function, Y, (the paradoxical combinator) is such that $YF = F$ but Y, which may yield many different results (or the set of all such results) is usually particularised in computing to fix, representing the least fixed point. This will be defined later.

4.2   Language Syntax

Syntax definitions (in the Backus Naur Form used in describing IAL, the International Algebraic Language, the precursor of Algol 60, in 1958, and subsequently manipulated as equations by Ginsberg [7]) provide only a description of structure,

Revolution, Evolution or Renaissance

there is no meaning! Take an example:

$$< expression > ::= < expression > + < term > \mid < term >$$

This says that an expression is either an expression followed by a plus sign followed by a term, or a term on its own. This gives rise to a sequence of 'terms' separated by '+' symbols. For this to be complete we must also give a definition of $< term >$.

Grammars are used to 'generate' languages – or, used 'backwards' – in the analysis of potential sentences. Hence we can regard the rewrite rules as equations; the 'Language generated' by the grammar being the minimal solution of this set of equations. So we may regard formal linguistics as a branch of mathematics.

With a simplification of notation using upper case letters for 'unknowns' and lower case for constants, we might have:

$$A \to xB|c$$
$$B \to yA|zB|d,$$

i.e.
$$A = xB|c$$
$$B = yA|zB|d$$

By simple manipulations within the regular algebra, where addition is 'or' and multiplication is 'concatenation of strings', we get

hence
$$B = yxB|yc|zB|d$$
$$B = yxB + yc + zB + d$$
$$= (yx + z)B + (yc + d)$$

This leads to $L(B) = \{(yx + z)^n(yc + d) : n \geq 0\}$
and $\quad L(A) = xL(B) \cup \{c\}$
$\quad\quad\quad\quad = \{x(yx + z)^n(yc + d) : n \geq 0, c\}$

where $L(A)$ is the set of all sentences, the language, generated using the rules to expand A into sequences of x, y, z, c, and d, and similarly for $L(B)$. There is an alternative approach, applicable when the grammar is more complex and regular manipulations are not possible. Here we can express the set of equations as a single *matrix* equation over sets of strings. (Here we have followed the convention of replacing $\emptyset$ by 0 and $\{x\}$ by x etc..)

$$\begin{pmatrix} A \\ B \end{pmatrix} = \begin{pmatrix} 0 & x \\ y & z \end{pmatrix} \begin{pmatrix} A \\ B \end{pmatrix} + \begin{pmatrix} c \\ d \end{pmatrix}$$

So

$$\begin{pmatrix} A \\ B \end{pmatrix} = \mathcal{F} \begin{pmatrix} A \\ B \end{pmatrix} \quad \text{where} \quad \mathcal{F} : \begin{pmatrix} X \\ Y \end{pmatrix} \mapsto \begin{pmatrix} 0 & x \\ y & z \end{pmatrix} \begin{pmatrix} X \\ Y \end{pmatrix} + \begin{pmatrix} c \\ d \end{pmatrix}$$

This is then a fixed point equation, the solution of which is the vector

$$\begin{pmatrix} L(A) \\ L(B) \end{pmatrix}$$

Using a result to be proved later, we state that, starting with a suitable 'undefined' value which we denote by $\perp$, and consists of a column vector of empty sets of strings, that:

$$\begin{pmatrix} L(A) \\ L(B) \end{pmatrix} = \bigsqcup_{n=0}^{\infty} \mathcal{F}^n(\perp) \qquad \text{a supremum (proper definitions to follow)}$$

$$= \lim_{n \to 0} \mathcal{F}^n(\perp) \qquad \text{since } \mathcal{F} \text{ is increasing.}$$

$$\perp = \begin{pmatrix} 0 \\ 0 \end{pmatrix} \qquad \text{so } A_0 = \emptyset \text{ and } B_0 = \emptyset$$

$$\mathcal{F}(\perp) = \begin{pmatrix} c \\ d \end{pmatrix} \qquad A_1 = \{c\} \text{ and } B_1 = \{d\}$$

$$\mathcal{F}(\perp)^2 = \begin{pmatrix} xd \\ yc + zd \end{pmatrix} + \begin{pmatrix} c \\ d \end{pmatrix}$$

and so $A_2 = \{xd, c\}$, $B_2 = \{yc, zd, d\} \subseteq \{(yx + z)^n(yc + d) : n \geq 0\}$, etc.

This yields an infinite sequence of 'better' finite approximations to $L(A)$ and $L(B)$, and finite approximations are always sufficient – this is effectively the link between computability and countability.

4.3 Language Semantics

We are now going to use recursion equations, fixed points etc., to illustrate how the (denotational) semantics – the 'meaning' of a program, and hence of a programming language – may be specified. As stated before, the meaning of a program can be characterised by a function.

We ought first to dot some i's and cross some t's. In an attempt to put computation theory on a sound basis that relates more readily to the programs that people actually write rather than to the internal machine states that had hitherto been studied by automata theorists, Scott [16], presented a refined version of earlier work on domains. We outline the constructions.

Take a set S of data values and augment it with an element $\perp$, called *bottom*. Let this enlarged set $S \cup \{\perp\}$ be called D. Define a *approximation relation*, $\sqsubseteq$, on D

by $\bot \sqsubseteq x$ for all x in D. $(D, \sqsubseteq)$ is a *domain*. A set of elements of D, $\{x_1, \ldots, x_n, \ldots\}$ with $x_i \sqsubseteq x_{i+1}$ is a *chain*.

We can also have domains of functions (on domains, but there are extra restrictions!). In particular there is the 'undefined function' $\bot$ defined by $\lambda x.\bot$, and if f, g are functions $D \to D$, then $f \sqsubseteq g$ iff $f(x) \sqsubseteq g(x)$ for all x in D. Similarly, f is *monotonic* (increasing) if $f(x) \sqsubseteq f(y)$ whenever $x \sqsubseteq y$. For x and y in D we define their *supremum* (with respect to $\sqsubseteq$) as $x \sqcup y \triangleq \sup\{x, y\}$ and this can be extended over larger sets, in particular chains. Finally, a function f is *continuous* if $f(\bigsqcup X) = \bigsqcup f(X)$ for all chains X.

To get a big enough domain (including basic types, n-tuples, unions, functions etc.), we must impose certain restrictions so as not to fall foul of all the problems that sets have; see [16] for details. With f continuous and monotonic increasing we can justify that the least fixed point of f, $\mathbf{fix}(f)$, is given by:

$$\mathbf{fix}(f) \triangleq \bigsqcup_{n=0}^{\infty} f^n(\bot)$$

Proof: (This is the only rigorous development presented in this paper.) f is monotonic so $f^n(\bot) \sqsubseteq f^{n+1}(\bot)$ for all $n \geq 0$, so the set $\{f^n(\bot) : n \geq 0\}$ is a chain. f is continuous so

$$f(\bigsqcup_{n=0}^{\infty} f^n(\bot)) = \bigsqcup_{n=0}^{\infty} f(f^n(\bot)) = \bigsqcup_{n=0}^{\infty} f^n(\bot) = p \text{ (say)},$$ but

$\bot \sqsubseteq f(\bot) \sqsubseteq \ldots \sqsubseteq f^n(\bot)$ for all $n \geq 0$. So,

$$p = \bigsqcup_{n=0}^{\infty} f^n(\bot) = f(p).$$

Thus p is a fixed point of f. Suppose now that $\alpha = f(\alpha)$. But $\bot \sqsubseteq \alpha$, $f(\bot) \sqsubseteq f(\alpha) = \alpha$. By induction $f^n(\bot) \sqsubseteq \alpha$ for all $n \geq 0$. So $\bigsqcup_{n=0}^{\infty} f^n(\bot) \sqsubseteq \bigsqcup_{n=0}^{\infty} \alpha = \alpha$, i.e. $p \sqsubseteq \alpha$ and p is the *least* fixed point.

We are now in a position to investigate the functions that describe the meanings of common programming constructs. Let C be a function with signature: command $\to$ state $\to$ state , i.e. given a command it will describe how it should cause the current state to be transformed into the next.

The example chosen is not trivial but is easy to follow because it has a familiar pictorial representation. We look at the function which defines the meaning of the 'while B do C od' construct commonly found in block- structured programming languages. The 'od' is included so as to resolve problems of ambiguity in parsing complete programs. The function

$C [\![\text{ while } B \text{ do } C \text{ od }]\!]$

has the signature state $\to$ state.

Informally, and assuming readers are familiar with flowcharts, the meaning of this construct corresponds to the flowchart in Figure 2(a)

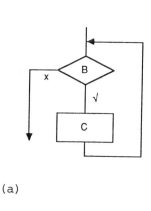 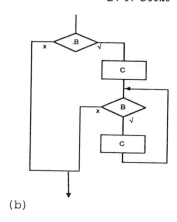

(a) (b)

Figure 2.

To indicate the motivation behind the definition we start with the following, *wrong* , definition.

(while B do C od)$(s) \triangleq$ if $B(s)$ then $(C$; while B do C od)(s) else s

if B yields 'true' then we do C and this is followed by the 'while B do C od' construct once more, otherwise we exit with the state s unchanged. We have effectively unrolled the flow chart to give Figure 2(b).

More properly, but ignoring the possibility of B and C giving rise to errors, we have

$C[\![$ while B do C od $]\!]s \triangleq$ if $\mathcal{B}[\![B]\!]s$ then $(C[\![B$ do C od $]\!])(C[\![(C)]\!])s$ else s

where $\mathcal{B}[\![B]\!]$ denotes the evaluation of the Boolean condition B and $C[\![C]\!]$ denotes the execution of the command C. If now we replace 'if P then Q else R' by $P \to Q, R'$ we arrive at the usual form of the definition:

$C[\![$ while B do C od $]\!]s \triangleq \mathcal{B}[\![B]\!]s \to (C[\![$ while B do C od $]\!])(C[\![C]\!])s, s$

Then

$C[\![while\ B\ do\ C\ od\ ]\!] = \lambda s.(\mathcal{B}[\![B]\!]s \to (C[\![$ while B do C od $]\!])(C[\![C]\!]s, s)$
$ = \lambda f.(\lambda s.(\mathcal{B}[\![B]\!]s \to f(C[\![C]\!])s, s))(C[\![$ while B do C od $]\!])$

In this equation $C[\![$ while B do C od $]\!]$ is merely the unknown, and can therefore be replaced by any name, say W. So,

$W = \lambda f.(\lambda s.(\mathcal{B}[\![B]\!]s \to f(C[\![C]\!]s, s))W$

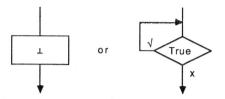

Figure 3.

Therefore $W = \text{fix}(\lambda f.(\lambda s.(\mathcal{B}[\![B]\!]s \to f(\mathcal{C}[\![C]\!])s,s)))$, and $W = \text{fix}X$

where $X = \lambda f.(\lambda s.(\mathcal{B}[\![B]\!]c \to f(\mathcal{C}[\![C]\!])s,s))$. X is increasing so, using the 'formula' for fix,

$W = \lim_{n \to \infty} X^n(\bot)$

The first few terms of the 'expansion' are as follows. The first is $X^0(\bot) = \bot$. This gives no results at all and can be represented as in Figure 3.

The next term is

$X^1(\bot) = \lambda f.(\lambda s.(\mathcal{B}[\![B]\!]s \to f(\mathcal{C}[\![C]\!])s,s))\bot$
$= \lambda s.(\mathcal{B}[\![B]\!]s \to \bot(\mathcal{C}[\![C]\!])s,s) \qquad = \lambda s.(\mathcal{B}[\![B]\!]s \to \bot,s)$

since $\bot(\mathcal{C}[\![C]\!])s = (\lambda x.\bot) \circ (\mathcal{C}[\![C]\!])s = (\lambda \mathcal{C}[\![C]\!]s.\bot) = \bot$. This models iteration of the loop zero times if B is false, otherwise it gives no answer (figure 4).

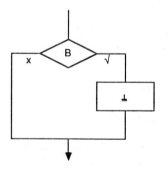

Figure 4.

Similarly, the next term is

$X^2(\bot) = (\lambda f.(\lambda s.(\mathcal{B}[\![B]\!]s \to f(\mathcal{C}[\![C]\!])s,s)))X^1(\bot)$
$= (\lambda s.(\mathcal{B}[\![B]\!]s \to X^1(\bot)(\mathcal{C}[\![C]\!])s,s))$

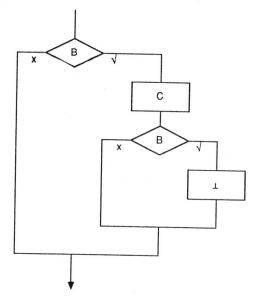

Figure 5.

$$= (\lambda s.(\mathcal{B}[\![B]\!]s \to (\lambda s.(\mathcal{B}[\![B]\!]s \to \bot, s))(\mathcal{C}[\![C]\!])s, s))$$
$$= (\lambda s.(\mathcal{B}[\![B]\!]s \to ((\mathcal{B}[\![B]\!](\mathcal{C}[\![C]\!])s \to \bot, (\mathcal{C}[\![C]\!])s))s)).$$

This corresponds to an immediate exit (if B is false), or a single iteration (if evaluation of B after C gives false), or else the result is undefined (Figure 5). And so on. Unfold as far as you need, but notice that in any (proper, terminating) situation a finite approximation to $\mathcal{C}[\![$ while B do C od $]\!]$ is always sufficient.

We have a finite mathematical representation of this program construct which represents infinitely many execution sequences. Such representations are important when considering the equivalence or approximation of programs, yet there are no really new mathematical concepts embodied within this key area of computing.

5 Is there anything new?

We have limited discussion to relatively simple ideas which can be appreciated, at least in principle by anyone who has a mathematical background and who has written a computer program. Are there any genuinely new pieces of mathematics which derive from computing? There are certainly some topics of great interest which we have not considered here, notably

- Combinators [19], but their origins can be traced to 1924
- Type Theory [1], in a new guise but really much older (1934?)
- Category Theory [2], but originated in 1945 [5]

- Temporal Logic, see [6] for an introduction, but the ideas are hinted at in 1903 in [14] or even earlier, see also [13] for a formulation of work relating back to ancient times.

As with other areas of applied mathematics, problems that arise in Computing have caused mathematicians to reorientate their research; but this is not new, and it is certainly not revolutionary. Most formal Computing is based on "seemingly useless" (not my sentiments) bits of pure mathematics which have been around for many years but *which are in danger of being removed from many mathematics degree course due to their abstract nature.*

6 Conclusion

In computing there is a move back to 'proper' mathematics – a move away from algorithms back to algebra. This means that the exercise of having to construct convoluted ways of mapping problems onto computers is gradually coming to an end and maybe computers are at last going to be machines which work in the way that mathematicians think (or thought) they ought to. Whether the mathematician is the client/customer or the provider of systems, he should be encouraged to stick to his mathematical principles.

There are 'new' techniques , but they are slow coming. When they arrive it will be debatable whether they will really be Mathematics applied to Computing Science or pure Computing Science. In any case, a proper (rigorous) background in pure mathematics will certainly be an adequate (even a 'good') preparation for both specifying and developing modern computing systems.

Maybe the evolution/revolution debate is merely dependent on the different historical standpoint (in a temporal context). A change seen in the context of 40 years of computing could be perceived as rather radical if viewed against a mathematical backdrop of several thousand years.

Computing is getting back to mathematics, but not to numbers *per se*. Maybe what we have is a renaissance in (a reappraisal of, a proper reformulation of) mathematics caused by (a proper appreciation of the foundations of) computing.

References

[1] Backhouse, R., Chisholm, P., Malcolm, G. and Saaman, E., 'Do-it-Yourself Type Theory', *Formal Aspects of Computing*, 1(1), 19-84, 1989.

[2] Barr, M., Wells, C., *Category theory for computer science*, Prentice Hall, (London), 1990.

[3] Benyon, M., Russ, S. 'The development and use of variables in mathematics and computer science', in *The mathematical revolution inspired by computing*, J.H. Johnson & M.J. Loomes (eds), Oxford University Press, 1991.

[4] Church, A., *The Calculi of Lambda Conversion*, Princeton University Press, 1941.

[5] Eilenberg, S. and Maclane, S., 'General Theory of Natural Equivalences', *Trans. Amer. Math. Soc.*,**58**, 231-294, 1945.

[6] Galton, A. (ed), *Temporal Logics and their Applications*, Academic Press, 1987.

[7] Ginsberg, S., *The mathematical theory of context-free languages*, McGraw Hill, 1986.

[8] Goguen, J. A., Tardo, J. J., 'An introduction to OBJ', *Proc. IEEE conference on Specifications of Reliable Software*, Cambridge, Mass, 170-189, 1979.

[9] Hinley, J.R. and Seldin, J.P., *Introduction to Combinators and λ-Calculus*, London Mathematics Society Student Text, No 1, 1986.

[10] Hyland, J. M. E., 'Computing and foundations', in *The Mathematical Revolution Inspired by Computing*, J.H.Johnson & M.J.Loomes (eds), Oxford University Press, 1991.

[11] Jouannaud, J-P.(ed), *Rewriting Techniques and Applications*', Academic Press, 1987.

[12] Jones, C.B., *Systematic Software Development Using VDM*, Printice-Hall, 1986.

[13] Prior, A.N., 'Diodoran Modalities', *Philosophical Quaterly*, 5, 205-213, 1955.

[14] Russell, B., *Principles of Mathematics*, George Allen and Unwin, 1903.

[15] Schönfinkel, M., 'Über die Bausteine der mathematischen Logik', *Math. Annalen*, **92**, 305-316, 1924.

[16] Scott, D.S., *Domains for Denotational Semantics*, Springer Lecture Notes in Computer Science, 140, 577-613, 1982.

[17] Slater, G., (ed) *et al*, *Essential mathematics for software engineers*, Peter Peregrinus, 1987.

[18] Stone, R. G., Cooke D. J., *Program Construction*, Cambridge Computer Science Texts, 22, Cambridge University Press, (Cambridge), 1987.

[19] Turner, D.A., 'A New Implementation Technique for Applicative Languages', *Software Practice & Experience*,9, 267-270, 1979.

The Superfluous Paradigm

Daniel I. A. Cohen
Hunter College, City University of New York
695 Park Avenue, New York 10022, U.S.A.

From our current vantage point, the influence on mathematics of the advent of the computer can be separated into five basic categories. The first of these is the development of new mathematics required to serve the discipline of Computer Science. There are branches of mathematics which appear to have been created for the purpose of building computers, for the purpose of producing algorithms, for the purpose of analysing the action of computers and for analysing the action of computer programs.

The second category is the creation of new mathematical results in branches which were already important for more classical reasons. This yield seems to be facilitated, expanded, or made possible for the first time through the application of gigantic computations, manipulations of fantastically complex systems of symbols, exploitation of the striking graphic capabilities for picturing perplexing structures, or simply exercising the phenomenal speed to produce innumerable repetitions of basic operations, as in projects of simulation or trial-and-error.

The third effect is upon those older branches of mathematics which appear to have been rendered obsolete by the calculational and manipulational abilities of the computer. This is a Darwinian pruning of the tree of mathematics.

Fourthly, there is a significant impact on the classroom through the employment of computer-aided teaching and educational software. Computer interactions can augment or replace human instruction with varying degrees of benefit.

Lastly, and most dastardly, there is the much ballyhooed extension of the notion of mathematical proof, far beyond the classical Greek paradigm of axioms and rules of inference, to include the so-called computer proofs, whose non-surveyability by human beings demands appeals to faith inimical to the enterprise of science, and yet allegedly yields results which are otherwise of necessity far beyond the powers of mortal man to obtain.

I would like to discuss these five areas individually and demonstrate that each contains significantly less than meets the eye.

As for the possibility that the computer has opened new branches of mathematical research, let me reprise the mathematical history of this century from the watershed address of David Hilbert delivered before the International Congress of Mathematicians in 1900 [1]. By that date it was already understood that mathematics is as much a collection of algorithms as it is the accumulation of technical interrelations stored in the form of theorems. The entire project of Linear Algebra, was seen clearly by Hilbert as one algorithm, to wit, the resolution of any set of linear equations in a predictable number of steps. Similarly, all of the results in Differential Calculus fit together to form one algorithm – the constructive proof that the derivative of any function in closed form is again a function in closed form. To Hilbert, mathematics was not properly a potpourri of fragmented correlations, but an intentionally structured pursuit of discovering which questions about mathematical objects can be answered, and of manufacturing reliable procedures to implement the conclusion. Hilbert presented 23 problems which he predicted would be the major focus of mathematical attention in the 20th century. Despite the fact that he failed to anticipate the invention of the high speed calculating machine, the topics he did enumerate describe accurately the progress made in this century, including most of those advances forming the subject now known as Computer Science.

Hilbert wanted to see mathematics given a logical foundation and logic given a mathematical foundation. He prophesied the existence of (and personally sought) an axiom set so inclusive that all mathematical questions could be framed therein and once so formulated could be automatically resolved. Hilbert's vision of a universal algorithm to solve mathematical theorems required a unification of Logic, Set Theory and Number Theory. This project was initiated by Frege, rerouted by Russell, repaired by Whitehead, derailed by Gödel, restored by Zermelo, Frankel, Bernays and von Neumann, shaken by Church and finally demolished by Turing. Hence, to say that the interest in algorithmic methods in mathematics or the progress in logic was engendered by the computer is wrong way round. For these subjects it is more correct to observe the revolution in computing that was inspired by mathematics.

One may note that because of the economic exigencies of Academe, financial support for logicians had sorely ebbed until their connection to something as lucrative as computers was established and they found new sources of funding and a home in the newly formed departments of Computer Science. But this is hardly a change in mathematics as such.

The other branches of mathematics that have taken on an increased importance because of the existence of funding for Computer Science, such as Combinatorics, Operations Research, Cryptology, Graph Theory, Statistics and Probability, all have historically well identified independent justifications for existence and all are pursuing much the same goals they did before the invention of the computer. The fundamental nature of these subjects has not been affected by the existence of the computer; only the perception of the autonomy of these fields has suffered. One perfect case in point is the discipline of Game Theory. Nothing could be more perfectly positioned to be confused with computers - both inventions were heavily influenced by the same man, roughly at the same time and both employ the word "programming" as a term of art. However, the development of Game Theory has proceeded exactly along the lines originally conceived in its pre-computer days. The existence of computers has merely made its results applicable to larger problems, not to different problems. Certainly there is nothing revolutionary in this distinction.

The fashionable question of whether P equals NP appears on the surface to have been computer-inspired yet closer inspection casts severe doubts on its much touted practicability. When n becomes large enough so that n to the tenth is less than 2 to the n, it is already far beyond the computational capabilities of any machine present or future. The difference between exponential and polynomial growth, when confined to the size of problems that computers can handle, is decidedly uninteresting. The very concept of non-deterministic programming is uncompromisingly theoretical. This is the type of thinking non-applied mathematicians have been up to for centuries.

Correspondingly, Graph Theorists are interested in the travelling salesman problem for the same reasons they were interested in crossing the bridges of Königsberg. Modern Cryptology, on the other hand, passes all the tests for being a computer inspired field but by this very nature fails to embed itself properly into the domain of mathematics. It is a combination of rules of thumb, speculations, proprietary secrets, military secrets, commercial secrets, experimental data, experiential data and conjecture more like science or engineering than mathematics. If security in the large can be construed to be a mathematical issue then it is such without appeal to the existence of the computer.

Parenthetically we might admit that the computer has introduced some very interesting legal issues into mathematics, e.g., the prioritization of algorithms, the patentability of computer software and firmware, the implications of copyright protection for programs, as well as the moral and jurisprudential aspects of keeping mathematical discoveries trade secrets à la Karmarkar. All these are social issues and do not constitute a revolution properly within mathematics.

The second ramified aspect of the influence of the computer on mathematics is as a facilitator, catalyst or adjutant for pure research into classical areas. It is true that the symbolic algebraic manipulation packages can aid in the discovery of new identities, the solution of differential equations and so on but these capabilities have already been demonstrated by gifted mathematicians of the past. When Lagrange wished to show that every integer could be written as the sum of four squares he knew that he could reduce the problem to consideration of primes alone by means of an algebraic identity demonstrating that the product of two such sums is again a sum of four squares. He needed an eight variable polynomial and he found one. Gauss and Euler similarly pulled beautiful and not obvious identities from the air as their work required. As for analysis by computer graphics, it is true that some discoveries, such as the new surfaces of constant curvature, have in fact been based in large measure on computer displays. However, Lobachevsky, Bolyai, Gauss and Poincaré seemed to be able to imagine surfaces of constant negative curvature without the need for printed circuits.

The moral seems to be that nothing actually substantively revolutionary is being discovered, only that it is being discovered by more unlikely sources. A mediocre mathematician with a computer might be able to simulate the creative powers of a top notch mathematician with pencil and paper. If this is true it is much less a revolution in the nature of mathematics than in the nature of society. It is also mildly distasteful to unnecessarily offend those mathematicians who have made computer-aided discoveries by suggesting that they would never have been able to achieve as much on their own.

It is also hard to be sure that there is any real value in the computer's predilection for producing vast quantities of case by case data. Kepler's laws were extracted from crude hand-produced data, Gauss conjectured the prime number theorem from hand-produced data, Goldbach arrived at his conjecture exclusively from hand-produced data. How could one test whether or not computer produced data has actually led to more, or more profound, conjectures. The claim that such capacity only adds to the tool chest of mathematicians without diminishing any other tools and therefore must be beneficial, may miss an important drawback of computer dependency – supplanting creative efforts by programming time. I have often seen it happen that a problem that could have been solved by abstract mental reasoning was instead resolved by computer calculation thereby precluding the possibility of discovering deep abstract truth which would provide a deeper understanding. As an example of programming replacing thinking I once asked a first year Computer Science class whether there was any square less than a million that ended in the digit seven; all wrote programs and therefore all evaded making a mathematical discovery. Half of the mathematicians that have ever been born are alive today. Somewhere out there there should be a Gauss and an Euler. Perhaps instead of having developed their facility with algebraic and numerical manipulations and their geometric intuition they have been writing code. Wouldn't it be horrible if this were the true revolution in mathematics inspired by the computer?

This brings us properly to the third category, those old branches of mathematics that have been made obsolete by electronic equipment. We can observe that, without exception, anything which can now be done by the computer that was, formally done by a human was never properly part of mathematics. The fact that computers can now integrate into closed form every function which can be so integrated has made obsolete the burdensome drill in techniques of integration which used to plague Calculus classes with integration by parts, partial fractions, and trigonometric substitutions. Just because these were taught in the mathematics department does not make them part of mathematics. These are all merely lemmas for the grand algorithm of integration. Once the theory has been established the practice is properly within the purview of engineering. That part of the subject which was mathematics – the constructive existence theorem – is still alive and necessary (necessary even for the production of the computer algorithms). That part which was never more than engineering - the bag of tricks and shortcuts in execution – has thankfully been superseded. This is not an effect within mathematics at all, nor is it revolutionary for engineering. All other supplanted aspects of mathematics can be disposed of with similar crocodile tears.

The fourth field of impact listed above, the changes in mathematical education, yields to an attack analogous to the three arguments already presented. We plead in the alternative (1) that the revision of mathematical education was already in progress before computers were introduced, for example, the early introduction of set theory, and logic and the like; (2) that mathematical education already had all the aspects of computer- assistance even before the invention of the machine -

the visual aids of analytic geometry, the computational aids of tables of integrals and logarithms and the aid to symbolic manipulation gained from centuries of textbooks, etc., rendering computer-aided instruction more a change in degree than in substance, and (3) that whatever revolution there is in mathematical education it is a revolution in education and not in mathematics. We can also reassert our caveat that if there is a revolution we are not willing to presume it is beneficial without specific proof.

This brings us to the fifth horseman of the computer onslaught on mathematics, the alleged new paradigm of mathematical proof. There are several examples: perhaps the most annoying is the work on the Four Colour Problem. In 1976, Appel and Haken announced that they had solved the Four Colour Problem by a computer examination of nearly two thousand cases. In the analysis of each case the program only announced whether or not the procedure terminated successfully. The entire output from the machine was a sequence of hundreds of yeses. This must be distinguished from a program which produces a quantity as output which can subsequently be verified by humans as being the correct answer. Furthermore, the procedure employed by the machine to analyse each case, of necessity involved billions of logical inferences; this means that even though a human can duplicate by hand any small subset of the machine's deliberations there is not even a remote chance that, in an entire lifetime, a human could trace the program's run on even one case. Nor is the task of tracing the algorithm severable so that it can be worked in parallel by a number of individuals. The program, in the form constructed by them, is by its fundamental nature unsurveyable. The result of the computer calculation then either has to be accepted as truth on faith or rejected as nothing more than some sort of experimental evidence akin to that found in laboratory sciences. Mathematics then jumps from the Kantian status of *a priori* truth to an epistemological *a posteriori* condition.

If we encounter in a mathematical paper a seventh degree polynomial with integer coefficients and a list of seven integers claimed to be its roots we do not care whether these numbers were found with the aid of a computer. We can verify for ourselves that they are indeed the roots; whether they were discovered through lucky guesses, calculation or plagiarism is mathematically irrelevant. The circumstance is completely different if a computer tells us that the one zillionth digit in the decimal expansion of pi is a 7. In this case there is no known independent verification algorithm and we are obliged to duplicate the entire calculation ourselves to be sure of the result. An examination of the program might help to convince us of the truth of the proposition but we must still then exercise blind faith to be confident that the computer executed the program without electronic flaw. The bitter truth is that computers are not perfectly reliable; they do make errors. We might then increase our belief that 7 was indeed the one zillionth digit of pi by re-running the program a number of times, perhaps even on a variety of different machines, perhaps even rewriting the algorithm in other ways, or in other languages. The more times the same answer appears the more confident we might become. But this is all scientific evidence, the discovery of truth through the paradigm of inductive reasoning, not the seminal deductive method inherited from the Greeks. Convictions derived in this manner might be valid but they are not mathematics. Such a result is still unproven, and should be so considered.

Two arguments can be raised in support of this new method of proof. The first is to ask what all the fuss is about. Why not admit that this evidence is convincing? This position mistakes the goal of mathematics. Our pursuit is not the accumulation of facts about the world or even facts about mathematical objects. The mission of mathematics is understanding. No matter how many different computers or how many different algorithms or how many different programs come out with the answer 7 it still does not increase our understanding of why the one zillionth digit of pi is 7. A proof must provide this. The Appel and Haken work on the Four Colour Problem amounts to a confirmation that a map-maker with only four paint pots will not be driven out of business. This is not really what mathematicians were worried about in the first place. The real thrill of mathematics is to show that as a feat of pure reasoning it can be understood why four colours suffice. Admitting the computer shenanigans of Appel and Haken to the ranks of mathematics would only leave us intellectually unfulfilled. Kepler gleaned the laws of planetary motion from the data of Brahe, and this was science par excellence, but the real mathematical achievement was Newton's demonstration that the elliptical orbits were a necessary consequence of an inverse square law of gravitation. Newton did not increase our belief that the planets moved in conic sections. What he contributed was not a new fact but an explanation of an already believed fact [2]. If mathematicians were primarily interested in new results the literature would not be as rife as it is with additional proofs of already established theorems. Each new proof adds more understanding not more information.

The second proffered justification for accepting the work of Appel and Haken recognises the possibility that some results do not admit any humanly surveyable proof. This, of course, is unsubstantiated supposition. The Four Colour Problem is not a result which requires an unsurveyable computer proof, as my own work in this field shows [3]. Nor was there anything about the nature of the problem that would suggest that it might.

If mathematics continues to be a subject of interest to human beings over the next million years there will undoubtedly be theorems built upon theorems which in turn are built upon theorems, to the extent that it may exceed several hundred years to move from first principles to the outermost layers. This may then mean that there will be humanly uncheckable proofs that are locally built out of ordinary mathematics. However, if this crisis arises it will have nothing to do with the existence of computers. If anything, computers may stave off this dilemma by enabling us to organise the mathematical results in a more accessible way based on data-base principles. But this idle meditation is all irrelevant to the topic at hand.

References

[1] Cohen, D. I. A., Miller, V. S., 'On the four color problem', *Actes* 16e, Seminaire

Lotharingien, IRMA, (Strassbourg), 5–62, 1987.

[2] Hilbert, D., 'Mathematiche Probleme', *Archiv f. Math u. Phys. 3. Reihe,* Bd. 1, S.44–63; S.213–237, (1901)

[3] Newton, I., *Principia,* S. Pepys, Reg. Soc. Praeses, (London), 1686.